Metal Forming Processes

Metal forming processes include bulk forming and sheet metal forming with numerous applications. This book covers some of the latest developments aspects of these processes such as numerical simulations to achieve optimum combination and to get insight into process capability. Implementation of new technologies to improve performance based on Computer Numerical Control (CNC) technologies are discussed, including use of CAD/CAM/CAE techniques to enhance precision in manufacturing. Applications of AI/ML, the Internet of Things (IoT), and the role of tribological aspects in green engineering are included to suit Industry 4.0.

The Main Features:

- Covers latest developments in various sheet metal forming processes
- Discusses improvements in numerical simulation with various material models
- Proposes improvements by optimum combination of process parameters
- Includes finite element simulation of processes and formability
- Presents a review on techniques to produce ultra-fine-grained materials

This book is aimed at graduate students, engineers, and researchers in sheet metal forming, materials processing and their applications, finite element analysis, manufacturing, and production engineering.

Metal Forming Processes

Developments in Experimental and Numerical Approaches

Edited by

Kakandikar Ganesh Marotrao

Anupam Agrawal

D. Ravi Kumar

CRC Press
Taylor & Francis Group
Boca Raton London New York

CRC Press is an imprint of the
Taylor & Francis Group, an **informa** business

First edition published 2023
by CRC Press
6000 Broken Sound Parkway NW, Suite 300, Boca Raton, FL 33487-2742

and by CRC Press
2 Park Square, Milton Park, Abingdon, Oxon, OX14 4RN

ISBN: 978-1-032-12890-0 (hbk)
ISBN: 978-1-032-12891-7 (pbk)
ISBN: 978-1-003-22670-3 (ebk)

DOI: 10.1201/9781003226703

Typeset in Times
by SPi Technologies India Pvt Ltd (Straive)

Contents

Preface

Metal forming processes play an extremely important role in mass production of many parts for various applications. Among them, sheet metal forming is vital to produce a large number of simple-to-complex parts for aircrafts, automobiles, appliances, and many more. For successful stamping operations to produce intricate parts, design process is often complicated, expensive and time consuming because forming dies and other tools need to be built for each trial. The continuously increasing costs of material, energy, and manpower demand that the new processes be developed, and tooling be designed with the minimum possible defects and lead times. Therefore, research in the last few decades in sheet metal forming has focused on the development of methods to process new and lightweight materials, characterization of materials before and after processing, innovative methods to optimize process and design variables to obtain the desired combination of microstructures, mechanical properties, formability, and investigations into the processing of materials over a wide range of temperatures (cryogenic to recrystallization temperature). We are delighted to present this book covering the latest topics in most of these important areas.

This collection aims to provide information on state-of-the-art research in the field of sheet metal forming and summarizes the advances and innovations made in recent times. The book might serve as a useful reference for students, research scholars, engineers, scientists, researchers, and academicians working in the broad areas of manufacturing, materials processing, metal forming, and materials science.

The book is divided into 14 chapters and each chapter is contributed by a group of authors who have been actively involved in research and have published many journal articles in their respective fields. Chapters 1–4 cover some recent advances in sheet metal forming processes such as microforming of hot stamping steel and shock tube based forming. Advanced characterization of sheets subjected to cryorolling, texture, and tribological behavior of sheets are presented in Chapters 5–7. Formability characterization in terms of hole expansion ratio and forming limit curves for different sheet materials under different conditions is presented in Chapters 8–10. Incremental forming of sheets has gained significant importance for low-volume productions and hence recent advances in various aspects of incremental forming are covered in Chapters 11–14.

We express our thanks to all the authors who have spent enormous time and effort to contribute chapters to this book and all the others who have helped us publish this book. We hope that this book will be of great use to all those who are striving for advancing the knowledge of metal forming science and technology.

Kakandikar Ganesh Marotrao

Anupam Agarwal

D. Ravi Kumar

Contributors

Anupam Agrawal
Department of Mechanical Engineering
Indian Institute of Technology
Ropar, India

Saibal Kanchan Barik
Department of Mechanical Engineering
IIT Guwahati
Guwahati, India

Vishal Bhojak
Mechanical Engineering Department
Malaviya National Institute of Technology Jaipur
Jaipur, India

Kandarp Changela
Saint Gobain Research India, Indian Institute of
    Technology Madras Research Park
Chennai, India

Sukanta Das
Department of Mechanical Engineering
IIT Guwahati
Guwahati, India

S. Deb
Indian Institute of Technology
Madras, Chennai, India

Kishore Debnath
Department of Mechanical Engineering
National Institute of Technology Meghalaya
Meghalaya, India

R. Ganesh Narayanan
Department of Mechanical Engineering
IIT Guwahati
Guwahati, India

Amit Kumar Gupta
Birla Institute of Technology and Science
Pilani Hyderabad Campus, India

K. Hariharan
Department of Mechanical Engineering
Indian Institute of Technology Madras
Chennai, India

Jinesh Kumar Jain
Mechanical Engineering Department
Malaviya National Institute of Technology
    Jaipur
Jaipur, India

Nitin Kotkunde
Birla Institute of Technology and Science
Pilani Hyderabad Campus, India

Gautam Kumar
Mechanical Engineering Department
National Institute of Technology Patna
Patna, India

Narinder Kumar
Department of Mechanical Engineering
Indian Institute of Technology
Ropar, India

Gauri Mahalle
Birla Institute of Technology and Science
Pilani Hyderabad Campus, India

Kuntal Maji
Mechanical Engineering Department
National Institute of Technology Patna
Patna, India

Amrut Mulay
Mechanical Engineering Department
Sardar Vallabhbhai National Institute of
    Technology
Surat, India

Murugabalaji V.
Department of Production Engineering
National Institute of Technology
    Tiruchirappalli
Tamil Nadu, India

K. Narasimhan
Department of Metallurgical Engineering and
    Materials Science
IIT Bombay
Mumbai, Maharashtra, India

S. K. Panigrahi
Indian Institute of Technology
Madras, Chennai, India

Surajit Kumar Paul
Mechanical Engineering Department
Indian Institute of Technology Patna
Patna, India

D. Ravi Kumar
Department of Mechanical Engineering
Indian Institute of Technology Delhi,
    Hauz Khas
New Delhi, India

Matruprasad Rout
Department of Production Engineering
National Institute of Technology Tiruchirappalli
Tamil Nadu, India

Niranjan Sahoo
Department of Mechanical Engineering
IIT Guwahati
Guwahati, India

Parnika Shrivastava
Department of Mechanical Engineering
National Institute of Technology
Hamirpur, India

Amarjeet Kumar Singh
Department of Metallurgical Engineering and
    Materials Science
IIT Bombay
Mumbai, Maharashtra, India

Swadesh Kumar Singh
Gokaraju Rangaraju Institute of Engineering
    and Technology
Hyderabad, India

Puneet Tandon
Department of Mechanical Engineering
PDPM Indian Institute of Information
    Technology Design and Manufacturing
    Jabalpur
Jabalpur, India

Sameer Vadhera
Mechanical Engineering Department
Sardar Vallabhbhai National Institute of
    Technology
Surat, India

Notes on the Editors

Professor Dr. Kakandikar Ganesh Marotrao is Professor and Head of School - Mechanical Engineering at Dr. Vishwanath Karad MIT World Peace University, Pune. He completed a Ph.D. in Mechanical Engineering from Swami Ramanand Teerth Marathwada University, Nanded in 2014.

- He has 22 years of experience in Teaching, Research and Administration.
- He has authored four books published internationally by CRC Press, Taylor and Francis Group and contributed book chapters published by Wiley, Springer, and Elsevier.
- 80+ publications are at his credit in national/international journals and conferences.
- He is a reviewer of many reputed journals.
- He has chaired sessions in various national and international conferences and also worked on their advisory committees.
- His publications have been widely cited by researchers.
- He is Fellow of The Institutions of Engineers [India]. He is professional member of various society's including International Society on Multiple Criteria Decision Making, International Society for Structural and Multidisciplinary Optimization and Sheet Metal Forming Research Association.
- He has executed many research projects with industries especially on optimization of automotive components.
- He has guided several M.Tech. theses and adjudicated many at other universities.
- He is recipient of many academic and industry awards including:
  1. Young Scientist International Travel Grant by Government of India in 2013 to visit Federal Institute of Technology, Zurich, Switzerland.
  2. Indo Global Engineering Excellence Award 2015 from Indo Global Chamber of Commerce, Industries and Agriculture for Research Work in Engineering,
  3. Cooper Engineering Prize for Best Paper in Mechanical Engineering for paper entitled "Thinning Optimization in Punch Plate using Flower Pollination Algorithm at Annual Technical Paper Meet 2015 of The Institution of Engineers (India) at Pune.
  4. Appreciation Award in ANSYS Hall of Fame 2017.
  5. Rajarambapu Patil National Award for Promising Engineering Teacher 2020 from Indian Society for Technical Education, New Delhi.
  6. Teacher of the year award 2021 in Technology category from Academisthan Foundation Mumbai.
- He regularly writes columns in India print media on education systems and reforms.
- He is also contributing for quality technical education as Accessor for NAAC.
- He is presently executing a Research Project funded by Defense Research and Development Organization, New Delhi.
- He is presently executing a Techno Societal Research Project funded by Dassault Systemes, India.

Dr. Anupam Agrawal received his Bachelor's degree in Mechanical Engineering from Madan Mohan Malviya Engineering College Gorakhpur, UP, Masters degree in Mechanical Engineering from Indian Institute of Technology Kharagpur, WB and Ph.D. in Mechanical Engineering from Indian Institute of Technology Kanpur, UP. He worked as Postdoctoral Research Fellow at International Center for Automotive Research (ICAR) at Clemson University, US, for two years. After his postdoctoral stint he joined as an Assistant Professor in 2010, and as Associate Professor in 2015, in the Department of Mechanical Engineering at IIT Ropar. Dr. Anupam Agrawal has been working in forming/micromachining/advanced manufacturing processes, CAD/CAM for more than 15 years. His current research interests are in the area of microfabrication using EDM, Laser Machining, and Mechanical Micromachining. He is also working on 3D printing of biomedical implants using suitable additive manufacturing processes along with micromachining processes. He has wide experience of micromanufacturing processes and also he has established a dedicated lab for performing micro-manufacturing processes at IIT Ropar. He has more than 70 publications in international journals, as well as international and national conferences.

Prof. D. Ravi Kumar is a Professor in the Department of Mechanical Engineering, Indian Institute of Technology Delhi, India. He has more than 25 years of experience in teaching and research. His research interests include Metal Forming, Plasticity and Materials Processing. His specific research has been focused mainly on Sheet Metal Forming and Finite Element Analysis. He has guided several Ph.D. scholars and Masters' students in their thesis work. He has published over 100 research papers in international journals and conferences worldwide. He was a Humboldt Research Fellow at the Institute for Metal Forming Technology (IFU), University of Stuttgart and University of Erlangen, Germany. He was a visiting faculty at the University of Waterloo, Canada. Before joining IIT Delhi, he was a researcher at Tata Steel (R&D), Jamshedpur. Dr. Ravi Kumar has received several awards and honors for his teaching and research. He is a member of the Editorial Board of Production Engineering Journal (Springer). He is also an active member of Sheet Metal Forming Research Association (SMFRA).

1 Artificial Neural Network (ANN) Based Formability Prediction Model for 22MnB5 Steel under Hot Stamping Conditions

Amarjeet Kumar Singh and K. Narasimhan
IIT Bombay, Mumbai, India

CONTENTS

1.1 INTRODUCTION

Due to new government policy, it is responsibility of industries and automobile to reduce CO_2 and greenhouse gases emitted. To reduce carbon emission automotive industries strive to decrease the weight of component by use of Advanced High Strength Steel (AHSS). However, AHSS material has lower formability and high springback at room temperature and therefore press hardening or hot stamping process was developed. In this process sheet specimen is heated above austenization temperature and held for sufficient time for homogenization after which the specimen is transferred to forming press where forming and quenching are done simultaneously in a single stage. In last few years [1] hot stamped components have increased drastically in automobile industries because of advantage associated with process. At room temperature, FLC is determined with help of in plain and out of plane method developed by Marciniak test [2] and Nakazima test [3] respectively. These processes are standardized for room temperature but not for elevated temperature applications [4]. Therefore, in this work a method is to determine formability under hot stamping conditions is developed.

Formability of 22MnB5 steel depends on temperature, strain rate, process, strain path, sheet thickness etc. H. Karbasian et al. [1] reviewed the hot stamping process and concluded that formability depends on cooling medium, temperature, transfer time of specimen from furnace to press, tribological conditions and temperature of deformation etc. Li et al. [5] investigated effect of sheet thickness and temperature on FLC of 22MnB5 steel. To study effect of sheet thickness of 1 mm and 1.4 mm

DOI: 10.1201/9781003226703-1

at 600°C and 700°C on 22MnB5 steel Nakazima test was performed and found that with increase in sheet thickness and temperature formability increased. Georgiadis et al. [6] conducted experiments using the Nakazima test for sheet thickness of 0.5 mm, 0.8 mm, 1.25 mm at elevated temperature. In this work experiments were performed under isothermal and non-isothermal conditions and it was found that under isothermal conditions formability increased with greater in sheet thickness but in the case of non-isothermal conditions an increase in sheet thickness led to a decrease in formability. Singh et al. [6–10] studied the effect of strain rate on 22MnB5 steel during hot stamping conditions at various strain rate 0.01/s, 0.1/s, 1/s at elevated temperature and concluded that in drawing region formability increased with increase in strain rate. Lee et al. [11] analyzed the effect of strain rate of 3/s and 9/s at elevated temperature and found similar effect of strain rate on formability of 22MnB5 steel. Hongzhou Li et al. [12] predict formability of 22MnB5 steel at different temperature and strain rates and concluded that with increase in temperature and strain rate formability increase however pointed out that effect of strain rate was negligible. Gerdooei et al. [13] studied effect of strain rate at room temperature on various materials and summarized that in most of materials increase in strain rate formability decreases. From the literature, effect of strain rate on formability is not conclusive therefore in this work effect of strain rate on formability analyzed. Min et al. [14] developed FLC of 22MnB5 steel using Marciniak-Kuczynski (MK) model and Storen & Rice's (SR) model and concluded that SR model had better prediction compared to MK model. Sarawagi et al. [9] compared various necking criteria that are commonly used in sheet metal forming and found that thickness gradient based necking criterion had good prediction capability then other necking criteria. In this work thickness gradient based necking criterion was used for necking analysis.

Lin et al. [15] reviewed different types of constitutive models used in sheet metal forming and classified these constitutive models in three groups namely phenomenological model, physical based constitutive model, and ANN based model. It was observed that phenomenological model had a smaller number of constants to determine constitutive behavior compared to the physical based model. Physical based models are more complex and determination of constant for such models is time consuming. However, ANN model gives better results compare to others and less time resource and time consuming. Therefore, in this work an ANN model is developed that can predict stress strain behavior along with the formability in a single network.

1.2   EXPERIMENT

22MnB5 steel is widely used in automobile industries for hot stamping process. Due to advantage associated with this material like higher strength to weight ratio after hot stamping process. Therefore, same material was used in this study.

1.2.1   Material Used

Uncoated 22MnB5 steel sheet under hot rolled condition with thickness of 1.8 mm received from JSW Ispat Steel Ltd was used for this study. The chemical composition (in percentage) and mechanical properties of received material are given in Table 1.1 and Table 1.2 respectively. This material consists of dual phases in as received condition. Initial microstructure of 22MnB5 steel consists of ferrite and perlite.

TABLE 1.1

Chemical Composition (in Percentage) of Uncoated 22MnB5 Steel as in the Received Condition

Al	B	Mn	Si	Cr	Ni	Cu	P	Ti
0.108	0.003	1.17	0.23	0.163	0.0055	0.063	0.013	0.018

TABLE 1.2
Mechanical Properties of Uncoated 22MnB5 Steel at Room Temperature as in the Received Condition

Tensile Strength	Yield Strength	Total Elongation
530MPa	352MPa	24.25%

1.2.2 THERMO-MECHANICAL TEST (GLEEBLE-3800 TEST)

Hot stamping or press hardening is a high temperature non-isothermal process as discuss in earlier section. Industrial hot stamping process cycle consists of heating to austenization, holding and forming and quenching based on this cycle previous work done by Singh et al [7] optimize this process for lab scale as given in Figure 1.1. During optimization of this process it may be noted that properties at end of lab scale and industrial process were same.

Elevated temperature tensile tests were performed using Gleeble-3800 in a closed chamber with strain rate control mode. Heating of specimen was done with help of resistance heating for monitoring temperature of specimen K-type thermocouple was welded on it. To prevent oxidation of specimen experiment was conducted in inert medium. Dimensions and geometry for elevated temperature tensile test is shown in Figure 1.2.

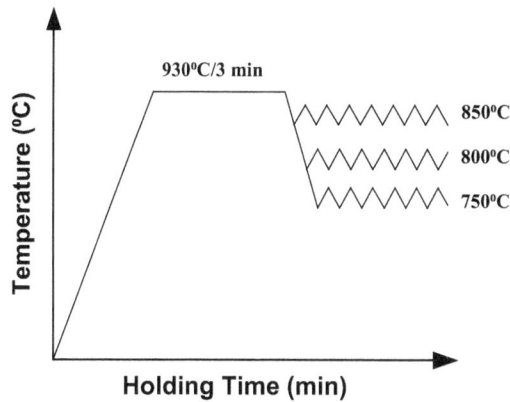

FIGURE 1.1 Temperatures Vs holding time used in Gleeble-3800 for lab scale experiment at various temperature and strain rates.

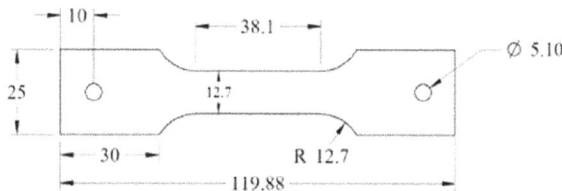

FIGURE 1.2 Uniaxial test specimen dimensions & geometry of tensile test (without notch).

1.3 RESULTS AND DISCUSSION

1.3.1 TENSILE TEST

Elevated temperature tensile test in temperature range of 750°C to 850°C and strain rate range of 0.01/s to 1/s are shown in Figure 1.3 and Figure 1.4 respectively. It was observed that with increase in temperature flow stress of material decreases due to softening at higher temperature. Whereas with increase in strain rate flow stress of 22MnB5 steel increases.

1.3.2 FORMING LIMIT DIAGRAM

Forming Limit Diagram (FLD) is used to determine formability of sheet metal component. It is graphical representation of fail and safe region of sheet metal during forming. It depends on temperature, strain rate, thickness of sheet, strain path etc. In this work temperature and strain rate effects were analyzed under hot stamping conditions. Gleeble-3800 was used to conduct experiment in order to determine FLD in drawing region. In order to predict limiting strain, thickness based necking criterion was used [4]. To determine formability in drawing region simple tensile geometry was modified in shape such that it can produce plane strain condition as shown in Figure 1.5. To measure strain circular shape grid with diameter size of 1.5 mm and 2 mm center to ends distance between neighbor grids were engraved on surface of sheet with the help of laser marking machine as shown in Figure 1.6. After deformation four types of distorted grid were observed namely safe grid (before necking), critical grid (at point of necking), cracked or failed, neighbor to critical grid. Calculation of major and minor strain for FLD were done with help of critical grid.

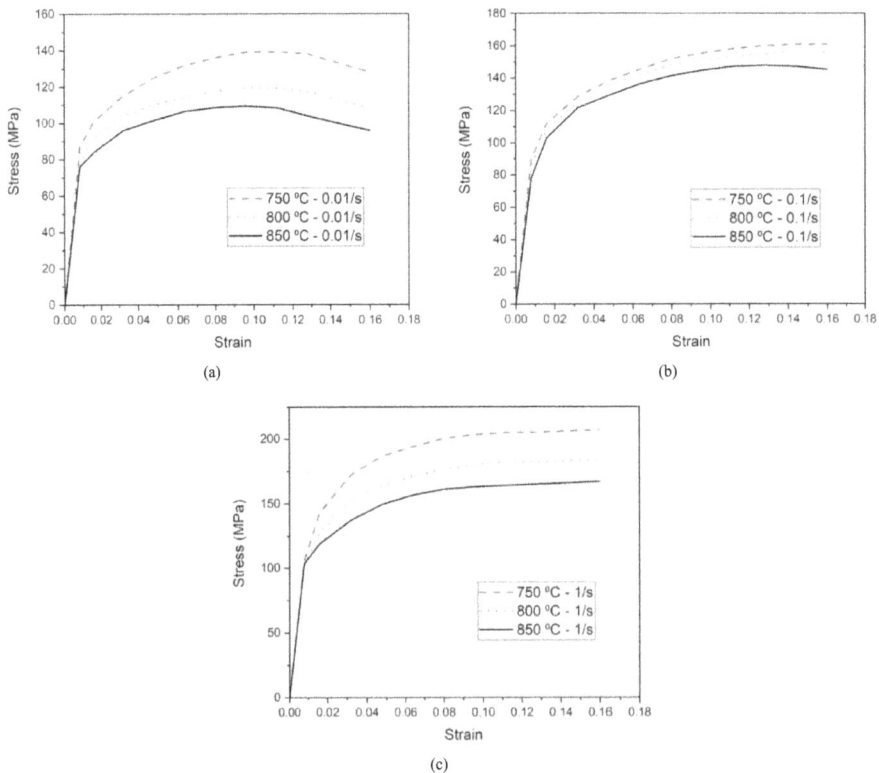

FIGURE 1.3 True Stress Vs True Strain at different temperature with (a) Strain rate of 0.01/s, (b) Strain rate of 0.1/s, and (c) Strain rate of 1/s.

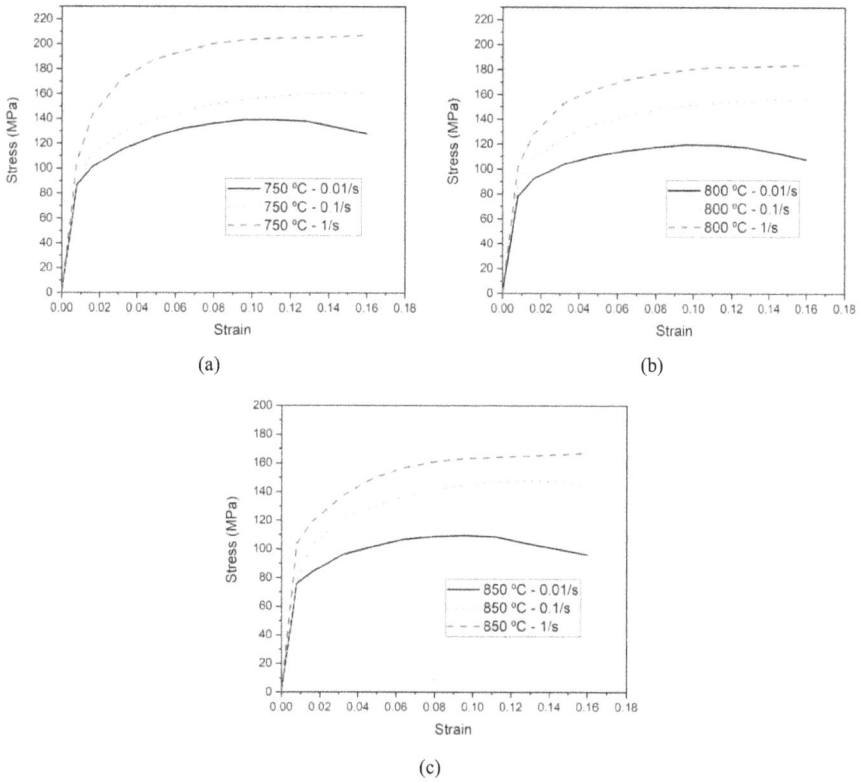

FIGURE 1.4 True Stress Vs True Strain at different strain rate with (a) Temperature 750°C, (b) Temperature 800°C, and (c) Temperature 850°C.

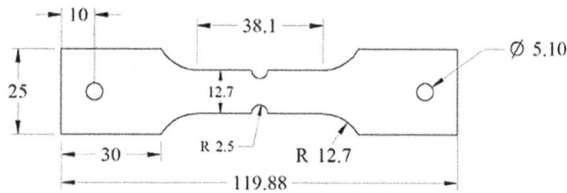

FIGURE 1.5 Uniaxial tensile specimen for determination of formability in drawing region (notch and without notch specimen) used for different experimental conditions.

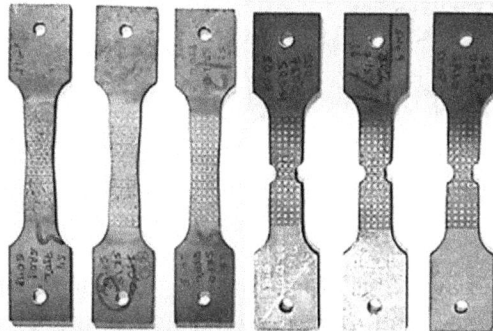

FIGURE 1.6 Tensile notch and without specimen after deforamtion at various experimental conditions.

Uniaxial deformation for notch and without notch specimen were done to point of necking using hot stamping cycle as given in the previous section. Measured major and minor strain at different temperature and strain rates shown in Figure 1.9. It was found that with increase in strain rate and temperature formability increase in drawing region. It was also found that flow stress and elongation for notch specimen was lower than without notch specimen because of stress concentration induced in notch specimen due to notch [10].

1.4 ANN MODEL

Development of constitutive model at elevated temperature and strain rates using physical based model is challenging and computationally time consuming. Whereas in case on ANN model it is easy to develop and takes a few seconds to predict the result. ANN model was developed to predict flow behavior and formability of 22MnB5 steel. Concept of ANN models are based on human biological neuron in which complex neuron connection was done to solve difficult problems. Neurons are basic building block for ANN models. There are various types of neuron are present in a network like input neuron, output neurons and hidden neurons etc. These neurons are interconnected different weight, bias and transfer functions to solve various problems.

MATLAB was used to generate ANN model in this work. To develop ANN model different stages are involved such as selection of data, normalization of all available data, training, validation and testing of model etc. In this work strain, temperature, strain rate, and strain ratio were selected as input datasets whereas stress, major strain, and minor strain as output datasets as shown in Figure 1.7. Architecture of neural network consists of sigmoid transfer function, Feed forward method, Levenberg-Marquardt training algorithm. Developed model consists of two hidden layers with five neurons in each hidden layer for optimization of model. To obtain optimal solution by ANN model mean square error between target datasets and output datasets should be close to 0.0001 as defined. To minimize mean square error weight and bias of each neuron was adjusted using training algorithm.

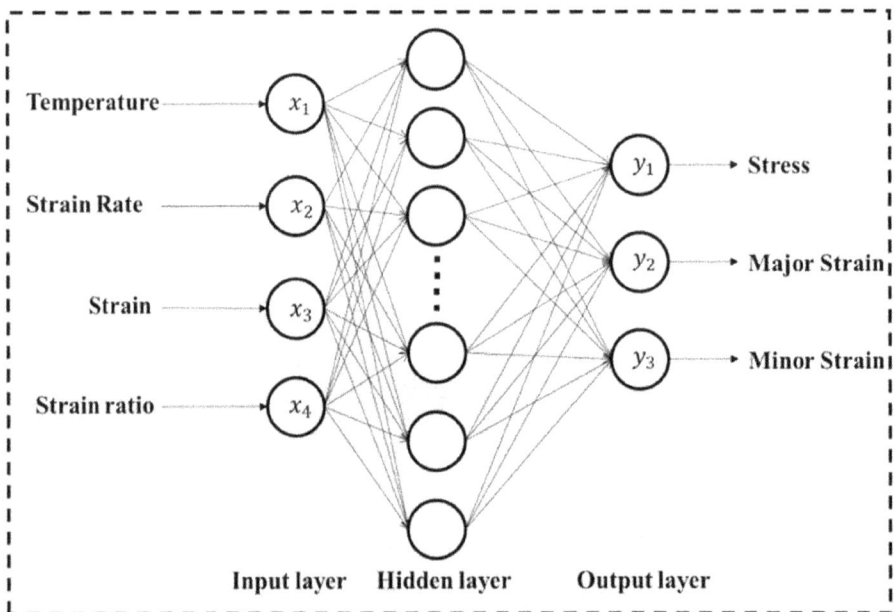

FIGURE 1.7   Architecture used in ANN Model.

Experimental and predicted flow behavior and FLD of 22MnB5 steel is shown in Figures 1.8 and 1.9 respectively at various temperatures and strain rates. It was found that develop ANN model was able to predict flow behavior and formability of material. Comparison of experimental and predicted value of stress and strains are shown in Figure 1.10. It was observed that R-values for stress and strains are 0.99. R-value indicate the correlation between measured and predicted values. R-value is defined in such a way that if it is equal to one then close relation between measured and predicted values but if it is zero then random relation observed.

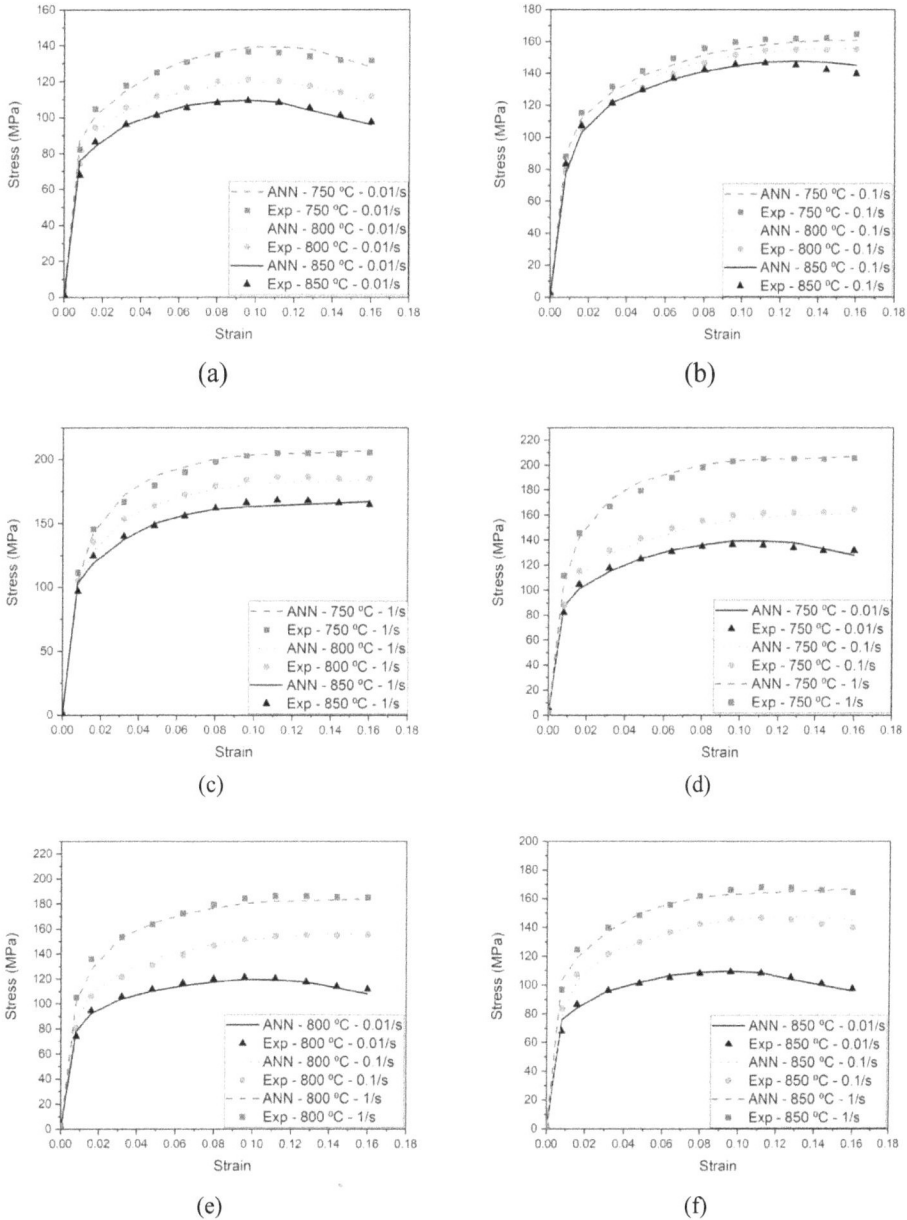

FIGURE 1.8 Experimental and ANN Model predicited true stress vs true strian at tempeature of 750°C to 850°C and strain rate of (a) 0.01/s, (b) 0.1/s, (c) 1/s, (d) at strain rates of 0.01 to 1/s and tempeature of 750°C, (e) 800°C, and (f) 850°C.

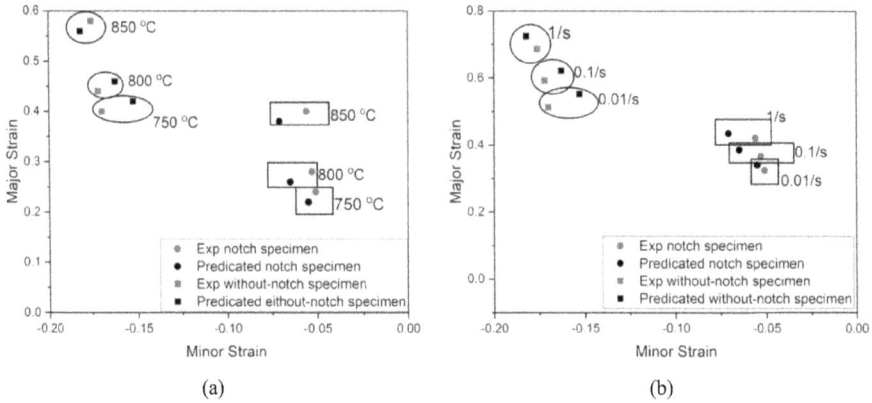

FIGURE 1.9 Experimental and ANN predicted FLC in Drawing region at (a) various elevated temperature range of 750°C to 850°C at strain rate of 0.1/s and (b) different strain rates of 0.01/s to 1/s at 800°C.

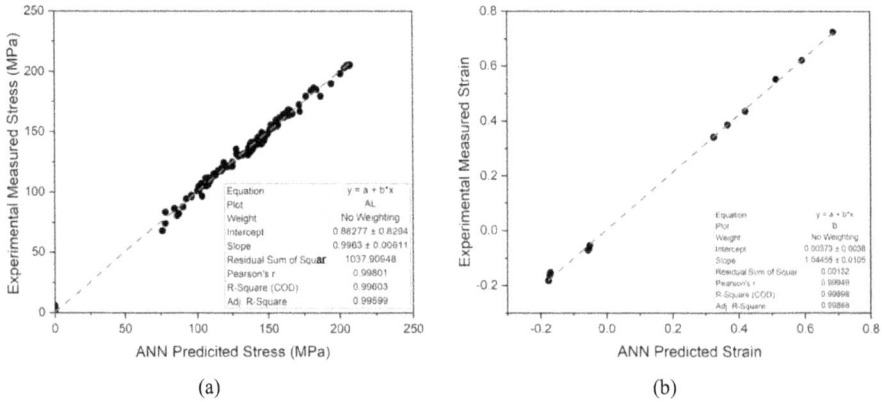

FIGURE 1.10 Comparision of experimental and ANN model predicted result for (a) Stess and (b) limiting strain.

1.5 CONCLUSION

Elevated temperature uniaxial deformation conducted using Gleeble-3800 at different temperature and strain rates under hot stamping conditions. Formability was measured in drawing region with help of new experimental procedure. An ANN model was developed that capture effect of strain, temperature, strain rate and strain ratio on flow stress and FLD of 22MnB5 steel. In future work it requires to develop an ANN model that can able to predict formability at any stain ratio and complex strain paths. Based on this study, following conclusions can made.

   i. Modified Gleeble experiment procedure can produce FLD in drawing region where as notch specimen can produce plane strain condition for different temperatures and strain rates under hot stamping conditions. It was found that with increase in temperature flow stress of 22MnB5 steel decrease whereas with increase in strain rate flow stress increases.
   ii. It was found that in drawing region of FLD formability increase with increase in strain rate and temperature.
   iii. Developed ANN model is able to predicted flow stress and formability of 22MnB5 steel rapidly and with good accuracy. Therefore, this method can be used for online monitoring.

REFERENCES

[1] H. Karbasian and A. E. Tekkaya, "A review on hot stamping," *J. Mater. Process. Technol.*, vol. 210, no. 15, pp. 2103–2118, 2010, doi:10.1016/j.jmatprotec.2010.07.019.

[2] Z. Marciniak and K. Kuczyński, "Limit strains in the processes of stretch-forming sheet metal," *Int. J. Mech. Sci.*, vol. 9, no. 9, pp. 609–620, 1967, doi:10.1016/0020-7403(67)90066-5.

[3] Dorel Banabic, *Sheet Metal Forming Processes Constitutive Modelling and Numerical Simulations*, 2010.

[4] A. K. Singh and K. Narasimhan, "Determination and predication of formability on 22MnB5 steel under hot stamping conditions using Gleeble," *Adv. Mater. Process. Technol.*, vol. 00, no. 00, pp. 1–13, 2021, doi:10.1080/2374068X.2021.1878709.

[5] J. Y. Li, J. Y. Min, K. Y. Qin, J. P. Lin, F. Q. Liu, and L. C. Liu, "Investigation on the effects of sheet thickness and deformation temperature on the forming limits of boron steel 22MnB5," *Key Eng. Mater.*, vol. 474–476, pp. 993–997, 2011, doi:10.4028/www.scientific.net/KEM.474-476.993

[6] G. Georgiadis, A. E. Tekkaya, P. Weigert, S. Horneber, and P. Aliaga Kuhnle, "Formability analysis of thin press hardening steel sheets under isothermal and non-isothermal conditions," *Int. J. Mater. Form.*, vol. 10, no. 3, pp. 405–419, 2017, doi:10.1007/s12289-016-1289-4.

[7] A. K. Singh and K. Narasimhan, "Prediction of necking & thinning behaviour during hot stamping conditions of 22MnB5 steel," pp. 4–10.

[8] P. S. Narayanasamy, A. K. Singh, and K. Narasimhan, "Effect of cooling system in hot stamping process," *IOP Conf. Ser. Mater. Sci. Eng.*, vol. 967, no. 1, 2020, doi:10.1088/1757-899X/967/1/012059.

[9] I. O. P. C. Series and M. Science, "Formability studies on 22MnB5 steel during hot stamping process conditions," 2018, doi:10.1088/1757-899X/418/1/012011.

[10] A. K. Singh and K. Narasimhan, "Effect of strain rate on formability of 22MnB5 steel during hot stamping process," *IOP Conf. Ser. Mater. Sci. Eng.*, vol. 1157, no. 1, p. 012022, 2021, doi:10.1088/1757-899x/1157/1/012022.

[11] R. S. Lee, Y. K. Lin, and T. W. Chien, "Experimental and theoretical studies on formability of 22MnB5 at elevated temperatures by Gleeble simulator," *Procedia Eng.*, vol. 81, no. October, pp. 1682–1688, 2014, doi:10.1016/j.proeng.2014.10.213.

[12] H. Li, X. Wu, and G. Li, "Prediction of forming limit diagrams for 22MnB5 in hot stamping process," *J. Mater. Eng. Perform.*, vol. 22, no. 8, pp. 2131–2140, 2013, doi:10.1007/s11665-013-0491-5.

[13] M. Gerdooei, "Strain-rate-dependent forming limit diagrams for sheet metals," no. February, 2016, doi:10.1243/09544054JEM1193.

[14] J. Min, J. Lin, J. Li, and W. Bao, "Investigation on hot forming limits of high strength steel 22MnB5," *Comput. Mater. Sci.*, vol. 49, no. 2, pp. 326–332, 2010, doi:10.1016/j.commatsci.2010.05.018.

[15] Y. C. Lin and X. M. Chen, "A critical review of experimental results and constitutive descriptions for metals and alloys in hot working," *Mater. Des.*, vol. 32, no. 4, pp. 1733–1759, 2011, doi:10.1016/j.matdes.2010.11.048.

2 Shock Tube Based Forming of Sheets

Saibal Kanchan Barik, R. Ganesh Narayanan, and Niranjan Sahoo
IIT Guwahati, Guwahati, India

CONTENTS

DOI: 10.1201/9781003226703-2

2.1 INTRODUCTION

The increasing regulations on fuel economy, exhaust gas emissions, and material recycling potential draws the attention of aerospace and automobile sectors towards the production of light-weight sheet structures. It is observed from the review of Joost (2012) that a potential weight saving of 10% can enhance the fuel efficiency by 6–8%. Thus, among all the sheet metals aluminium alloys are mostly preferable because of their specific engineering properties such as light-weight, good static and dynamic strength, superior corrosion resistance and recycling ability (Fridlyander et al. 2002). However, because of the poor formability at ambient temperature, the sheets can easily fracture at the round corners of the dies during the stamping process (Hsu et al. 2008). As a result, many manufacturing industries are focused on enhancing the formability of the aluminium alloys.

Though it is well known that a significant increase in the formability of the aluminium alloys can be achieved by carrying out the same operation at a higher temperature (Gu et al. 2019), it adds expense to the conventional stamping process. On the other hand, forming the sheets at higher strain rates also enhance the forming properties (Ahmed et al. 2017). The velocity gradient developed on the material during the deformation is minimized by the inertial forces generated during the high strain rate forming which permits the material to stretch further without strain localization. It results in larger deformation than the traditional quasi-static forming.

When the strain rate approaches a certain limit, almost all material exhibit strain sensitive behavior (Smerd et al. 2006). As a result, determining rate-dependent material properties at various strain rates is crucial. To extract material properties at lower and higher strain rates, the quasi-static tensile test and Split Hopkinson Pressure Bar (SHPB) test have traditionally been used for sheet metals. However, these tests are limited to deformation in the uniaxial direction. However, in actual forming condition, material deforms biaxially and it is always required to interpolate uniaxial test data with extended range of plastic strain (Koç et al. 2011). Thus, hydraulic bulge test has been used since so long as a convenient method for determining the biaxial material properties (Mahabunphachai and Koç 2010). Due to the biaxial state of stress-induced condition, the maximum achievable strain before the fracture is much higher than the tensile test. Further, the use of hydraulic bulge test is restricted to certain strain rate range. Thus, various high strain rate biaxial forming devices become more popular.

Thus, the present article is focused to give an overview about the different high strain rate forming devices available to study the biaxial forming behavior of the sheet metals. During this analysis, the important process parameters that affect the forming conditions are discussed in details. Further, the applicability of the shock tube facility during the characterization of the forming behavior of AA 5052-H32 sheet is analysed.

2.2 STATIC AND DYNAMIC BEHAVIOR OF SHEET METAL

Since the strain rate has an influence on both the plastic flow stresses and the ductility of materials, it also affects the identification of crucial process parameters such as forming forces, energies, and forming limits. Light-weight materials like Al and Mg alloys, which have surge in demand in automotive industries, show little dependency on strain rate and exhibit insignificant and sometimes negative values of strain rate sensitivity at quasi-static strain rates ($10^{-5}\,\mathrm{s}^{-1} \leq \bar{\varepsilon} \geq 10^{-1}\,\mathrm{s}^{-1}$) (Ghosh 1977). On the other hand, the flow stress of various aluminium alloys increases with strain rate, showing a positive SRS at high-strain rates (above $10^3\,\mathrm{s}^{-1}$) (Smerd et al. 2006). This positive strain rate sensitivity delays the onset of diffuse necking and hence total elongation and formability, can be improved at high-strain rates. Dorward and Hasse (1995) also highlighted that at high-strain rates, the hardening effects of Al alloys become prominent, resulting in a rise in flow stress and elongation.

Because of the limited range of velocity condition in particular machines, it has been impossible to accurately describe the mechanical behavior under a wide range of strain rates using one testing machine. According to the magnitudes of the strains, tensile experiments are divided into

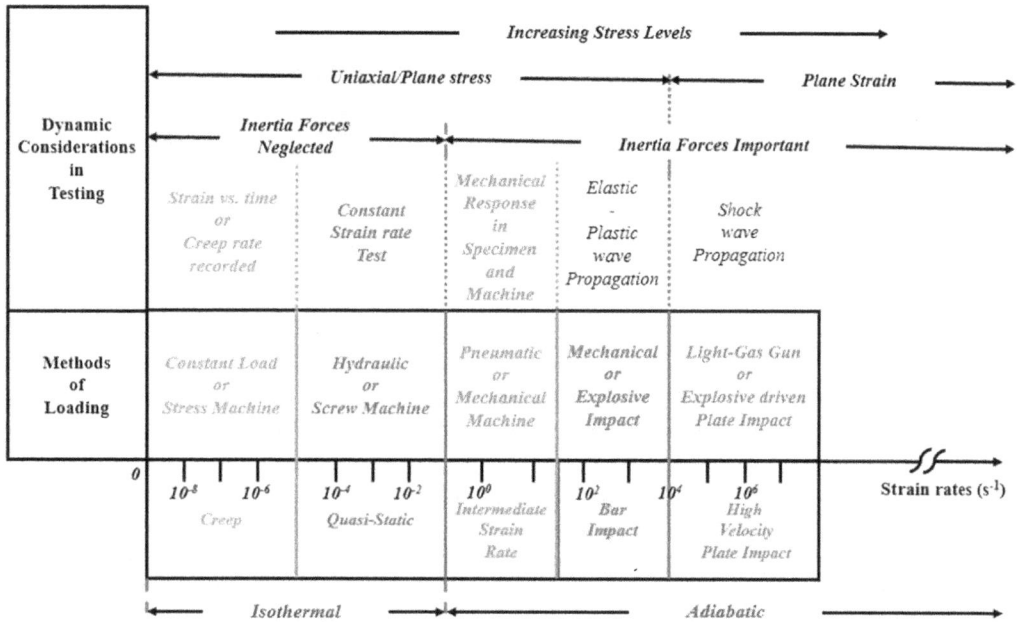

FIGURE 2.1 Classification of tensile experiments and loading methods with respect to the value of strain rates. (Reprinted with permission from Abd El-Aty et al. 2019, Elsevier).

quasi-static, static, and dynamic experiments. Based on the strain rate regime, different types of tensile experimental apparatus used nowadays are reported in Figure 2.1 (Abd El-Aty et al. 2019).

In quasi-static strain rate regime, generally hydraulic experiments are used to obtain the material properties. Traditionally, hydraulic bulge test has long been recognised as a convenient and practical method for determining the formability of sheet metal, as well as an effective way for determining the biaxial stress–strain relationships (Mahabunphachai and Koç 2010). However, because of the use of hydraulic fluid during the experiment, its application is restricted only in the lower ranges of strain rates. In addition, many changes to the existing experimental setup have been made in order to acquire rate-dependent forming conditions during biaxial forming.

2.3 DYNAMIC SHEET FORMING DEVICE

2.3.1 DROP HAMMER RIG

Broomhead and Grieve (1982) used a drop hammer rig during the bulge test to investigate the effect of strain rate on the strain to fracture of sheets under biaxial tension. With this setup, they determined the forming limit curves of low carbon steel for strain rates of up to 70 s^{-1}. Percy and Lim (1983) identified the forming limit diagram of the steel sheet with the same setup and reported an enhancement in the formability at the higher forming rate. Further, Pickett et al. (2004) used a similar apparatus to investigate the high strain rate response of high strength steels. However, the use of pressurized oil/water used in this process make the process complicated.

2.3.2 MODIFIED SPLIT HOPKINSON PRESSURE BAR

2.3.2.1 Working Principle

In the conventional SHPB apparatus, the specimen is placed between the so-called input and output bars, and the dynamic pressure loading is generated by a striker bar (Smerd et al. 2006).

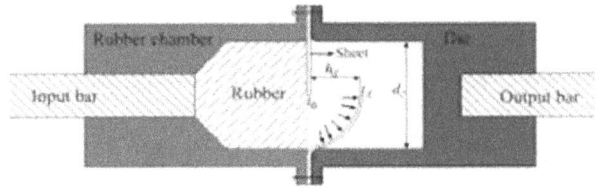

FIGURE 2.2 Modified SHPB setup used for dynamic bulging of sheet. (Reprinted with permission from Ramezani and Ripin 2010, Elsevier).

In a recent study, Grolleau et al. (2008) modified the conventional SHPB by using a movable bulge cell to perform dynamic bulge testing of aluminium sheets. In this work, the bulge cell was filled with a fluid and when the pressure wave transmitted through the fluid from the input bar, it caused the dynamic bulging of the sheet specimen. They conducted dynamic bulge studies on aluminium sheets for plastic strain rates of up to 500 s^{-1}. They looked into the experimental setup and measurement accuracy in depth and discovered that input bars made of low impedance materials must be required to obtain achievable pressure measurement accuracy. Further, an inverse analysis was utilised to determine the parameters of the rate-dependent material properties using a finite element model of the testing system.

In the further research, in order to eliminate the problem of the filling and removals of the fluid used in the bulge cell, Ramezani et al. (2010) performed the dynamic bulging of the copper sheet by using a polyurethane rubber inside the bulge cell (Figure 2.2). The use of hyper-elastic rubber as a pressure medium made the bulge test simple to perform. From the dynamic forming analysis of the copper sheet it was observed that the strength of the sheet increased with the increase in the strain rate. In an another study, Ramezani and Ripin (2010) developed a theoretical approach to attain pressure-strain curves during this high strain rate bulge forming process.

2.3.2.2 Advantages

The modified SHPB as a dynamic bulge testing device can be easily utilized in lab scale as well as in industries to characterize the material properties at different strain rates.

2.3.2.3 Limitations

The use of fluid as a pressure medium complicates the process. Further, in the current setup, Digital Image Correlation (DIC) methods would add benefits and eliminate the need for inverse modeling.

2.4 HIGH ENERGY RATE FORMING PROCESSES

During the high energy rate forming (HERF) operations, a large amount of energy is applied for a short period of time to deform the sheets to the final shape. Various HERF processes such as Electromagnetic Forming (EMF), Electrohydraulic Forming (EHF) and Explosive forming (EF) are utilized to investigate the biaxial forming behavior of the material at different strain rates. The HERF processes are not only used to encase safe forming of complex parts but also utilized to identify the limiting strain. All of the HERF processes increase the forming parameters and highlighted three distinct mechanisms that could account for it: (a) the inertial effect generated during the experiment diffuses the neck development and it leads to an increase in ductility, (b) the variation in the material constitutive behavior leads to increase in the formability, and (c) in closed die forming, the impact of the material with the die wall at high velocity causes the material to spread radially and it suppresses the necking and damage evolution. A detailed literature review on various HERF processes is given below.

2.4.1 ELECTROMAGNETIC FORMING

2.4.1.1 Working Principle

An electromagnetic forming (EMF) is a non-contact technology that uses pure electromagnetic interaction to impart significant forces to any electrically conductive work-piece (Imbert et al. 2005). EMF setup consists of a capacitor bank, a conductive coil, and the metallic sheet to be deformed (Figure 2.3). The forming coil, which is close to the metallic sheet is connected to the capacitor bank. By charging a large capacitor to a high voltage, it stores a significant amount of energy (typically between 5 and 200 kJ). The charge is switched over low inductance conductive bus work through a coil or actuator. Large currents run through the coil and the peak current is normally in the range of 10 to 1000 kA. The time to peak current is measured in microseconds. This produces a very strong transient magnetic field in the vicinity of the coil. The magnetic field induces eddy currents in any nearby conductive materials in the similar way the primary circuit of a transformer induces voltage and current in the secondary (Psyk et al. 2011). As a result, currents will be induced in any metallic sheet nearby and these will normally be opposite in direction to the primary current. According to Lenz's law, the currents in the coil and the metallic sheet move in the opposite directions. The Lorentz force governs the electromagnetic repulsion between oppositely flowing currents, which produces the deformation in the metallic sheet. This can eventually lead the sheet to deform plastically at a speed more than 100 m/s. (Oliveira et al. 2005).

2.4.1.2 Process Parameters and Their Influences

There has been a surge in interest in high-speed forming of light-weight materials in the recent years, which has accelerated the use of the EMF process in many industries. There are several process parameters which have influenced the dynamic forming behavior of the material are discussed below.

Al-Hassani et al. (1974) investigated the sheet forming by EMF of aluminium sheets using four types of spiral coils on a die having holes. During this analysis, the magnetic pressure distribution was anticipated by measuring the height of the dimples formed. Further, they estimated the magnetic pressure distribution during the process analytically. Padmanabhan (1997) studied the ways to prevent wrinkling and springback during EMF. The influence of various factors on the springback and wrinkling were investigated and observed that the increase in the magnetic pressure would reduce the wrinkling. Imbert et al. (2005) investigated the effect of tool-sheet interaction on the sheet formability. During this study, the sheet was formed freely and by using a conical die. The influence of tool-sheet interaction on the formability was studied by strain measurement. It is observed that employing a conical die increased the sheet's formability, particularly in locations where the die and sheet come into contact. Golovashchenko (2007) performed comparative analysis between the traditional forming limit diagram (FLD) obtained by stretch of aluminium sheet with the hemispherical punch to the results obtained by EMF. They reported a significant increase in the formability of the sheet during the high strain rate forming condition. It was noticed that the forming in a V-shape die or into a conical die further increased the formability of the material. Wang et al. (2017) studied

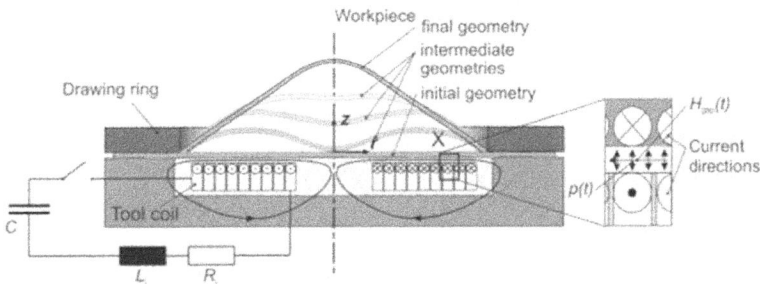

FIGURE 2.3 Electromagnetic sheet metal forming setup. (Reprinted with permission from Psyk et al. 2011, Elsevier).

the effect of EM bulging on the fatigue behavior of the AA 5052 sheets and reported that the bulged specimens exhibited enhanced fatigue strength as compared to the original Al alloy. From the TEM analysis, it was observed that the enhancement in the fatigue strength and increased resistance to fatigue crack propagation was caused due to strain hardening and dislocation shielding effect. Meng et al. (2011) performed warm and electromagnetic hybrid forming on Mg alloy sheets and studied the effects of voltage, capacity, and temperature on the bulge height. A rise in formability was evident during the rise in forming temperature. Moreover, it was noticed that with the constant discharge energy, the bulging height first decreased (<150°C) and then increased (>150°C) from room temperature to 230°C. Xu et al. (2013) studied the effect of process parameters on the formability of AZ31 Mg alloy during EMF and reported that the bulging height is directly proportional to the discharging energy which was tuned by discharge voltage and capacitance. The limit dome height was also compared with the quasi-static forming and a significant rise in formability was noted because of the inertial effect associated with EMF. Further, with the use of an Al driver sheet, the discharging energy was reduced nearly by 50% while forming the sheet to same dome height.

2.4.1.3   Advantages

EMF has many advantages that make this process an alternative technique of conventional forming processes. Among all the high-speed forming techniques EMF is one of the most suitable techniques of plastic deformation. The major advantages of EMF are listed below.

i.   During EMF, the sheet metal experiences improvement in mechanical properties in terms of reduction in springback and wrinkling. Stretching exceeds conventional process limits thus improving the ductility of the material.

ii. In EMF, the deformation occurs as a result of high pressure generated by the tool coil against the metal sheet for a fraction of second. As a result, rather than accommodating additional punch forces, fixture design should be focused on accelerating or decelerating the low-mass thin sheet. This simplifies the tooling and fixture design associated with EMF process.

iii. The process of forming becomes environmental friendly because of the no use of lubricants. This also makes the operation zone out of messy and stinky environment.

2.4.1.4   Limitations

Besides several advantages, there are some disadvantages of the EMF process:

i.   It is restricted to the materials with a high electrical conductivity. However, non-conducting or low-conductive material can be formed by the help of a driver sheet. Further, this adds to the operation's complexity.

ii. Due to large pressure involved in the process, the tool life is short.

iii. The EMF is relatively inefficient, as only around 20% of the charging energy is really used for plastic deformation.

iv. Because of the enormous currents and voltages involved in EMF operation, there are many safety risks involved in it.

2.4.1.5   Applications

The demand of sheet and tubular components in the current automotive, aerospace and nuclear industries require high productive, low-cost and robust forming processes. In such cases, the EMF can produce parts with complex shape with high accuracy.

2.4.2   Electrohydraulic Forming

2.4.2.1   Working Principle

Electrohydraulic forming (EHF) is one of the high-strain rate forming processes in which the electrical energy is converted into work without involving the magnetic field (Ahmed et al. 2017).

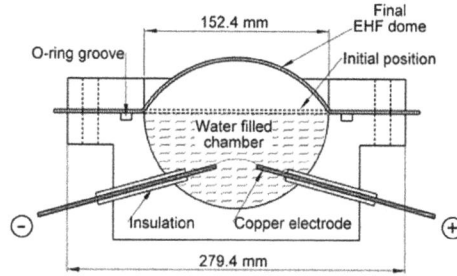

FIGURE 2.4 Schematic illustration of Electrohydraulic forming. (Reprinted with permission from Rohatgi et al. 2012, Elsevier).

The basic working principle of EHF and explosive forming (EF) is similar, with the exception that EHF uses an electrical discharge between the two electrodes as the driving force rather than a chemical explosion. The schematic representation of EHF device is shown in Figure 2.4. During the electrical discharge, the majority of the energy held in the capacitors is released in a short period of time (usually around 100 μs). As a result, the liquid surrounding the electrode's bridge wire vaporizes and results a high-velocity shock wave. When the shock wave propagates through the water in the chamber, it forces the sheet metal into the die (Rohatgi et al. 2012).

The EHF process can be carried out either in free condition, which is represented as electrohydraulic free forming (EHFF), or with the use of a die, which is called as electrohydraulic die forming (EHDF) (Rohatgi et al. 2012). During EHFF, the sheet decelerates at the end of the forming and it results in the decrease in the strain rates at the end of the operation. Further, in EHDF process, if the initial energy is insufficient to fully form the part, the blank speed might be affected and there would be no formability improvement. The main limitation in EHF is the lack of the availability of a bank of capacitors that store sufficient energy. The amount of energy required for a discharge highly dependent on the part's size and the material grade of the sheet metal. Most of the time, EHF is best suited for small to medium-sized sheets and tubes with relatively thin wall thickness (Golovashchenko et al. 2013).

2.4.2.2 Process Parameters and Their Influences

The EHF is more flexible manufacturing process than EMF, because it is the most economical and it does not require an expensive machine and complicated dies as in the case of EMF. Several consecutive electrical discharges in EHF can be utilised to completely fill the die of calibrate the springback without removing the die. There are several process parameters which affect the forming condition of sheet metal are discussed below.

Balanethiram and Daehn (1992) reported the increase in the formability of interstitial free steel when deformed at high strain rate using EHF. The forming outputs were compared with the FLD obtained from quasi static punch stretching test and observed a threefold improvement in in-plane failure strain without any substantial change in the material constitutive behavior. It was argued that the improvement is due to the effect of inertia in stabilizing neck growth and not due to change in material constitutive behavior. In another study, Balanethiram et al. (1994) investigated the improvement of formability in EHF as a hyperplastic phenomenon. Using conical dies with an apex angle of 90° in biaxial stretching, the formability of Al 6061-T4, oxygen-free high-conductive (OFHC) copper and interstitial free iron was instigated. The axisymmetric expansion of Al 6061-T4 ring was also investigated in order to observe the material's uniaxial behavior at high strain rates. Biaxial deformation showed a significant improvement in formability, but uniaxial deformation showed just slight improvement. Homberg et al. (2010) successfully achieved the forming of sharp edge contour of high strength steel using EHF process. During this analysis. They formed a smaller radius of 0.8 mm by a discharge energy of less than 6 kJ, whereas the minimum radius of curvature formed in the case of quasi-static hydroforming was 1.75 mm. Ahmed et al. (2017) used the EHF technique to test the formability of AA5052 alloys in terms of FLD and compared it to the traditional quasi-static forming results. The limit strains were confirmed to have increased by nearly 45–50% as a result of the inertial stabilization. Rohatgi et al. (2012) investigated the influence of open (free forming) and conical closed dies in EHF of Al and DP600 materials with the help of deformation history. It was discovered that conical closed die produces better results when compared to open die. Golovashchenko et al. (2013) demonstrated a proof of concept for electrohydraulic calibration to reduce springback of DP 980 and DP600 steels for obtaining actual geometry using an industrial tool that were optimized using numerical modelling. A stamped panel was clamped to a calibration die surface with a target shape, internal stress in the stamped component was relieved with a pulse of pressure. A pressure pulse from a newly built electrohydraulic chamber was used to calibrate the bent shape for springback.

2.4.2.3 Advantages
Major advantages of EHF are listed below.

  i. In EHF, several successive discharges can be utilised to fill the die completely or to calibrate springback without removing the die.
 ii. Unlike EMF, EHF can be used to deform material with lower conductivity.
iii. The high-velocity forming environment improves the formability of the material as compared to conventional forming processes. The absence the tool friction stretches the material towards larger limit strain.
 iv. The elimination of tooling and fixture design reduces the complexity of the process and it is reliable and well suited for high-volume production.

2.4.2.4 Limitations
Besides several advantages, there are some disadvantages of the EHF process:

  i. The volume of the equipment is too large and it requires large area to keep.
 ii. The high speed of the process requires more safety and skilled personnel.
iii. Higher thickness of the work piece is difficult to form.
 iv. The sheet profile is usually convex after forming.

2.4.2.5 Applications
From large body panels to miniature and fancy equipment with complicated profile in automotive and aerospace industries can be easily manufactured by EHF with higher accuracy.

2.4.3 Explosive Forming

2.4.3.1 Working Principle
Explosive Forming (EF) is likely the most widely used high-strain rate forming process, particularly in the aerospace industries (Mynors and Zhang 2002). The pressure required to shape the work-piece in EF is generated by an explosive charge submerged in a liquid medium, most often water (Figure 2.5). Pressure waves propagate through the liquid medium after detonation and reach the sheet surface, forming it into a die or into free space. EF is widely used in numerous industries to create complex parts from high strength steel and aluminium alloys, due to its excellent repeatability and ability to produce large and geometrically complex parts with good accuracy.

2.4.3.2 Process Parameters and Their Influences
Fengman et al. (2000) utilized the a non-die explosive forming technique to fabricate thin-wall spherical parts which was difficult to form in conventional forming procedure. The process showed

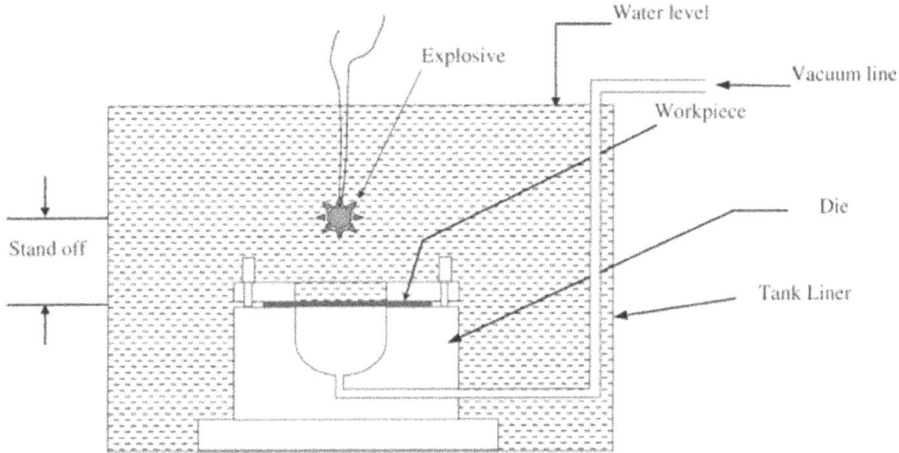

FIGURE 2.5 Schematic illustration of explosive forming. (Reprinted with permission from Mynors and Zhang 2002, Elsevier).

high dimensional accuracy with less wrinkling than conventional punch forming and did not require any special set of dies and forming devices. Dariani et al. (2009) studied the effect of forming velocity on FLDs and investigated experimentally for Al 6061-T6 and AISI 1045 sheets. Further, the FLDs obtained at low impact (50/s) as well as explosive free-forming (1000/s) were compared with the conventional FLDs at quasi-static condition. The results showed a substantial improvement in formability at high strain rate for aluminium sheets, whereas it was not considerable for steel sheets. In a study, Babaei et al. (2015) used impulse loading from the detonation of the oxygen and acetylene mixture at varied volume ratios to conduct experiments on a clamped circular mild steel sheet. For their experiments, they developed an analytical and empirical model to show how the mechanical properties of the plate and gas, the impulse of applied load, plate geometry, the velocity of sound in different gases and the strain-rate sensitivity affect the large deformation of circular plates in high rate energy forming. Further, Spranghers et al. (2012) analyzed the full field measurements of transient nature deformation of aluminium plates under free air blast loading conditions. 3D high-speed DIC system was used for the assessment of the dynamic response of the structure such as surface displacement and deformation accurately at the blast loading condition.

Besides several advantages of the EF process, the most common limitations observed are higher capital cost, complexity in instrumentation and difficulties in handling. In order to minimize these limitations and to study the biaxial dynamic properties of the material, the shock tube has been introduced in various research works which works on the same phenomenon (Kumar et al. 2012).

2.4.4 SHOCK TUBE BASED FORMING

2.4.4.1 Working Principle

A shock tube is a device that creates a gas flow condition that is difficult to achieve with conventional gas flow devices. It is a laboratory apparatus that can produce high velocity, pressure, and temperature conditions for a short interval of time, allowing it to be used in the aerodynamic and thermo-chemical engineering. It is made up of a long rigid cylinder that is split into two sections: a high-pressure driver section and a low-pressure driven section. A diaphragm made of a metal sheet or by the layers of thin Mylar sheets separates the two portions. The diaphragm ruptures when the pressure difference between the driver and driven sections reaches a threshold value, causing a rapid release of gas and it creates a shock wave inside the tube. It propagates with a Mach number greater than one along the driven section. The schematic illustration of the pressure variation inside the shock tube is represented in Figure 2.6.

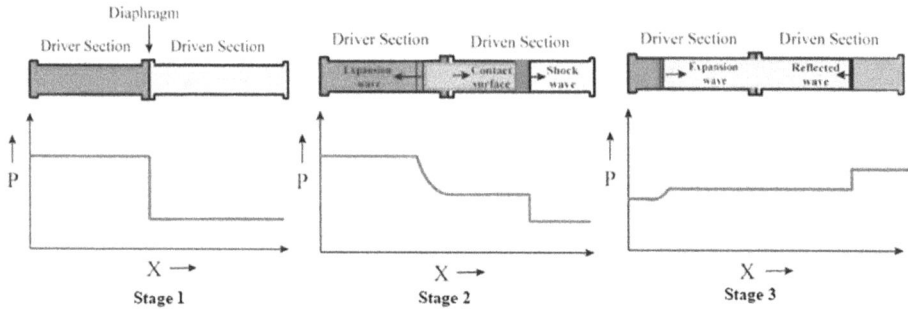

FIGURE 2.6 Working principle of shock tube.

The magnitude of shock Mach number (M_s) is decided by various factors. The main factor is the thickness of the diaphragm. In most of the studies, constant layers of Mylar sheets have been considered to generate the required magnitude of Mach number repetitively (Barik et al. 2021). The speed of the shock wave is also influenced by the gas used in the driver section. For example, during an experiment, for a constant bursting pressure (21.87 bar), the heavier gas such as nitrogen (N_2) travels slowly down the tube resulting in lower Mach number ($M_s = 1.86$), whereas lighter gas such as helium (He) allows for higher Mach number ($M_s = 2.5$). Applying vacuum prior to bursting across the driven section also influences the shock wave to travel faster than the normal speed. The propagation of the shock wave generates a high-pressure and high-temperature gas behind it. The equations describing the change in pressure and temperature with respect to the speed of the shock wave are well described in the work of Kumar et al. (2012). When the shock wave travels down the tube and imparts on an end wall, it is reflected back by producing a higher pressure and temperature gas behind the reflected shock wave.

In the shock tube, the high-pressure field zone is developed for a short interval of time after the incident shock wave is reflected back. The shock tube's end wall can be replaced with a sheet metal and this advantage may be used to perform typical bulge tests at various strain rates. This is the same as performing a bulge test with forced air rather than a hydraulic fluid Justusson et al. (2013). The schematic illustration of the shock tube facility as a dynamic bulging device used by Barik et al. (2020a) is represented in Figure 2.7. Generally, hydraulic bulge tests are confined to low strain rates,

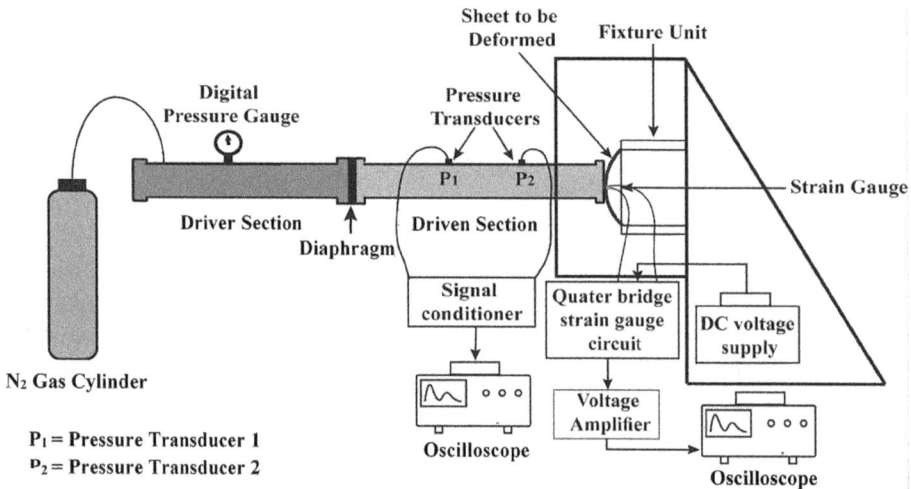

FIGURE 2.7 Schematic representation of the shock tube facility. (Reprinted with permission from Barik et al. 2020a, Elsevier).

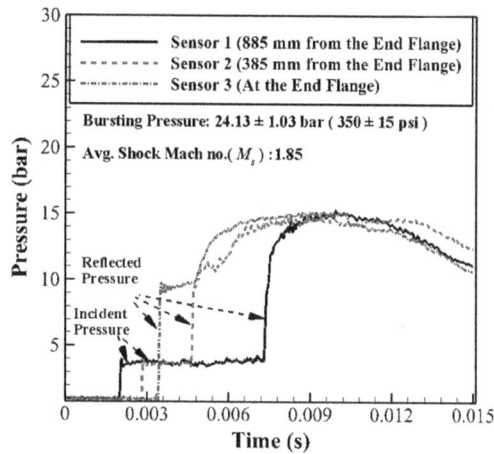

FIGURE 2.8 Pressure signals obtained for a shock tube experiment for a bursting pressure 24.13 bar. (Reprinted with permission from Barik et al. 2020a, Elsevier).

whereas in shock tube, it can be used to investigate the forming behavior of sheet metals at intermediate and high strain rates by varying the diaphragm thickness or the driver gas.

In shock tube experimental facility, the busting pressure is measured by a digital pressure gauge. Pressure transducers are mounted on the shock tube facility to measure the incident pressure, reflected pressure and shock Mach number developed during the experiment. A typical pressure signal captured by Barik et al. (2020a) during an experiment with a bursting pressure of 24.13 bar is represented in Figure 2.8. In order to determine the transient variation of forming parameters, high-speed 3D digital image correlation (DIC) system has been used in many cases (Kumar et al. 2012). In some studies, strain gauge has also been used to measure the strain rate and the results are well validated with DIC results (Louar et al. 2015).

2.4.4.2 Process Parameters and Their Influences

In many recent studies, the shock tube facility has been utilized to study the dynamic behavior of sheets and composite materials. The major process parameter is the bursting pressure of the driver gas, which is mainly decided by the diaphragm used during the experiment. The effect of impulsive loading on the dynamic forming behavior of different materials are discussed below.

Kumar et al. (2012) conducted both the experimental and numerical investigation on the effect of plate curvature on blast response of aluminium panels using a shock tube. A 3D digital image correlation (DIC) technique coupled with high-speed photography was used to instigate out-of-plane deformation, velocity and the in-plane strain generated during experiment. The transient response of the panels during impact loading was evaluated by a dynamic computational simulation by incorporating fluid-structure interaction. Stoffel et al. (2001) utilized a shock tube during the study of the dynamic behavior of the steel sheet. Further, a material model was developed by considering the elastic-plastic behavior, isotropic and kinematic hardening, and strain rate sensitivity of the material and the evolution of deflection, stresses, and plastic zones are predicted and validated with the experimental results. Justusson et al. (2013) retrieved the biaxial rate-dependent mechanical properties of thin aluminium sheets using a shock tube. The DIC technique is used to record the out of plane deflection and strain fields. the rate dependent constitutive properties of the material were determined using an inverse modeling technique in conjunction with the finite element (FE) simulation.

The intra-granular misorientation on aluminium sheet while studying the effect of shock wave based deformation at the microstructural level was observed by Ray et al. (2015). During this study, the improvement in hardness indicated the absence of recovery and strain hardening. Together with this, Bisht et al. (2019) used the same shock wave assisted deformation process to understand the microstructural response of FCC metals and also elucidated the role of stacking fault energy on the evolution of microstructure and texture. Barik et al. (2020a) used the shock tube facility to investigate the forming behavior of pre-strained AA 5052-H32 sheets. Further, in another study Barik et al. (2020b), the strain evolution, effective stress and strain distribution of AA 5052-H32 sheets obtained from the shock tube experiments are validated with the FE simulation results.

2.4.4.3 Advantages
Major advantages of shock tube based forming are listed below.

   i. Uniformity in loading and ease of handling make the shock tube based forming useful to study the dynamic behavior of sheet metals.
   ii. It can be utilized to study the rate-dependent forming behavior of the material from intermediate strain rate range to high strain rates.
   iii. As the pressure energy developed because of rapid motion of shock wave is responsible for the deformation, any material, both conductive and non-conductive can be deformed at a high strain rate.
   iv. The inertial effect developed due to impulsive loading condition stretches the sheet metal significantly without strain localization and the stretched material has higher formability than the conventional forming.
   v. The process is environmental friendly and can be used in laboratory scale as a dynamic bulging device by following safety precaution.

2.4.4.4 Limitations

   i. The repeatability of shock tube experiments is not so accurate because of the use of Mylar sheets as the diaphragm. However, this error can be eliminated by mounting a quick opening valve instead of Mylar sheets in the diaphragm section.
   ii. Deforming sheets up to failure is based on the capacity of the shock tube facility.
   iii. The bursting of diaphragm creates large noise during experiment.

2.4.4.5 Applications
The shock tube facility can be used in lab scale as well as in industries to study the dynamic behavior of sheet metals. Further, addition of a closed end die at the end of the shock tube can be useful to fabricate an end product with low cost and high accuracy.

2.4.4.6 Forming Behavior and Failure Response Analysis using a Shock Tube
It is observed from the working principle of the shock tube that the shock wave increases the pressure and temperature of the gas behind it and induces a mass motion. In a recent study, Barik et al. (2019) have used the impulsive induced gas to move a hemispherical end nylon bar and performed impact loading on AA 5052-H32 sheets. The schematic of the modified shock tube used by the authors is represented in Figure 2.9. The pressure energy of the induced gas has been used to drive the nylon bar towards the end of the shock tube at a high velocity. It is explored as a dynamic Erichsen testing device where a high-velocity hemispherical end striker is utilized to deform the sheet at different strain rates. This device has been used to study the forming behavior with in the safe limit and beyond the safe limit for better assessment of the dynamic behavior of the material. The deformed samples at different velocities are shown in Figure 2.10.

Further, AA 5052-H32 sheet has been stretched to failure in order to study the increase in the forming limits during the high-velocity shock tube experiments. The distribution of failure strains for both high-speed forming and quasi-static forming have been compared and observed a considerable improvement in the formability during shock tube based forming (Figure 2.11). The increase in the formability is due to the inertial forces involved with the process, which help the material to avoid sharp local velocity gradient on the material surface, allowing it to stretch further without strain localization.

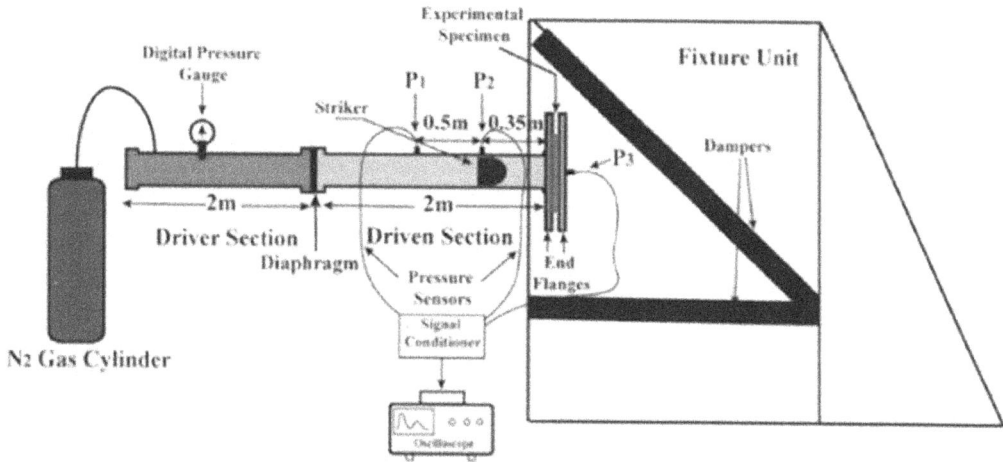

FIGURE 2.9 Shock tube facility used as a dynamic Erichsen testing device. (Reprinted with permission from Barik et al. 2019, Springer).

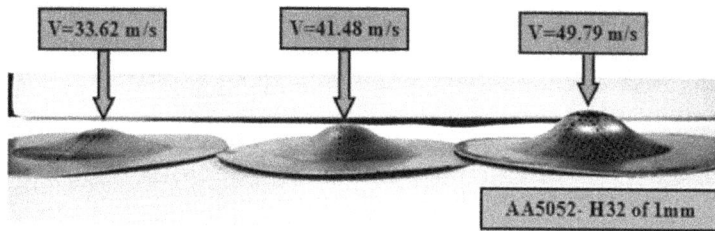

FIGURE 2.10 AA 5052-H32 sheets deformed under different velocity conditions. (Reprinted with permission from Barik et al. 2019, Springer).

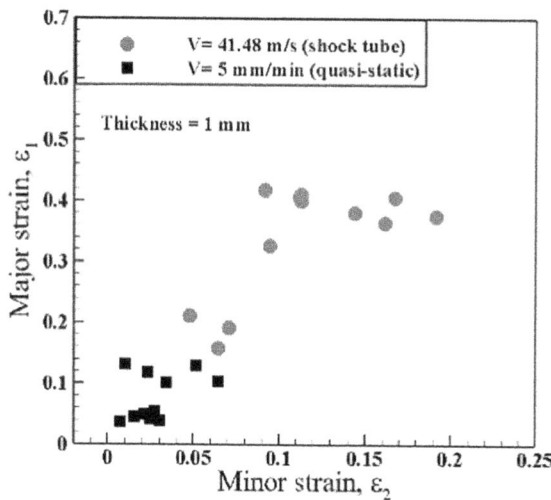

FIGURE 2.11 Comparison of limit strains in quasi-static and shock tube based experiments. (Reprinted with permission from Barik et al. 2019, Springer).

REFERENCES

Abd El-Aty, A., Xu, Y., Zhang, S. H., Ha, S., Ma, Y., & Chen, D. (2019). Impact of high strain rate deformation on the mechanical behavior, fracture mechanisms and anisotropic response of 2060 Al-Cu-Li alloy. *Journal of Advanced Research*, *18*, 19–37.

Ahmed, M., Kumar, D. R., & Nabi, M. (2017). Enhancement of formability of AA5052 alloy sheets by electro-hydraulic forming process. *Journal of Materials Engineering and Performance*, *26*(1), 439–452.

Al-Hassani, S. T. S., Duncan, J. L., & Johnson, W. (1974). On the parameters of the magnetic forming process. *Journal of Mechanical Engineering Science*, *16*(1), 1–9.

Babaei, H., Mostofi, T. M., & Sadraei, S. H. (2015). Effect of gas detonation on response of circular plate-experimental and theoretical. *Structural Engineering and Mechanics*, *56*(4), 535–548.

Balanethiram, V. S., & Daehn, G. S. (1992). Enhanced formability of interstitial free iron at high strain rates. *Scripta Metallurgica et Materiala*, *27*(12), 1783–1788.

Balanethiram, V. S., Hu, X., Altynova, M., & Daehn, G. S. (1994). Hyperplasticity: enhanced formability at high rates. *Journal of Materials Processing Technology*, *45*(1–4), 595–600.

Barik, S. K., Narayanan, R. G., & Sahoo, N. (2021). Failure strain and fracture prediction during shock tube impact forming of AA 5052-H32 sheet. *Journal of Engineering Materials and Technology*, *143*(3), p. 031009.

Barik, S. K., Narayanan, R. G., & Sahoo, N. (2019). Experimental investigation on the forming of AA 5052-H32 sheet using a rigid-body-based impact in a shock tube. In *Advances in Forming, Machining and Automation* (pp. 79–90). Springer.

Barik, S. K., Narayanan, R. G., & Sahoo, N. (2020a). Forming response of AA5052--H32 sheet deformed using a shock tube. *Transactions of Nonferrous Metals Society of China*, *30*(3), 603–618.

Barik, S. K., Narayanan, R. G., & Sahoo, N. (2020b). Prediction of forming of AA 5052-H32 sheets under impact loading and experimental validation. *Journal of Materials Engineering and Performance*, *29*(6), 3941–3960.

Bisht, A., Kumar, L., Subburaj, J., Jagadeesh, G., & Suwas, S. (2019). Effect of stacking fault energy on the evolution of microstructure and texture during blast assisted deformation of FCC materials. *Journal of Materials Processing Technology*, *271*, 568–583.

Broomhead, P., & Grieve, R. J. (1982). The effect of strain rate on the strain to fracture of a sheet steel under biaxial tensile stress conditions. *Transactions of ASME: The Journal of Engineering Materials and Technology*, *104*(1), 102–106.

Dariani, B. M., Liaghat, G. H., & Gerdooei, M. (2009). Experimental investigation of sheet metal formability under various strain rates. *Proceedings of the Institution of Mechanical Engineers, Part B: Journal of Engineering Manufacture*, *223*(6), 703–712.

Dorward, R. C., & Hasse, K. R. (1995). Strain rate effects on tensile deformation of 2024-0 and 7075-0 aluminum alloy sheet. *Journal of Materials Engineering and Performance*, *4*(2), 216–220.

Fengman, H., Zheng, T., Ning, W., & Zhiyong, H. (2000). Explosive forming of thin-wall semi-spherical parts. *Materials Letters*, *45*(2), 133–137.

Fridlyander, I. N., Sister, V. G., Grushko, O. E., Berstenev, V. V., Sheveleva, L. M., & Ivanova, L. A. (2002). Aluminum alloys: promising materials in the automotive industry. *Metal Science and Heat Treatment*, *44*(9–10), 365–370.

Ghosh, A. K. (1977). The influence of strain hardening and strain-rate sensitivity on sheet metal forming. *Journal of Engineering Materials and Technology*, *99*(3): 264–274.

Golovashchenko, S. F. (2007). Material formability and coil design in electromagnetic forming. *Journal of Materials Engineering and Performance*, *16*(3), 314–320.

Golovashchenko, S. F., Gillard, A. J., & Mamutov, A. V. (2013). Formability of dual phase steels in electrohydraulic forming. *Journal of Materials Processing Technology*, *213*(7), 1191–1212.

Grolleau, V., Gary, G., & Mohr, D. (2008). Biaxial testing of sheet materials at high strain rates using viscoelastic bars. *Experimental Mechanics*, *48*(3), 293–306.

Gu, R., Liu, Q., Chen, S., Wang, W., & Wei, X. (2019). Study on high-temperature mechanical properties and forming limit diagram of 7075 aluminum alloy sheet in hot stamping. *Journal of Materials Engineering and Performance*, *28*(12), 7259–7272.

Homberg, W., Beerwald, C., & Pröbsting, A. (2010). Investigation of the electrohydraulic forming process with respect to the design of sharp edged contours. *4th International Conference on High Speed Forming*, 58–64.

Hsu, E., Carsley, J. E., & Verma, R. (2008). Development of forming limit diagrams of aluminum and magnesium sheet alloys at elevated temperatures. *Journal of Materials Engineering and Performance*, *17*(3), 288–296.

Imbert, J. M., Winkler, S. L., Worswick, M. J., Oliveira, D. A., & Golovashchenko, S. (2005). The effect of tool--sheet interaction on damage evolution in electromagnetic forming of aluminum alloy sheet. *Journal of Engineering Materials and Technology*, *127*(1), 145–153.

Joost, W. J. (2012). Reducing vehicle weight and improving US energy efficiency using integrated computational materials engineering. *Jom*, *64*(9), 1032–1038.

Justusson, B., Pankow, M., Heinrich, C., Rudolph, M., & Waas, A. M. (2013). Use of a shock tube to determine the bi-axial yield of an aluminum alloy under high rates. *International Journal of Impact Engineering*, *58*, 55–65.

Koç, M., Billur, E., & Cora, Ö. N. (2011). An experimental study on the comparative assessment of hydraulic bulge test analysis methods. *Materials and Design*, *32*(1), 272–281.

Kumar, P., LeBlanc, J., Stargel, D. S., & Shukla, A. (2012). Effect of plate curvature on blast response of aluminum panels. *International Journal of Impact Engineering*, *46*, 74–85.

Louar, M. A., Belkassem, B., Ousji, H., Spranghers, K., Kakogiannis, D., Pyl, L., & Vantomme, J. (2015). Explosive driven shock tube loading of aluminium plates: experimental study. *International Journal of Impact Engineering*, *86*, 111–123.

Mahabunphachai, S., & Koç, M. (2010). Investigations on forming of aluminum 5052 and 6061 sheet alloys at warm temperatures. *Materials and Design*, *31*(5), 2422–2434.

Meng, Z., Huang, S., Hu, J., Huang, W., & Xia, Z. (2011). Effects of process parameters on warm and electromagnetic hybrid forming of magnesium alloy sheets. *Journal of Materials Processing Technology*, *211*(5), 863–867.

Mynors, D. J., & Zhang, B. (2002). Applications and capabilities of explosive forming. *Journal of Materials Processing Technology*, *125*, 1–25.

Oliveira, D. A., Worswick, M. J., Finn, M., & Newman, D. (2005). Electromagnetic forming of aluminum alloy sheet: free-form and cavity fill experiments and model. *Journal of Materials Processing Technology*, *170*(1–2), 350–362.

Padmanabhan, M. (1997). Wrinkling and Springback in Electromagnetic Sheet Metal Forming and electromagnetic Ring Compression. (Electronic Thesis or Dissertation). Retrieved from https://etd.ohiolink.edu/

Percy, J. H., & Lim, S. G. (1983). Forming limits for mild steel sheet at high forming rates. *CIRP Annals*, *32*(1), 177–180.

Pickett, A. K., Pyttel, T., Payen, F., Lauro, F., Petrinic, N., Werner, H., & Christlein, J. (2004). Failure prediction for advanced crashworthiness of transportation vehicles. *International Journal of Impact Engineering*, *30*(7), 853–872.

Psyk, V., Risch, D., Kinsey, B. L., Tekkaya, A. E., & Kleiner, M. (2011). Electromagnetic forming—a review. *Journal of Materials Processing Technology*, *211*(5), 787–829.

Ramezani, M., & Ripin, Z. M. (2010). Combined experimental and numerical analysis of bulge test at high strain rates using split Hopkinson pressure bar apparatus. *Journal of Materials Processing Technology*, *210*(8), 1061–1069.

Ramezani, M., Ripin, Z. M., & Ahmad, R. (2010). Plastic bulging of sheet metals at high strain rates. *The International Journal of Advanced Manufacturing Technology*, *48*(9–12), 847–858.

Ray, N., Jagadeesh, G., & Suwas, S. (2015). Response of shock wave deformation in AA5086 aluminum alloy. *Materials Science and Engineering A*, *622*, 219–227.

Rohatgi, A., Stephens, E. V., Davies, R. W., Smith, M. T., & Soulami, A. (2012a). Journal of Materials Processing Technology Electro-hydraulic forming of sheet metals: free-forming vs. conical-die forming. *Journal of Materials Processing Technology*, *212*(5), 1070–1079.

Rohatgi, A., Stephens, E. V., Davies, R. W., Smith, M. T., Soulami, A., & Ahzi, S. (2012b). Electro-hydraulic forming of sheet metals: free-forming vs. conical-die forming. *Journal of Materials Processing Technology*, *212*(5), 1070–1079.

Smerd, R., Winkler, S., Salisbury, C., Worswick, M., Lloyd, D., & Finn, M. (2006). High strain rate tensile testing of automotive aluminum alloy sheet. *International Journal of Impact Engineering*, *32*(1–4), 541–560.

Spranghers, K., Vasilakos, I., Lecompte, D., Sol, H., & Vantomme, J. (2012). Full-field deformation measurements of aluminum plates under free air blast loading. *Experimental Mechanics*, *52*(9), 1371–1384.

Stoffel, M., Schmidt, R., & Weichert, D. (2001). Shock wave-loaded plates. *International Journal of Solids and Structures*, *38*(42–43), 7659–7680.

Wang, D. zhen, Li, N., Han, X. Tao, Li, L., & Liu, L. (2017). Effect of electromagnetic bulging on fatigue behavior of 5052 aluminum alloy. *Transactions of Nonferrous Metals Society of China (English Edition)*, *27*(6), 1224–1232.

Xu, J. R., Yu, H. P., & Li, C. F. (2013). Effects of process parameters on electromagnetic forming of AZ31 magnesium alloy sheets at room temperature. *The International Journal of Advanced Manufacturing Technology*, *66*(9–12), 1591–1602.

3 Bending of Sheet Metals
Challenges and Recent Developments

S. Deb and S. K. Panigrahi
Indian Institute of Technology, Madras, Chennai, India

CONTENTS

3.1   INTRODUCTION

Application of bending in the sheet metal forming industry is extensive. Automobile car body panel (indoor and outdoor), door and window frames in construction industry; bridges, ladders, stairs, scaffolders; bend profiles in defense sector are the few examples. The through thickness variation of stress is the common characteristics of all the bending dominated process. During bending, the inner radius of the sheet experience compression while the outer radius experience tension. In general, the bending processes are simple but can lead to several complexities in terms of determination of bend forming limit and accurate prediction of springback. The performance of the sheet metal under bending depends on the tooling, the process parameters and the characteristics of the sheet itself.

Sheet metals are produced by running slab of metal passing through series of rollers. The final thickness of the sheets depends on the percentage of rolling reduction, number of passes, the velocity of rollers and rolling temperature. Large number of sheets today is produced by hot-rolling

DOI: 10.1201/9781003226703-3

process which is usually performed above the recrystallization temperature. Hot rolling removes the porosities and the cavities of the block, breaks the grains, and results in formation of new grains. The sheet after hot rolling tends to shrink as it cools down. More dimensional control is achieved during cold rolling which actually started with hot rolling but the final stages are performed at near room temperature. Apart from dimensional stability, cold-rolled sheets have better surface finish, higher strength and hardness as compared to hot-rolled sheet.

The sheets metal products are the result of the secondary processing of the hot- or cold-rolled sheets, this includes blanking, piercing, bending, stamping, deep drawing, hole extrusion etc. Bending along a line to produce channel with U and V cross-section is one of the most common sheet metal forming process in industry. Compared to the bulk metal forming process, these processes are simple and can be performed at room temperature which eliminates the possible grain grown during bending. Moreover, loss of material is minimal during bending. In bulk forming, such as rolling, extrusion, drawing, forging etc., the intention is to reduce the thickness or the cross-sectional area of the specimen. But in bending, the deformation is limited to the region of bend line while the other part of the sheet remains almost untouched [1].

The focus of the current chapter is on bending and bending dominated forming processes. Unlike stamping and deep drawing, the through thickness variation of stress is the common characteristics of all the bending dominated process. So the formability under bending is different from that of uniaxial and biaxial tension and the bending limit cannot be determined from the standard Forming Limit Diagram (FLD). The failure in bending is marked as the appearance of crack on the surface under tension. The formation of cracks depends on the materials conditions as well as the tooling and the state of stress. The state of stress varies among different bending processes such as pure bending, three-point or four-point bending, roll forming, profile bending etc. As a result, a universal parameter to represent formability limit in bending is missing and the study of bending limit is based on the evaluation of various parameters. The present chapter deals with the various aspects of bending and at the end of the chapter following questions will be answered:

- Application of bending.
- Methods to the determine the bend forming limit and corresponding limitations.
- Effect of microstructure and mechanical properties on bend forming limit.
- Effect of sheet geometry and tooling on bend forming limit.
- Defects in bend products.
- Recent advancement in bending.

3.2 THEORY OF BENDING

In this section, simple theory of bending will be discussed briefly. Figure 3.1 demonstrates a common bending scenario where a sheet having a thickness of t is bend to an angle of θ and a curvature ρ by applying a moment M.

Under the applied moment, the line CD_0 changes to CD ($l_s = \rho\theta$) and the line AB_0 changes to AB ($l = \theta(\rho + y) = l_s(1 + \frac{y}{\rho})$) as shown in Figure 3.1. The strain at the fibre AB can be written as:

$$\varepsilon_1 = \ln\frac{l}{l_0} = \ln\frac{l_s}{l_0} + \ln\left(1 + \frac{y}{\rho}\right) = \varepsilon_a + \varepsilon_b \tag{3.1}$$

where ε_a is the axial strain at the mid-surface and ε_b is the bending strain.

$$\varepsilon_b = \ln\left(1 + \frac{y}{\rho}\right) \approx \frac{y}{\rho} \tag{3.2}$$

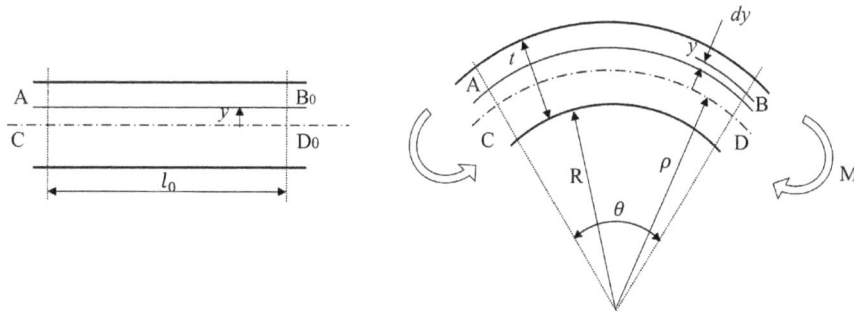

FIGURE 3.1 Schematic representation of bending.

As the material is not deforming along the width direction, the plane strain condition can be assumed as:

$$\varepsilon_1 = \frac{\sqrt{3}}{2}\bar{\varepsilon}; \ \varepsilon_2 = 0; \ \varepsilon_3 = -\varepsilon_1 \tag{3.3}$$

$$\sigma_1 = \frac{2}{\sqrt{3}}\bar{\sigma}; \sigma_2 = \frac{\sigma_1}{2}; \sigma_3 = 0 \tag{3.4}$$

where $\bar{\varepsilon}$ and $\bar{\sigma}$ are the effective strain and effective stress respectively. The bending moment can be calculated as (Figure 3.1):

$$M = \int_{-t/2}^{t/2} \sigma_1 y \, dy \tag{3.5}$$

The transition of the stress from compressive at the inner radius to tensile in the outer radius is the characteristics of bending as indicated in Equation 3.2. For radius below the neutral axis, "y" is negative representing compression zone while the for radius above neutral axis "y" is positive representing tensile zone and at "$y = 0$", the bending strain is "zero".

3.3   VARIOUS BENDING PROCESSES

Bending is performed extensively in the automotive, aircraft, and defense sectors. Figure 3.2–3.5 highlight various bending processes used in industry. The corresponding schematics provide the information regarding tooling, tool movement and direction of force required for bending. Most commonly, bending is employed to develop V profile using V-die bending. Other commonly used bending processes are Air bending, U profile bending, flanging and hemming of the edge. During the V-die bending, a flat sheet is bend against a V-die by a punch until the desired bend angle is achieved which is fixed by the angle of the V-die (Figure 3.2(a)). On the other hand, Air bending is a three-point bending (Figure 3.2(c)) where the sheet is not pressed against any die (Figure 3.2(d)). The accuracy of Air bending is less as compared to V-die, but re-tooling is not necessary in order to achieve different bend angle which in case of V-die is fixed by the die. The U bending is very similar to the V-die bending where a U-die is used instead of a V-die to produce a U shape from a flat sheet (Figure 3.2(b)). The Flanging, Rotary bending, Beading, Hemming, and Seaming are the edge bending processes (Figure 3.3). In flanging the edge of the sheet is usually bend up to 90°

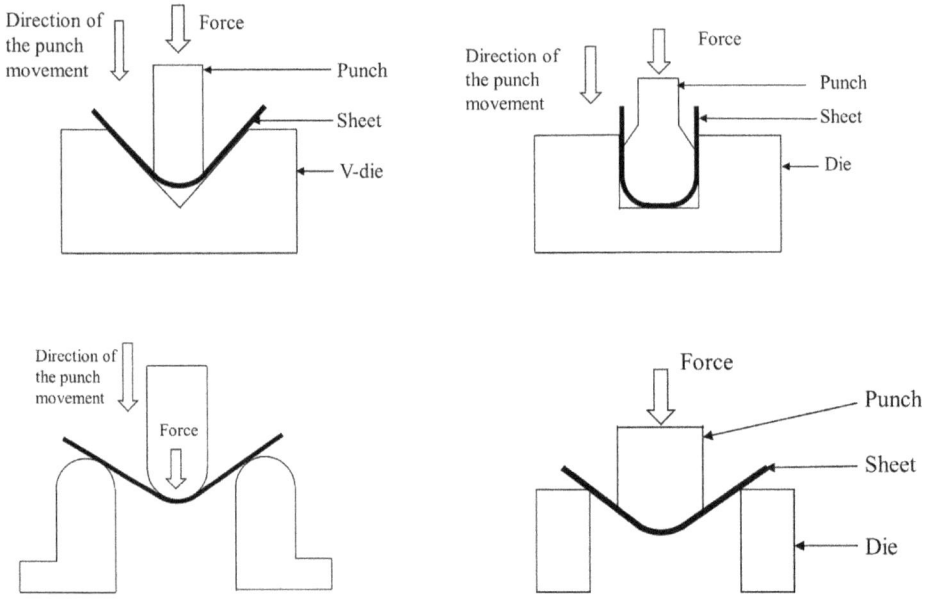

FIGURE 3.2   Schematic representation of various bending processes.

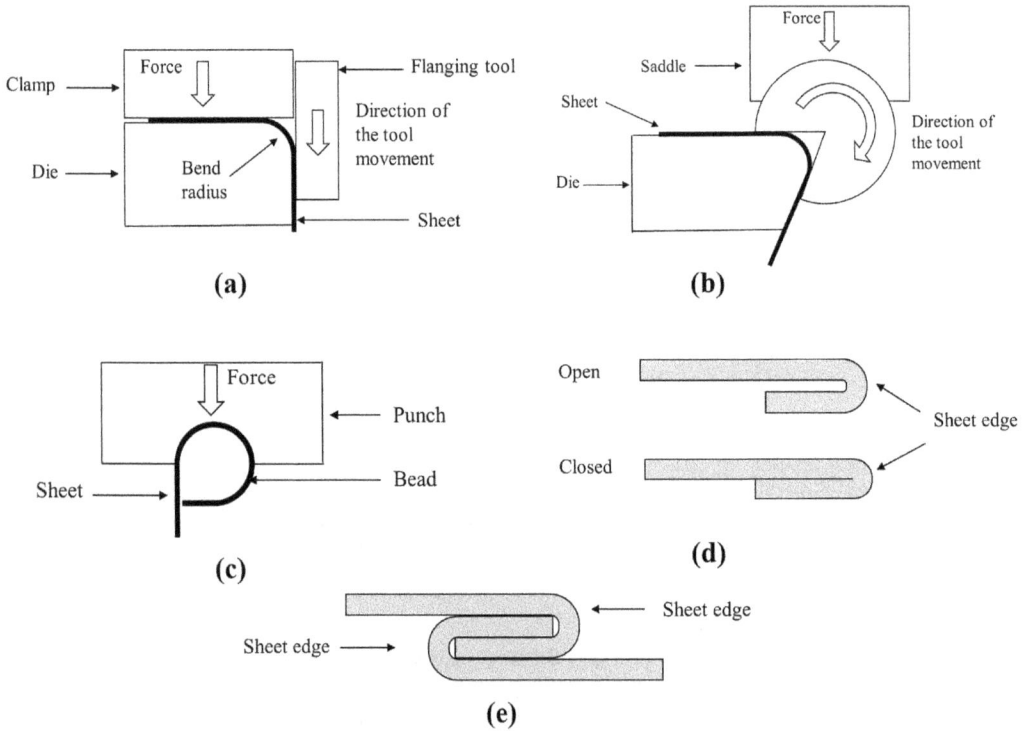

FIGURE 3.3   Schematic representation of (a) Flanging/Edge bending, (b) Rotary bending, (c) Beading, (d) Hemming, and (e) Seaming.

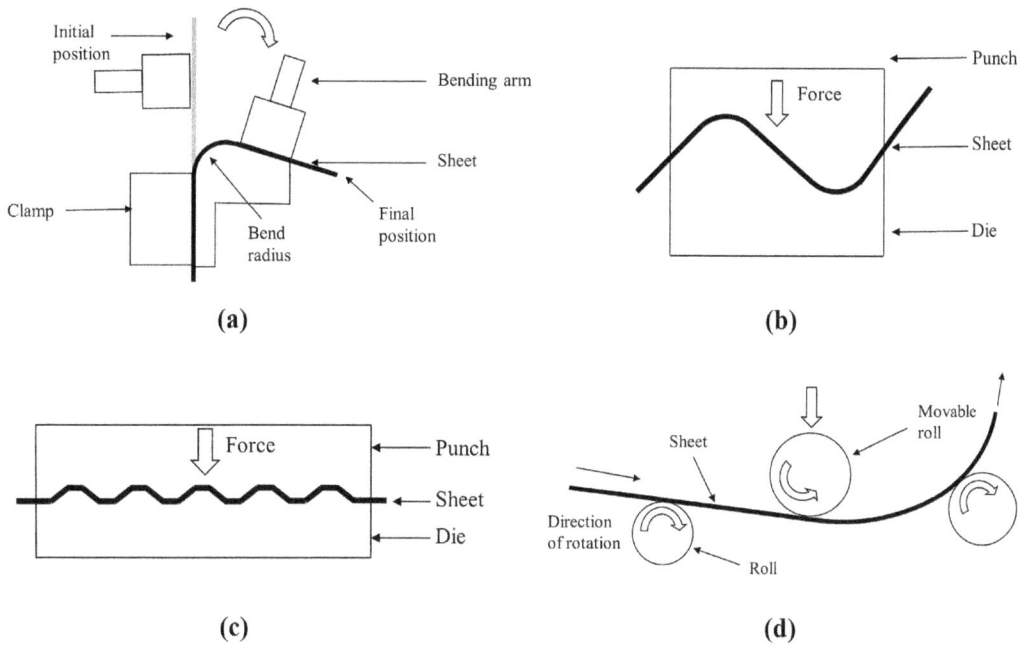

FIGURE 3.4 Schematic representation of (a) Wrap bending, (b) Offset bending, (c) Corrugating, and (d) Roll bending.

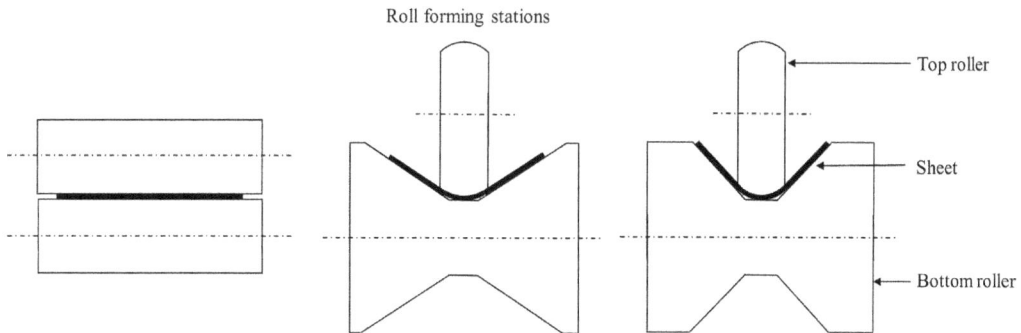

FIGURE 3.5 Schematic representation of roll forming.

whereas during hemming the edge is bend up to 180°. Other widely used bending operations are Wrap bending, Offset bending, Corrugating, and Roll bending (Figure 3.4). While these processes are generally performed in a single step, a multistep deformation is the key of roll forming process where the sheet incrementally bent along the parallel, straight and longitudinal bend line through successive sets of rollers (Figure 3.5) [2]. This is a continuous process which leads to high productivity. The process is versatile and suitable of producing complex cross-sectional profiles from flat sheets. Due to its incremental nature of deformation mode, the roll forming process is suitable for materials that exhibit high strength combined with low tensile ductility, such as Ultra-High Strength Steels (UHSS) [3] and Titanium sheet [4]. Recent studies have shown that cryo-rolled ultrafine grained Aluminium (Al) and nano composites with very low uniform elongation of less than 5% can be roll formed to profile shapes that are common for automotive structural components [5, 6]. Though conventional roll forming is restricted to the longitudinal components with constant cross-section, the Flexible roll forming process allows the manufacture of components with a non-uniform cross-section [7]. The web wrapping and wrinkling of the sheets are often observed and are the most critical issue associated with the Flexible roll formed parts [8]. It should be noted that though the processes are different in terms of tooling, state of stress, and output, but for most of the processes, the deformation zone is restricted to a very small region of the sheet. Also, the through thickness variation of strain is the common characteristic of all these processes.

3.4 MEASUREMENT OF BENDABILITY

The ability of the sheet metal to acquire different shapes is determined through its "Formability". Hereafter, the term formability will be used only for sheet metal formability as the bulk formability is not the interest of the current chapter. Commonly, the mechanical properties of a sheet are esti-mated through uniaxial tensile test. The mechanical properties namely strength and ductility are the function of its microstructural features such as grain size, grain boundaries, presence of coherent or incoherent particles; size, shape, and distribution of the second phases [9]. In reality, uniaxial tension situation hardly occurs during sheet metal forming. So, the applicability of uniaxial tensile test data for the prediction of formability is debatable. In general, the uniform tensile elongation provides the indication of materials hardening capacity and has been linked to its formability. On the other hand, the total tensile elongation is based on the sample geometry and not a fundamental property of a material [10]. The limit of the uniform elongation is the onset of diffused neck during uniaxial tension. According to Considere criterion, a diffused neck forms when the strain hardening rate ($d\sigma/d\varepsilon$) becomes same as the true stress (σ) of the material [1]. For materials that follows power law ($\sigma = K\varepsilon^n$), the uniform elongation equals to its strain hardening coefficient (n). The uniform elongation can underestimate the formability when the grain size is small. It has found that fine or ultrafine grained materials with very low uniform elongation often possess very high formability as a result of their positive strain rate sensitivity (m) [6, 11].

The most common way to represent the formability of a sheet metal is through Forming Limit Diagram (FLD) [12, 13]. FLD represents a range of strain path starting from uniaxial tension to biaxial stretching and the forming limit is marked as the onset of necking. The minimum value of the major strain at the point of zero minor should be theoretically equals to the n. The onset of necking is retarded on the left hand side due to the negative minor strain whereas on the right hand side due to geometrical constraints [1]. Therefore, the loading scenario is important while determining the formability. Another parameter that often is used to determine formability is the Lankford coeffi-cient (r- value). r- value is widely used to determine the deep drawability of a material and is related to the materials resistance to thinning and planner isotropy. r- value is determined from the ratio of strain in the width (ε_w) and the thickness direction (ε_t) measured during tensile test. The r- value ($=\varepsilon_w/\varepsilon_t$) varies along the direction of the sheet and an average normal anisotropy, $\bar{r}$ is estimated from there testing direction, $0°$, $45°$ and $90°$ ($\bar{r} = (r_0 + 2r_{45} + r_{90})/4$). A planner anisotropy, Δr is measured from the variation of the r- value along the three direction, as $\Delta r = (r_0 - 2r_{45} + r_{90})/2$. A high $\bar{r}$ indicates higher resistance to thinning, thereby improved formability whereas a low Δr is related to more homogeneous property of the sheet along its plane [1].

Neither FLD nor Lankford coefficient can represent the bend forming limit or bendability. Bendability is usually represented through the determination of Minimum Bend Radius (MBR). MBR is the limiting bend radius value beyond which the tensile surface of the bend will fracture. MBR depends on the microstructure and mechanical properties, but it cannot be classified as a mate-rial property as MBR varies with the geometry of the sheet and the bending process itself. MBR is generally represented as multiple of the thickness of the sheets, such as $3t$ or $5t$. MBR can be evalu-ated from the Equation 3.6:

$$MBR = \frac{t}{2}\left(\frac{1}{\varepsilon_{major}} - 1\right)$$
(3.6)

where, t is the thickness of the material and ε_{major} is the maximum bending strain at the surface under tension prior to cracking. Lower value of MBR represents tighter bend radius and better bendability for a material.

The bendability can be determined through various bend tests such as free bending, wrap bending, three-point and four-point bending, V-die bending, tight radius bending. In this section, the discussion will be confined to pure bending condition and V-die bending which is widely used and tight radius bend test which is more efficient.

3.4.1 Pure Bend Test

Pure bending condition is commonly found in four-point bend test between the two inner rollers (B and C) as shown in Figure 3.6(a). The process is suitable for small deformation. After large deformation, the contact point between tool and the workpiece changes, the pure bending condition is lost and additional shear force, normal force and friction are prevalent. Several attempts have been made to manufacture pure bending setup [14–18]. A schematic of such setup is shown in Figure 3.6(b). The bending moment on the sheet is produced by the rotation of the clamp [18]. The load and the crosshead displacement can be recorded and from that the moment-curvature can be computed. Figure 3.6(c) shows an out of plane pure bend tester which can be used in universal testing machine (UTM) [19, 20]. Here a LVDT based curvature gauge can be used for more accurate curvature measurement [21].

3.4.2 V-Bend Test

The most widely used technique to measure the bendability is through V-bend test. Figure 3.7(a) represent a schematic of the test apparatus which can be attached to an UTM. The bend angle is same as the angle of the V-die and the bend radius is equal to the radius of the punch. The test is performed by slowly applying load on the punch and the sheet under test is bend into the shape of V-die. Clearly, the bendability determined from V-bend test is based on the punch radius and angle of the V-die. For a fixed bend angle, a series of tool can be used to determine the bendability which usually represented as ratio of MBR to thickness of the sheet [22]. The direct measurement of the bend angle under loading is often not possible during V-die bending. For this purpose, photographs can be taken and imported into CAD software to analyze the bend angle [4]. Kinking often occurs during V-die bending especially for mild steel [19, 20]. The Figure 3.7(b) shows a zoomed view of the bend region. A gap between the sheet and the punch, evident from the Figure, is due to the instability that arises as a result of Kinking. The details of the Kinking phenomenon are explained later. Kinking lead to discrepancy between the punch radius and the actual bend radius and underestimation of the bend fracture limit (Figure 3.7(b)).

3.4.3 Tight Radius Bend Test

The tight radius bend test is a modified form of three-point bend test. A schematic representation of the test setup is shown in Figure 3.8. The bend test samples were supported by two rollers at one side and a sharp tip tool from the other side [23]. The rollers are usually suspended using roller bearings which helps in smooth rotation of the rolls when the force is applied on the sample by the punch during testing [24]. The test is performed according to VDA238-100 standard. The fracture is identified by setting a threshold value for the bend force i.e., fracture occurs when the bending force is reduced by 30 N or 60 N after reaching the maximum. The benefit of tight radius bend over V-die bending is that the contact between the tool radius and the sheet is never lost during the bending. Also, the secondary bending that usually occurs during pure bending is also eliminated [25]. However, the marking the threshold force or maximum force at the onset on fracture may over- or underestimate the bendability, especially when the material is ductile [23, 25]. It is therefore advisable to use direct

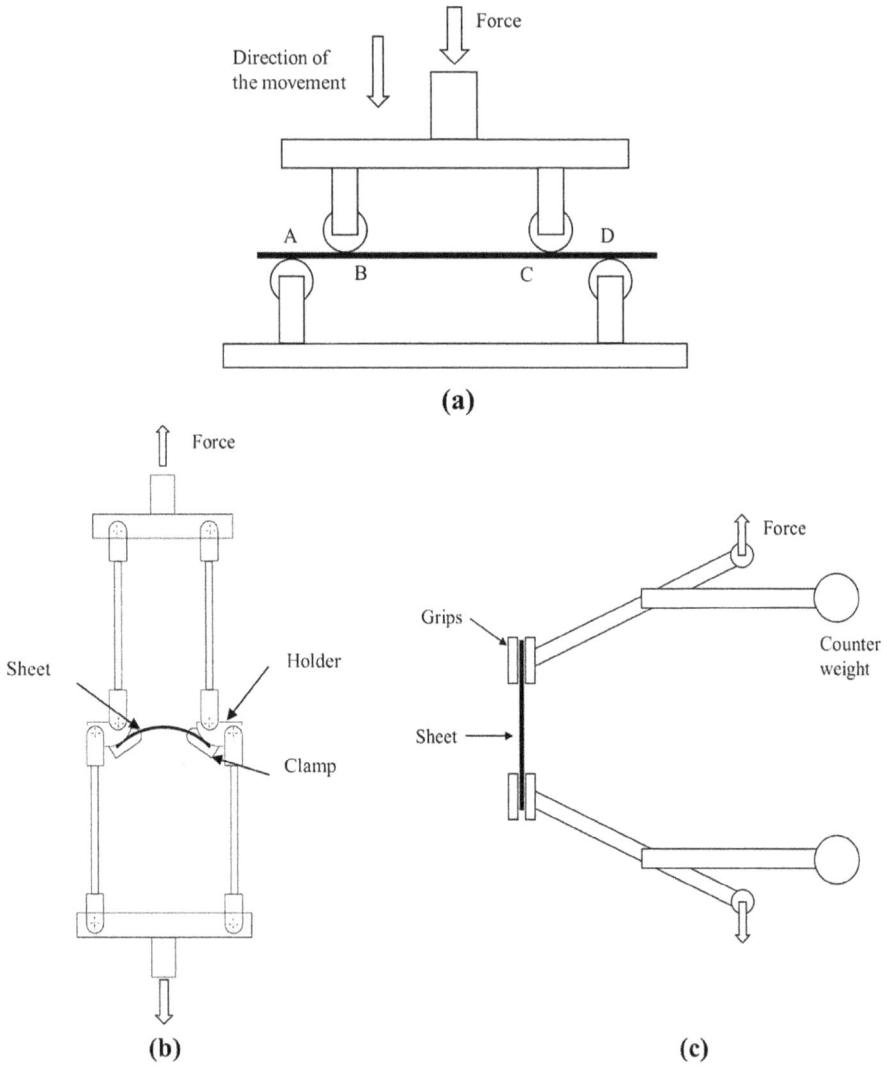

FIGURE 3.6   (a) 4-Point bend tester and (b, c) other pure bend testers are shown schematically.

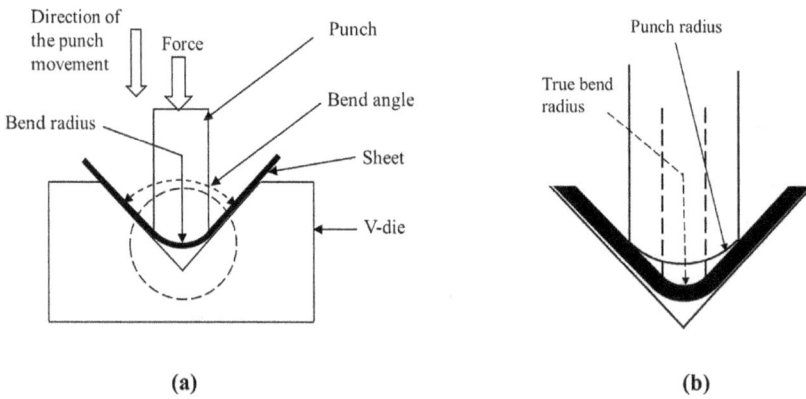

FIGURE 3.7   Schematic representation of (a) V-die bending and (b) A zooming view of the bend region representing Kinking.

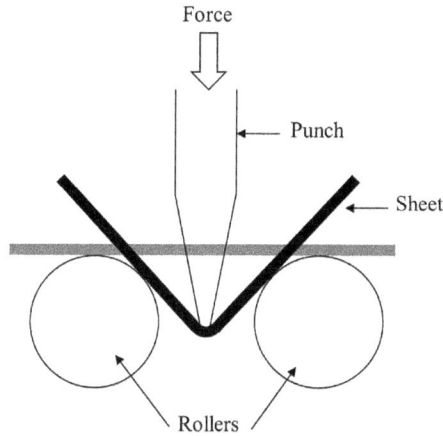

FIGURE 3.8 Schematic representation of the tight radius bend test.

or in-situ strain measurement technique, such as Digital Image Correlation (DIC) [23]. The failure strain in bending is determined from the appearance of crack from the DIC images [6]. Care should be taken when the crack is determined from DIC as the sample centre experience plane strain condition i.e., bending, while the edge of the sample may experience uniaxial tension [25]. Another way to measure the onset of crack is by attaching a piezoelectric acoustic sensor with the sharp tip tool [26]. The onset cracking can be identified from the peak of the acoustic emission.

3.5 EFFECT OF MECHANICAL PROPERTIES ON BENDABILITY

A definite relationship between bendability and the mechanical properties has not been established so far. In general, the bendability reduces with increase in the strength of the materials and closely related to the fracture strain. As mentioned before, the total tensile elongation is not the true indication of a material property. But the maximum reduction in area is directly associated with the fracture and can be used as "true ductility" of a material [10]. The most effective analytical model has been proposed by Datsko and Yang [27] to establish a relation between the bendability of a material with its true ductility as:

$$\frac{R}{t} = \frac{50}{A_r} - 1 \tag{3.7}$$

Or

$$\frac{R}{t} = \frac{60}{A_r} - 1 \tag{3.8}$$

where, R is the bend radius, t is the thickness of the sheet and A_r is the percentage reduction in area in tensile test. Equation (3.7) was derived by not considering neutral layer shift and by considering neutral layer shift, Equation (3.8) was obtained. Here, the bendability of a material is found to depend only on percentage reduction in area i.e., the true ductility of the material. This was experimentally supported by many researchers [28]. The above mentioned model assumes that the tensile fracture strain is same at the fracture strain of the tensile surface during bending. This model can predict the bendability of monolithic alloys with considerable accuracy, but does not consider the effect of microstructure. The anisotropy of the sheet, presence of second phases; shape, size and distribution of secondary particles and the residual stresses have significant effect on bendability. This will be discussed in the next section [29].

3.6   EFFECT OF MICROSTRUCTURE ON BENDABILITY

In a previous section, the relationship between mechanical properties with the bendability was discussed. In the current section, the effect of microstructure on the bend fracture limit will be discussed. It is well established that the microstructure and mechanical properties are influenced by thermo-mechanical treatment. For example, severe plastic deformation (SPD) processes, thermomechanical processing routes and cold working routes can result in grain refinement but the degree of grain refinement and nature of grain boundary depends on the strain path, state of stress, strain rate and temperature. Agehardenable and non-agehardenable alloys consist of coherent and non-coherent precipitates. The size, shape and distribution of those precipitate are based on the alloys chemistry as well as on the nature, sequence and duration of heat treatment. The precipitation morphology influences the bendability significantly. A comprehensive theory on the effect of microstructural constituents on bendability is not well established. This section will highlight the aspect of each constituent on the bendability on case to case basis with examples.

3.6.1   EFFECT OF GRAIN SIZE, SHAPE AND BOUNDARIES ON THE BENDABILITY

The grain size of material reduces when it is subjected to severe plastic deformation (SPD). The reduced grain size causes reduction in bendability and increases the requirement for higher force for bending due to grain boundary strengthening [30]. In addition, small grain size provides very little space for dislocation accommodation resulting in low strain hardening during bending. The grain size can be increased by annealing. During annealing, deformed material undergoes recovery and/or recrystallization followed by grain growth [31]. Larger grains provide more scope of strain hardening thereby improves the bendability [6]. Moreover, the changes in the nature of grain boundaries effect the bendability. The improvement of fraction of high angle grain boundary in annealed material over the deformed material has a positive impact on bendability [6].

Cold rolling process results in elongation of the grains along the direction rolling as shown in Figure 3.9. The distance between grain boundaries along the rolling direction (RD) is higher as compared to the transverse direction (TD). The tensile strength of the material along the TD is higher as compared to RD due to higher grain boundary strengthening. When the bend axis is parallel to RD (Figure 3.9(a)), the tension in the bend region is in the direction of TD. As the strength of the material along the TD is higher, the force required for bending will be higher. In other words, the bendability will be lower or the MBR will be high when the bend axis in parallel to RD. So, a better bendability and lower MBR can be achieved in cold-rolled material when the bend axis is parallel to TD (Figure 3.9(b)) [30].

During bending the grain boundaries which are located at the surface promote localized deformation and formation of slip band. The grain rotation and deformation at the outer surfaces appear as depression followed by formation of grooves at the surface. These grooves then grow parallel to the bend axis [28]. This is the main mechanism observed in case of ductile material. The strain localization points are based on the density of grain boundary on the surface. The formation of grooves is found to be related to the initial surface roughness. Finer surface finish can improve the bendability of a material [32].

3.6.2   EFFECT OF RESIDUAL STRESS ON BENDABILITY

Sheets are produced by rolling to a metallic block. A block passes through two rolls and under the compressive force the thickness reduces. At the surface of the metal which is in contact with the roller, shear stress is prevalent as a result of friction between the rollers and the sheet. So during

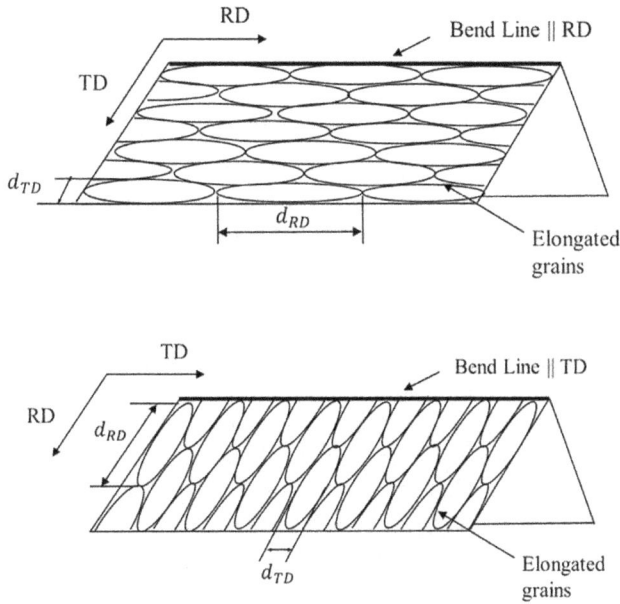

FIGURE 3.9   Orientation of bend line with respect to the rolling direction.

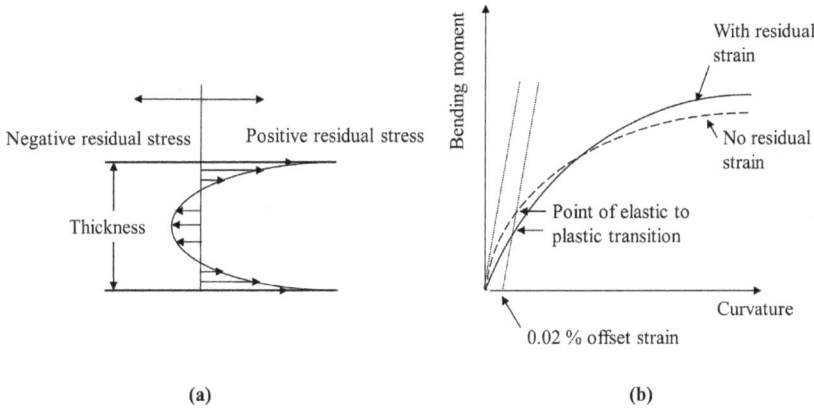

FIGURE 3.10   (a) Variation of through thickness residual strain and (b) typical bending moment vs curvature plot.

rolling, residual stress is induced in the material and the amount of residual stress or stored energy depends on the temperature, strain rate and percentage of rolling reduction at which rolling is performed [33–35]. A direct comparison between cryo-rolled (CR) and room-temperature-rolled (RTR) Al alloys shows that necking at outer surface occurs at lower bend radius for RTR alloy [5]. This is due to the dislocation densed or work hardened microstructure of CR materials which have very little scope for uniform elongation and lead to early necking [36]. The tendency of localized necking i.e., kinking is higher when the microstructure is more work hardened. For a work hardened material, during uniaxial tensile test, the stress-curve drops immediately after peak stress is achieved. This indicates an increased tendency towards strain localization during bending [5]. A through thickness variation in residual stress is observed for small rolling reduction where the residual stress is higher at the surface as compared to the core of the sheet (Figure 3.10(a)). A typical bending moment vs curvature curve is shown in Figure 3.10(b). This lowers the bending moment at which elastic to plastic transition occurs in bending, but improves the resistance to bending moment at higher curvature [37].

3.6.3 Effect of Crystallographic Texture on Bendability

The crystallographic texture is the distribution of orientation of grains in a polycrystalline material. If the distribution is random, the material is said to have weak texture whereas if a preferred orientation of grain is prevalent, then the texture is termed as a strong texture. The crystallographic texture itself is a function of alloy chemistry, deformation condition, and post-deformation heat treatment. The crystallographic orientation of grains influences several mechanical properties of a material. It should be noted that the effect of texture varies based on whether the rolling direction is perpendicular or parallel to the bend axis. For the former case, the Copper and Brass texture can significantly reduce the formation of surface undulations and shear bands during bending of rolled Al 6xxx alloys. They also reduce the crack propagation. On the other hand, microstructure with S, Goss, and rotated Goss texture promotes formation of shear band and intergranular cracking. The grains having Cube texture can accommodate higher deformation thereby restrict the formation of shear band. But higher number of Cube grains may result in strain localization during bending [38]. Moreover, the study on single crystal Al alloys has showed that the excellent bendability is observed in the material with Cube orientation regardless of the bending direction. But for Goss textured material the bendability is poor for bend line parallel to RD and superior for bend line parallel to TD [39]. For ultra-high strength steel (UHSS), the subsurface microstructure has strong effect on bendability. The formation of shear band is accelerated with $\{112\}\langle111\rangle\alpha$ texture component. Bendability reduces with increase in $\{112\}\langle111\rangle\alpha$ component for the subsurface microstructure having upper bainite with martesite-austenite island and the effect is more detrimental when the bend axis is perpendicular to the rolling direction. On the other hand, the softer granular bainite and ferrite subsurface microstructure have almost no effect on reducing bendability with increase in $\{112\}\langle111\rangle\alpha$ [22].

3.6.4 Effect of Precipitates/Second Phase Particles

The formation of precipitates or presence of other second phase particles may improve the strength of the alloys but have detrimental effect on ductility, formability, and bendability [6, 36]. The presence of these secondary phases lead to void formation at the particle-matrix interface as a result of strain localization. When large particles are formed at the grain boundary, the fraction of alloying element in matrix reduces near the grain boundary. This lead to the reduction in strength of the material adjacent to the grain boundary as compared to the grain interior. Moreover, the presence of large particles reduces the cohesive bond between two adjacent grains. This can promote grain boundary fracture during bending [40]. With the increased amount of particles at the grain boundary, de-cohesion of boundary becomes dominant fracture mechanism under bending for agehardenable alloys. In addition to this, ductile fracture of grain boundary can occur at the later stage of the bending due to the strain localization and can be viewed on the fractured surface in the form of microdimples. The void formation can also occur due to the breakage of particles under high strain [28]. While the large particle results in formation of crack, the growth and propagation is further influenced by the shape of the small precipitates present in the grain boundary [29]. Small amount of nano particle addition can have detrimental effect on bending while have almost no effect of tensile elongation [6].

3.7 DEFECTS OR INSTABILITIES DURING BENDING

3.7.1 Kinking

Kinking is a result of inhomogeneous deformation during bending of sheet. Figure 3.7(b) shows a schematic of the kinking phenomenon during V-die bending. When kinking occurs the bend radius

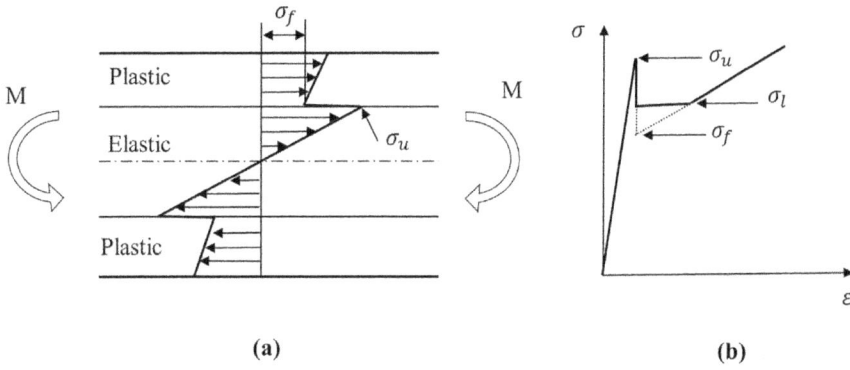

FIGURE 3.11 (a) Through thickness distribution of the stress during bending and (b) tensile stress-strain respond of aged mild steel.

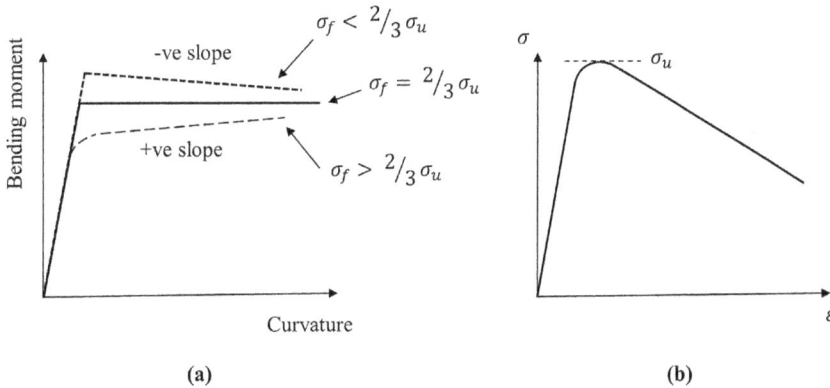

FIGURE 3.12 (a) Moment-curvature curve during bending and (b) Tensile stress-strain curve for typical work hardened material.

become smaller than the punch radius. The tendency of this localized deformation is more when a larger tool radius is employed [5]. Kinking occurs when instability arises in the bending moment. The through thickness distribution of the stress during bending of aged mild steel can be represented as indicated in Figure 3.11(a) and the tensile stress-strain response of the same material is shown in Figure 3.11(b) schematically. The sheet deform elastically till the outer fiber of the sheet reaches the upper yield point, σ_u. With further bending, the outer layer starts to deform plastically while the core of the sheet remains elastic. While the plastic flow stress (σ_f) at the outer layer reduces from the upper yield point, the max stress at the elastic core still remains as same as the upper yield point. When the plastic flow stress in the outer layer falls below 2/3 of upper yield stress, the moment required for further bending becomes lower than that required for elastic-to-plastic bending [41, 42]. This causes an instability in the bend region and results in kink. In moment-curvature curve, this point of instability is marked as zero or negative slope (Figure 3.12(a)) [43]. In some materials, such as cold-rolled Al, the moment-curvature curve shows a smooth transition from elastic to plastic bend, but still results in formation of kinks. This is related to the work hardened microstructure resulting low uniform tensile elongation and the stress drops suddenly after reaching at maximum tensile stress (Figure 3.12(b)) [5]. The kinking can be eliminated when the bend region is supported by tool from both the side the sheets, as observed during roll forming.

3.7.2 SPRINGBACK

Springback is a common defect is sheet metal forming. It is the result of elastic recovery. Springback in bending depends on various factors: microstructural feature of the material, the thickness of the sheet, the geometry of the tool and the parameters of the bending process. The sheet under bending deforms first elastically and then plastically. After the loading is removed or the bending moment is eliminated the elastic strain recovers. This lead to change in final shape of the bend. So, the prediction of springback is necessary in order to redesign the tool to compensate for springback. The sheet thickness and bend angle has most severe effect on springback. In general, the increase in sheet thickness can result in reduction in springback [44, 45]. When same material with different thickness bend at same radius, the total strain (elastic and plastic) is high for the sheets with higher thickness. So, the relative elastic strain compared to total strain is low for thicker sheets. Once the loading is removed the elastic recovery in both the sheets will be similar in absolute term but actually become relatively higher in thinner sheets as compared to thicker sheets. Also, higher the elastic modulus of sheet, lower will be the elastic strain thereby low springback. Further, the springback is directly proportional to the normal anisotropy of the sheet and inversely proportional to the strain hardening exponent [46]. The springback is found to change with the process parameters. The higher blank holder force reduces springback by restricting the sliding of the sheet, thereby increasing the tension in the sheet [47]. For a same bend angle, the deformation length is small for smaller bend radius and will result in more plastic deformation per unit length. Higher the plastic deformation, the contribution of the elastic deformation is low in the total strain; thereby low springback [48]. One of the way to eliminated springback is over-bending the sheet than the required amount by modifying the bending process or the tool or both accordingly. Once the load is eliminated the elastic recovery will cause the sheet to attain the desired shape [49]. Another way is to perform the bending process incrementally rather than going for single step bending. A multistep bending can reduce the springback as compared to single-step bending due to the accumulation of plastic strain in the bend region in consecutive passes [50]. In practice, this phenomenon is observed in roll forming of sheet where a reduced springback is observed as compared to conventional bending for same bend angle and bend radius [51]. In addition, annealing the material through local heating is an attractive way to eliminated springback. Also, soften the material locally by heating such as using laser only changes the microstructure at the bend region while the remaining material can hold its original strength [30].

3.8 RECENT DEVELOPMENTS

3.8.1 BENDING OF MAGNESIUM SHEET

Magnesium (Mg) has HCP crystal structure which results in tension-compression yield asymmetry [52]. In a rolled Mg sheet, for majority of grain the c-axis becomes parallel to the normal direction. When tension is applied along RD or compression along TD, the material deforms through slip dominated slip-twin deformation. On the other hand, when compression is applied along RD or tension along TD, twinning becomes the dominant deformation mechanism along with slip [53]. There are several ways to reduce tension-compression yield asymmetry, however these factors are not the focus/purpose of current discussion. The response of Mg sheet under bending is the interest of this section. Due to HCP crystal structure and limited number of slip systems at room temperature, the ductility and formability of Mg is very limited. Therefore, development of Mg sheet itself is a challenge and only few studies have been performed to examine its bendability. Moreover, the tension-compression asymmetry under bending has been rarely addressed. The bending of Mg sheet has shown that with increase in curvature, the thickness of the sheet starts to increase, reaches a maximum value and then decreases. This change in thickness is closely related to the position of the neutral axis. The neutral axis shifts towards the outer surface due to the yield asymmetry of Mg and lead to increase in compression zone at the cross-section of sheet. Therefore, under the influence of the asymmetric work hardening, the thickness of the Mg sheet increases during bending [54].

The shifting of neutral layer in Mg sheet during bending can be reduced by increasing the deformation temperature as a result of reduced tension-compression asymmetry at higher temperature [55]. Pre-straining of the Mg sheets prior to bending also reduces the neutral layer shift [56]. Due to the neutral layer shift coupled with the increasing thickness and tension-compression yield asymmetry, the prediction of springback in Mg sheet bending is difficult.

3.8.2 MICROBENDING

The increasing demand of the micro parts in electronic device, medical device, and fine chemistry increases the attention towards microforming processes. The deformation behavior of a material changes when the forming scale reduces from macro level to micro/meso level as a result of size effect. The grains at the surface are less constraint as compared to the internal grains and have low flow stress. When the geometry of the sheet becomes comparable to the grain size, the contribution from the surface layer increases; therefore, the overall strength of the sheet reduces. The springback in microbending generally reduces with increase in foil thickness irrespective of the microbending process [57–59]. This is similar to the conventional bending processes. But, for the same foil thickness, the springback increases with increasing grain size for three-point microbending [58]. On the other hand, for micro V-die bending springback increases with reducing grain size [57]. Moreover, for micro U-bending, no conclusive trend was observed between springback and grain size [59]. It should be noted that the number of grains present at the sheet/foil thickness has the most influence on the bend force and springback [60]. In microbending process, the number of grain along the thickness direction is a primary parameter that determines the contribution of the surface grain on the deformation behavior. Surface grains are weaker than the inner grains and its influence is pivotal in case of microbending. The microbending process depends on the strain gradient, thickness, and the grain size [61]. When the number of grains along the thickness reduces, the contribution of the surface grain increases. The rotation of the surface grains becomes restricted when the grain size becomes similar to or higher than the thickness of the sheet [62]. This can lead to reduced bendability of the foil during microbending process.

3.8.3 ELECTROMAGNETIC BENDING

In electromagnetic bending, the deformation is carried out with the help of electromagnetic force. There is no contact between the punch and the sheet and no requirement of lubrication. This process is a high speed forming process and is suitable for materials with low bendability such as Al and Mg. The overall springback is also reduced [63]. But the bending of sheet using electromagnetic forming may cause saddle like defect along the width direction where the middle region of the sheet will experience lower deformation as compared to the edges (Figure 3.13(a)). This is the result of

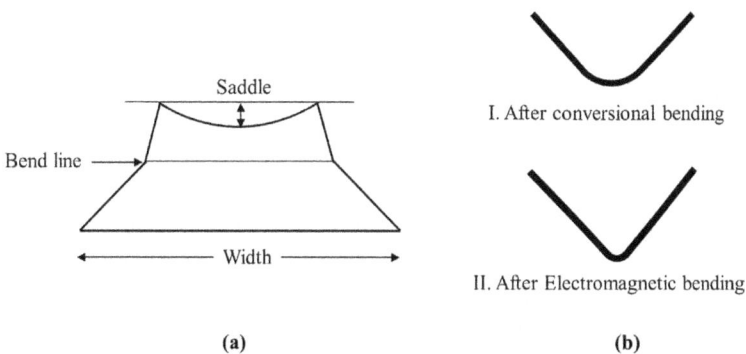

FIGURE 3.13 (a) Saddle defect in electromagnetic V-die forming and (b) sharpening of bending after electromagnetic bending.

the non-uniform distribution of the electromagnetic forces [64]. The height of the saddle decreases with the reduction in voltage as a result of more uniform force distribution during bending. On the other hand, the high voltage increases the forming force and collision velocity and helps in achieving smaller bend radius. For this reason, electromagnetic bending is often used to sharpen the preformed bend radius. The material having low bendability is first deformed through conventional bending and then electromagnetic bending is applied to sharpen the bend further (Figure 3.13(b)) which is difficult to achieve using conventional tooling [65].

3.9   CONCLUSIONS

The current chapter has discussed about the bending of sheet metal and related phenomenon. The application of bending is spread from automobile, aerospace, defense sector to various structural components. According to the theory of bending, the through thickness variation of strain is the unique characteristic of bending. The bend forming limit cannot be determined form uniaxial tensile test or forming limit diagram. Pure bend test, V-die bend test and newly developed tight radius bend test often used to determine the minimum bend radius (MBR). There is no definite relationship between materials' tensile properties with MBR, through it is closely related to the percentage reduction in area at the fracture during tensile test. MBR depends on various factors such as grain size, shape, and orientation; residual stress; crystallographic texture; size, shape, and distribution of the second phase particles; tooling; state of stress; strain and strain rate. The major defects generally occur during bending are springback and kinking. Majority of bending work is done for Aluminum and steel using conventional bending processes. Recently, the bending of Magnesium sheets, microbending, and electromagnetic pulse bending are gaining momentum.

REFERENCES

[1] Marciniak Z, Duncan JL, Hu SJ. *Mechanics of Sheet Metal Forming*. Butterworth-Heinemann, 2002.

[2] Abeyrathna B, Rolfe B, Hodgson P, Weiss M. An extension of the flower pattern diagram for roll forming. *Int J Adv Manuf Technol* 2016;83:1683–1695. https://doi.org/10.1007/s00170-015-7667-0

[3] Deole AD, Barnett MR, Weiss M. The numerical prediction of ductile fracture of martensitic steel in roll forming. *Int J Solids Struct* 2018;144–145:20–31. https://doi.org/10.1016/j.ijsolstr.2018.04.011

[4] Badr OM, Rolfe B, Hodgson P, Weiss M. Forming of high strength titanium sheet at room temperature. *Mater Des* 2015;66:618–626. https://doi.org/10.1016/j.matdes.2014.03.008

[5] Marnette J, Weiss M, Hodgson PD. Roll-formability of cryo-rolled ultrafine aluminium sheet. *Mater Des* 2014;63:471–478. http://dx.doi.org/10.1016/j.matdes.2014.06.036

[6] Deb S, Panigrahi SK, Weiss M. Understanding material behaviour of ultrafine-grained aluminium nano-composite sheets with emphasis on stretch and bending deformation. *J Mater Process Technol* 2021;293:117082. https://doi.org/10.1016/j.jmatprotec.2021.117082

[7] Abeyrathna B, Abvabi A, Rolfe B, Taube R, Weiss M. Numerical analysis of the flexible roll forming of an automotive component from high strength steel. *IOP Conf Ser Mater Sci Eng* 2016;159. https://doi.org/10.1088/1757-899X/159/1/012005

[8] Jiao J, Rolfe B, Mendiguren J, Weiss M. An analytical approach to predict web-warping and longitudinal strain in flexible roll formed sections of variable width. *Int J Mech Sci* 2015;90:228–238. https://doi.org/10.1016/j.ijmecsci.2014.11.010

[9] Deiter GE. *Mechanical Metallurgy*. 1988.

[10] Cockcroft MC, Latham DJ. *Ductility and Workability of Metals*. 1968:33–39.

[11] Ghosh AK. The influence of strain hardening and strain-rate sensitivity on sheet metal forming. *J Eng Mater Technol Trans ASME* 1977;99:264–274. https://doi.org/10.1115/1.3443530

[12] Goodwin GM. Application of strain analysis to sheet metal forming problems in the press shop. *Met Ital* 1968;60:767–774.

[13] Keeler SP. Determination of forming limits in automotive stampings. *Sheet Met Ind* 1965;42:683–691.

[14] Naseem S, Perdahcıoğlu ES, Geijselaers HJM, van den Boogaard AH. A new in-plane bending test to determine flow curves for materials with low uniform elongation. *Exp Mech* 2020;60:1225–1238. https://doi.org/10.1007/s11340-020-00621-5

[15] Perduijn AB, Hoogenboom SM. The pure bending of sheet. *J Mater Process Tech* 1995;51:274–295. https://doi.org/10.1016/0924-0136(94)01596-S

[16] Ben Zineb T, Sedrakian A, Billoet JL. An original pure bending device with large displacements and rotations for static and fatigue tests of composite structures. *Compos Part B Eng* 2003;34:447–458. https://doi.org/10.1016/S1359-8368(03)00017-9

[17] Arnold G, Calloch S, Dureisseix D, Billardon R. A pure bending machine to identify the mechanical behaviour of thin sheets. *6th Int ESAFORM Conf Mater Form* 2003:1–4.

[18] Duncan JL, Ding SC, Jiang WL. Moment-curvature measurement in thin sheet - Part I: Equipment. *Int J Mech Sci* 1999;41:249–260. https://doi.org/10.1016/s0020-7403(98)00031-9

[19] Weiss M, Hemmerich E, Hodgson PD, Rolfe B. Effect of thermal treatment on the bending properties of pre-strained carbon steel. *Mater Sci Eng A* 2011;528:4528–4536. https://doi.org/10.1016/j.msea.2011.02.059

[20] Hemmerich E, Rolfe B, Hodgson PD, Weiss M. The effect of pre-strain on the material behaviour and the Bauschinger effect in the bending of hot rolled and aged steel. *Mater Sci Eng A* 2011;528:3302–3309. https://doi.org/10.1016/j.msea.2010.12.035

[21] Badr OM, Rolfe B, Zhang P, Weiss M. Applying a new constitutive model to analyse the springback behaviour of titanium in bending and roll forming. *Int J Mech Sci* 2017;128–129:389–400. https://doi.org/10.1016/j.ijmecsci.2017.05.025

[22] Kaijalainen AJ, Liimatainen M, Kesti V, Heikkala J, Liimatainen T, Porter DA. Influence of composition and hot rolling on the subsurface microstructure and bendability of ultrahigh-strength strip. *Metall Mater Trans A Phys Metall Mater Sci* 2016;47:4175–4188. https://doi.org/10.1007/s11661-016-3574-8

[23] Cheong K, Omer K, Butcher C, George R, Dykeman J. Evaluation of the VDA 238-100 tight radius bending test using digital image correlation strain measurement. *J Phys Conf Ser* 2017;896. https://doi.org/10.1088/1742-6596/896/1/012075

[24] Pathak N, Adrien J, Butcher C, Maire E, Worswick M. Experimental stress state-dependent void nucleation behavior for advanced high strength steels. *Int J Mech Sci* 2020;179:105661. https://doi.org/10.1016/j.ijmecsci.2020.105661

[25] Kupke A, Barnett M, Luckey G, Weiss M. Determination of the bendability of ductile materials. *IOP Conf Ser Mater Sci Eng* 2018;418. https://doi.org/10.1088/1757-899X/418/1/012077

[26] Wagner L, Larour P, Dolzer D, Leomann F, Suppan C. Experimental issues in the instrumented 3 point bending VDA238-100 test. *IOP Conf Ser Mater Sci Eng* 2020;967. https://doi.org/10.1088/1757-899X/967/1/012079

[27] Datsko J, Yang CT. Correlation of bendability of materials with their tensile properties. *J Manuf Sci Eng Trans ASME* 1960;82:309–312. https://doi.org/10.1115/1.3664236

[28] Sarkar J, Kutty TRG, Wilkinson DS, Embury JD, Lloyd DJ. Tensile properties and bendability of T4 treated AA6111 aluminum alloys. *Mater Sci Eng A* 2004;369:258–266. https://doi.org/10.1016/j.msea.2003.11.022

[29] Castany P, Diologent F, Rossoll A, Despois JF, Bezençon C, Mortensen A. Influence of quench rate and microstructure on bendability of AA6016 aluminum alloys. *Mater Sci Eng A* 2013;559:558–565. https://doi.org/10.1016/j.msea.2012.08.141

[30] Maier V, Hausöl T, Schmidt CW, Böhm W, Nguyen H, Merklein M, et al. Tailored heat treated accumulative roll bonded aluminum blanks: Microstructure and mechanical behavior. *Metall Mater Trans A Phys Metall Mater Sci* 2012;43:3097–3107. https://doi.org/10.1007/s11661-012-1151-3

[31] Deb S, Panigrahi SK, Weiss M. The effect of annealing treatment on the evolution of the microstructure, the mechanical properties and the texture of nano SiC reinforced aluminium matrix alloys with ultrafine grained structure. *Mater Charact* 2019;154:80–93. https://doi.org/10.1016/j.matchar.2019.05.023

[32] Mattei L, Daniel D, Guiglionda G, Klöcker H, Driver J. Strain localization and damage mechanisms during bending of AA6016 sheet. *Mater Sci Eng A* 2013;559:812–821. https://doi.org/10.1016/j.msea.2012.09.028

[33] Panigrahi SK, Jayaganthan R. Development of ultrafine grained Al-Mg-Si alloy with enhanced strength and ductility. *J Alloys Compd* 2009;470:285–288. https://doi.org/10.1016/j.jallcom.2008.02.028

[34] Panigrahi SK, Jayaganthan R. Effect of rolling temperature on microstructure and mechanical properties of 6063 Al alloy. *Mater Sci Eng A* 2008;492:300–305. https://doi.org/10.1016/j.msea.2008.03.029

[35] Panigrahi SK, Jayaganthan R. A study on the mechanical properties of cryorolled Al-Mg-Si alloy. *Mater Sci Eng A* 2008;480:299–305. https://doi.org/10.1016/j.msea.2007.07.024

[36] Deb S, Panigrahi SK, Weiss M. Development of bulk ultrafine grained Al-SiC nano composite sheets by a SPD based hybrid process: Experimental and theoretical studies. *Mater Sci Eng A* 2018;738:323–334. https://doi.org/10.1016/j.msea.2018.09.101

[37] Weiss M, Rolfe B, Hodgson PD, Yang C. Effect of residual stress on the bending of aluminium. *J Mater Process Technol* 2012;212:877–883. https://doi.org/10.1016/j.jmatprotec.2011.11.008

[38] Muhammad W, Brahme AP, Ali U, Hirsch J, Engler O, Aretz H, et al. Bendability enhancement of an age-hardenable aluminum alloy: Part II—multiscale numerical modeling of shear banding and fracture. *Mater Sci Eng A* 2019;754:161–177. https://doi.org/10.1016/j.msea.2019.03.050

[39] Ikawa S, Asano M, Kuroda M, Yoshida K. Effects of crystal orientation on bendability of aluminum alloy sheet. *Mater Sci Eng A* 2011;528:4050–4054. https://doi.org/10.1016/j.msea.2011.01.048

[40] Davidkov A, Petrov RH, De Smet P, Schepers B, Kestens LAI. Microstructure controlled bending response in AA6016 Al alloys. *Mater Sci Eng A* 2011;528:7068–7076. https://doi.org/10.1016/j.msea.2011.05.055

[41] Duncan JL, Sue-Chu M, Wang XJ. *Instability in Plastic Bending of Thin Sheet Metal*. vol. 2. Pergamon Press Ltd., 1984. https://doi.org/10.1016/b978-1-4832-8372-2.50079-8

[42] Duncan JL, Ding SC, Jiang WL. Moment-curvature measurement in thin sheet - Part II: Yielding and kinking in aged steel sheet. *Int J Mech Sci* 1999;41:261–267. https://doi.org/10.1016/s0020-7403(98)00032-0

[43] Ding SC, Duncan JL. Instability in bending-under-tension of aged steel sheet. *Int J Mech Sci* 2004;46:1471–1480. https://doi.org/10.1016/j.ijmecsci.2004.09.008

[44] Bakhshi-Jooybari M, Rahmani B, Daeezadeh V, Gorji A. The study of spring-back of CK67 steel sheet in V-die and U-die bending processes. *Mater Des* 2009;30:2410–2419. https://doi.org/10.1016/j.matdes.2008.10.018

[45] Thipprakmas S, Phanitwong W. Process parameter design of spring-back and spring-go in V-bending process using Taguchi technique. *Mater Des* 2011;32:4430–4436. https://doi.org/10.1016/j.matdes.2011.03.069

[46] Leu DK. A simplified approach for evaluating bendability and springback in plastic bending of anisotropic sheet metals. *J Mater Process Technol* 1997;66:9–17. https://doi.org/10.1016/S0924-0136(96)02453-3

[47] Ouakdi EH, Louahdi R, Khirani D, Tabourot L. Evaluation of springback under the effect of holding force and die radius in a stretch bending test. *Mater Des* 2012;35:106–112. https://doi.org/10.1016/j.matdes.2011.09.003

[48] Buang MS, Abdullah SA, Saedon J. Effect of die and punch radius on springback of stainless. *J Mech Eng Sci* 2015;8:1322–1331

[49] Tan Z, Persson B, Magnusson C. An empiric model for controlling springback in V-die bending of sheet metals. *J Mater Process Tech* 1992;34:449–455. https://doi.org/10.1016/0924-0136(92)90140-N

[50] Badr OM, Rolfe B, Hodgson P, Weiss M. Experimental study into the correlation between the incremental forming and the nature of springback in automotive steels. *Int J Mater Prod Technol* 2013;47:150–161. https://doi.org/10.1504/IJMPT.2013.058961

[51] Weiss M, Marnette J, Wolfram P, Larrañaga J, Hodgson P. Comparison of bending of automotive steels in roll forming and in a V-die. *Key Eng Mater* 2012;504–506:797–802. https://doi.org/10.4028/www.scientific.net/KEM.504-506.797

[52] Sahoo SK, Sahoo BN, Panigrahi SK. Effect of in-situ sub-micron sized TiB2 reinforcement on microstructure and mechanical properties in ZE41 magnesium matrix composites. *Mater Sci Eng A* 2020;773:138883. https://doi.org/10.1016/j.msea.2019.138883

[53] Sahoo BN, Panigrahi SK. A study on the combined effect of in-situ (TiC-TiB2) reinforcement and aging treatment on the yield asymmetry of magnesium matrix composite. *J Alloys Compd* 2018;737:575–589. https://doi.org/10.1016/j.jallcom.2017.12.027

[54] Ahn K. Plastic bending of sheet metal with tension/compression asymmetry. *Int J Solids Struct* 2020;204–205:65–80. https://doi.org/10.1016/j.ijsolstr.2020.05.022

[55] Wang L, Huang G, Zhang H, Wang Y, Yin L. Evolution of springback and neutral layer of AZ31B magnesium alloy V-bending under warm forming conditions. *J Mater Process Technol* 2013;213:844–850. https://doi.org/10.1016/j.jmatprotec.2013.01.005

[56] Wang L, Huang G, Han T, Mostaed E, Pan F, Vedani M. Effect of twinning and detwinning on the spring-back and shift of neutral layer in AZ31 magnesium alloy sheets during V-bend. *Mater Des* 2015;68:80–87. https://doi.org/10.1016/j.matdes.2014.12.017

[57] Xu Z, Peng L, Bao E. Size effect affected springback in micro/meso scale bending process: Experiments and numerical modeling. *J Mater Process Technol* 2018;252:407–420. https://doi.org/10.1016/j.jmatprotec.2017.08.040

[58] Liu JG, Fu MW, Lu J, Chan WL. Influence of size effect on the springback of sheet metal foils in micro-bending. *Comput Mater Sci* 2011;50:2604–2614. https://doi.org/10.1016/j.commatsci.2011.04.002

[59] Wang J, Fu M, Ran J. Analysis of the size effect on springback behavior in micro-scaled U-bending process of sheet metals. *Adv Eng Mater* 2014;16:421–432. https://doi.org/10.1002/adem.201300275

[60] Diehl A, Engel U, Geiger M. Mechanical properties and bending behaviour of metal foils. *Proc Inst Mech Eng Part B J Eng Manuf* 2008;222:83–91. https://doi.org/10.1243/09544054JEM838

[61] Wang Z, Li S, Wang X, Cui R, Zhang W. Modeling of surface layer and strain gradient hardening effects on micro-bending of non-oriented silicon steel sheet. *Mater Sci Eng A* 2018;711:498–507. https://doi.org/10.1016/j.msea.2017.08.085

[62] Jiang Z, Zhao J, Xie H. Practice of micro bending. *Microforming Technol* 2017:417–430. https://doi.org/10.1016/b978-0-12-811212-0.00019-4

[63] Xiao W, Huang L, Li J, Su H, Feng F, Ma F. Investigation of springback during electromagnetic-assisted bending of aluminium alloy sheet. *Int J Adv Manuf Technol* 2019;105:375–394. https://doi.org/10.1007/s00170-019-04161-8

[64] Xiong W, Wang W, Wan M, Li X. Geometric issues in V-bending electromagnetic forming process of 2024-T3 aluminum alloy. *J Manuf Process* 2015;19:171–182. https://doi.org/10.1016/j.jmapro.2015.06.015

[65] Imbert J, Worswick M. Electromagnetic reduction of a pre-formed radius on AA 5754 sheet. *J Mater Process Technol* 2011;211:896–908. https://doi.org/10.1016/j.jmatprotec.2010.07.021

4 Friction Stir Welding (FSW) and Friction Stir Spot Welding of Dissimilar Sheet Materials

Sukanta Das and R. Ganesh Narayanan
IIT Guwahati, Guwahati, India

CONTENTS

4.1 INTRODUCTION

Friction stir welding (FSW) technology is considered a boon for the manufacturing industries in modern time. The joining technique is accepted as a solid-state joining technique because the joining is established at the temperature below the melting point of base materials to be joined. Conventional FSW is one most common joining technique of the FSW family. FSW was first introduced by The Welding Institute (TWI) in 1991 [1] and this technique can join a large variety of ferrous and non-ferrous alloys and polymers. The methodology of FSW is straightforward, as it uses a non-consumable rotating tool. This non-consumable rotating tool consists of a pin and a shoulder. The rotating tool employs three major roles during FSW, heating of the base plates or sheet, movement of the plasticized material, and the forging pressure. The pin allows the material to be plastically deformed and the shoulder provides the necessary frictional heat. As shown in Figure 4.1, the tool is initially plunged up to a certain thickness of base materials. Due to the relative motion of the tool, there generates a plasticized region around the tool. During plunging, the initial plastic deformation of the material starts. The shoulder constricts the movement of the plasticized material from the vicinity of the tool. As the rotating tool reached the desired thickness of the faying surfaces, it is traversed along the joint, allowing the plasticized material to coalesce at the rear of the tool to form a joint at the solid-state. FSW can join lightweight materials which cannot be joined with fusion welding processes. It has low distortion and shrinkage, good mechanical properties, followed by no arc and porosity [2–5].

The microstructure evolution of the weld zone is divided into four zones, Base material (BM), Heat-affected zone (HAZ), Thermo-mechanically affected zone (TMAZ), and nugget [6]. All four zones have been depicted in Figure 4.2.

DOI: 10.1201/9781003226703-4

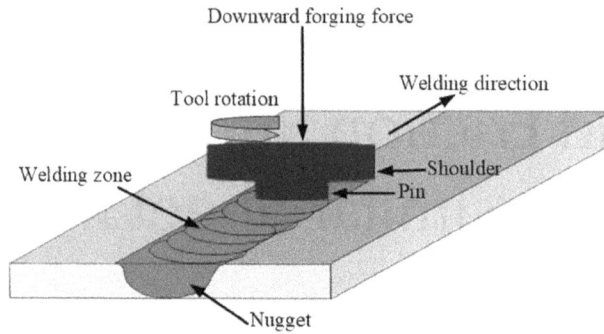

FIGURE 4.1 Schematic of FSW process.

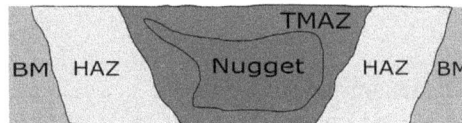

FIGURE 4.2 Microstructure evolution of weld made by FSW.

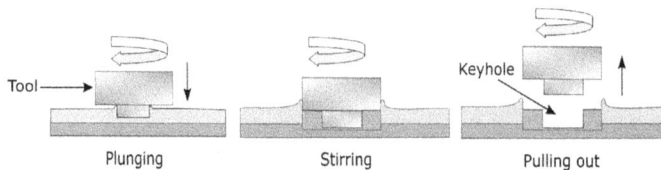

FIGURE 4.3 Schematic of FSSW process.

BM: This zone does not experience any plastic deformation. This zone may or may not experience a heat cycle during welding. The microstructure of this zone is the same as the base material.

HAZ: This zone is near the weld nugget and is affected by the heat cycle, altering the grain structure and orientation, but does not experience any deformation.

TMAZ: The material in this region is plastically deformed and experiences heat cycles during welding, which changes the microstructure evolution. Significant plastic strain can be observed in this region without any dynamic recrystallization.

The nugget zone is the heavily recrystallized zone in the weld zone, also called the stirred zone. The size of grains in this zone is smaller as compared to other weld zones. For example, Xie et al. [7] reported that the grain size at the nugget zone reduced from 9 to 3.5 μm.

Friction stir spot welding (FSSW) is another variant of the friction stir-based joining technique. The method is similar to the conventional FSW process in tooling, fixturing and clamping. The only difference is that here the tool is not traversed along the joint. The Mazda Corporation first introduces the FSSW process for the joining of automotive sheet materials. As shown in Figure 4.3, the FSSW process starts with plunging into the workpiece. The rubbing action between the rotating tool and the workpiece generates a huge frictional heat. Due to this frictional heat, the material around the periphery of the tool get softens and stirred around the tool allowing the sheets to be joined without any melting. After a short dwell time, after dwelling the rotating tool is retracted from the joint, leaving a keyhole [8–10]. The automotive and transportation industries are very much dependent on spot joining processes. FSSW process is considered vital for other spot joining processes like RSW, self-pierce riveting (SPR), and clinching. Earlier resistance spot welding process is used to join aluminium sheets or plates in a lap fashion. However, heavy electric power and erosion of the tip of the electrode hindered the success of this process. In addition, in SPR, the cost of the rivets and the limited range of joint configuration [11]. FSSW has many advantages over other spot joining processes. Since the traversing part is not there, this process is rapid, does not require special tooling, and can join a vast amount of lightweight materials [12–14].

Friction stir technologies have a significant impact on industrial innovations. FSW and FSSW are standard joining processes in the aviation and automotive industries. Since its invention in 1991, The Boeing company has been using this solid-state joining process to fabricate Delta IV and Delta II satellite launch rockets. They reported that they attain a reduced manufacturing time and a reduced cost of 60% [15]. Automotive industries take vast advantages from the FSW and FSSW of aluminium alloy. Rear axles, bumper beams, rear doors, and air suspension systems are typical examples of this process. In the shipbuilding industry, friction stir welded aluminium extrusions are used to manufacture hollow aluminium panels for the deep freezing of fish on boats [16]. In 2003, for the first time, was first used the Mazda group use the conventional FSSW to fabricate rear door panels in Mazda RX-8. It then joined the aluminium trunk lid to the steel bolt retainers in the 2005 Mazda MX-5 sports car. Toyota also used it for decklid and hood for Prius hybrid vehicles [17, 18].

Nowadays, every industry wants their fabricated product to weigh less and at the same time, it delivers better strength with no harmful emissions to the environment. In the automotive, aerospace and shipbuilding industries, hybrid structures fabricated from aluminium and steel alloys is very much appreciated. It allows both the alloys in the same structure. The hybrid structure consists of steel alloy with high strength, good creep resistance, and high corrosion resistance joined with low density, high thermal conductivity, corrosion resistance and low specific weight aluminium alloy [19]. Honda Motor Co. uses aluminium and steel alloy to fabricate the engine cradles. In cryogenic industries, cryogenic liquids are stored in a chamber fabricated from aluminium alloy and stainless steel. Nuclear industries use ion-sensitive chambers fabricated from aluminium and steel alloys [20]. Aluminium and copper dissimilar components also have great potential in the manufacturing and electrical industries. Copper poses improved mechanical, thermal and electrical properties. Along with copper, aluminium also has good electrical and thermal conductivities, making both alloys for electrical and thermal applications. Aluminium and copper joints have found their applications in the fabrication of bus bars, refrigeration tubes, tube sheets, and tubes of heat exchangers [21].

4.2 DISSIMILAR JOINING OF ALUMINIUM AND STEEL GRADES

The fusion welding process is one of the common joining processes in manufacturing industries. Before FSW technologies came into existence, fusion welding processes like arc welding, gas welding, TIG, and MIG are the only welding processes. Fusion welding of lightweight aluminium alloys is considered difficult due to high thermal and electrical conductivity, high thermal expansion coefficient, and low stiffness, which allows the aluminium alloy to consume more heat during welding, leading to distortion solidification defects [22]. Fusion welding of dissimilar alloys is complicated because of their vast differences in mechanical, metallurgical, and chemical properties. Nevertheless, due to the increase in demand for dissimilar or hybrid components (Al-Steel), which makes the structures more efficient, lightweight, cost-effective, and environment-friendly, researchers are always keen to introduce a feasible and suitable joining process to join the dissimilar alloys. However, fusion welding of dissimilar alloys triggers defects in the weld zone. Hydrogen-induced porosity is the most common type of defect due to hydrogen in the weld zone when the workpiece is not adequately cleaned, the presence of moisture in the electrode, the formation of an oxide layer, and shielding gas [23]. A considerable amount of hydrogen porosity in weld leads to reduced strength, ductility, and fatigue life. Inclusion defects are also occurring during the welding of aluminium alloys. This type of defect occurs due to unwanted materials in gases, thin films, and solid particles. The solidification defects also occur when the residual stresses exceed the strength of the solidifying weld metal [24, 25]. Moreover, high heat input during fusion welding processes leads to the formation of brittle intermetallic compounds (IMCs), which leads to crack formation, which ultimately leads to unwanted residual stresses. As shown in Figure 4.4, the appearance of the Fe_xAl_y phases is unavoidable during the fusion welding [26, 27].

The formation of the IMCs during the dissimilar joining cannot be avoided. To understand the different phases and the formation of IMCs, the binary phase diagram of the Fe and Al is shown in Figure 4.5. The main reasons which triggers the formation of the IMCs at the joint interface are the negligible solubility of Fe in aluminium which is in the range 0.01–0.022% between 225 and 600°C, excessive heat generation, differences in the chemical potential of Al and Steel, nucleation of phase

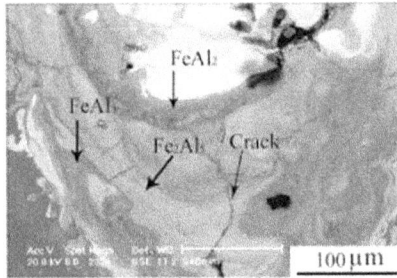

FIGURE 4.4 Intermetallic compounds formed during pulsed Nd: YAG laser welding of low carbon steel and AA5754. (Reprinted with permission from Torkamany et al. [26], Elsevier).

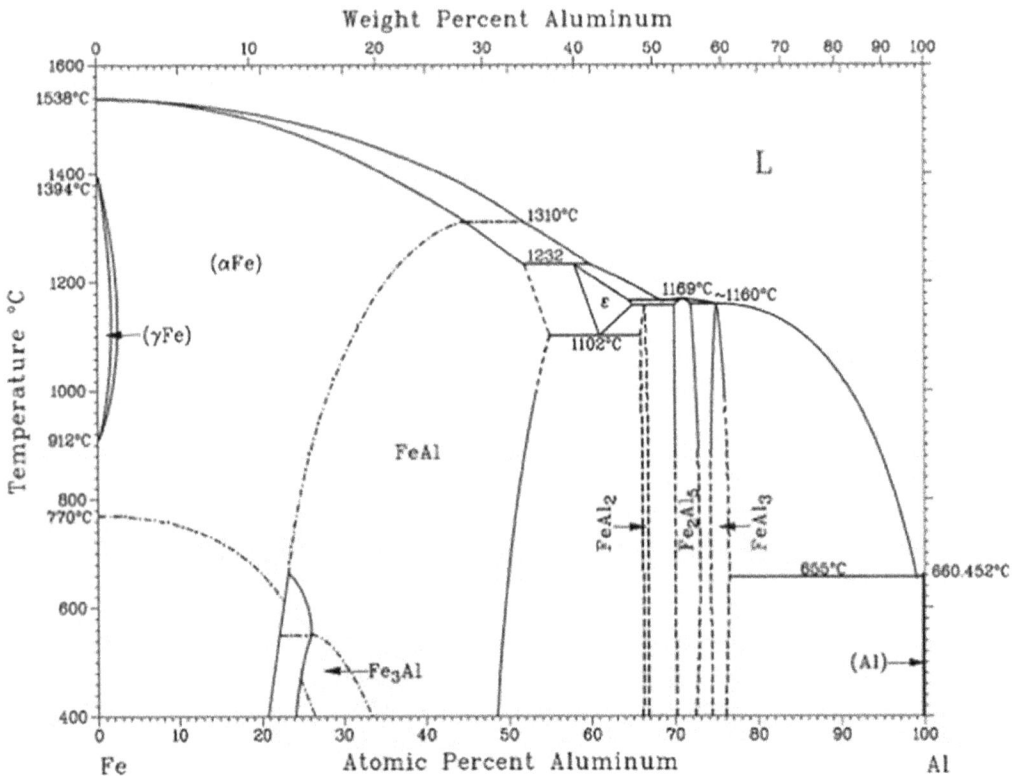

FIGURE 4.5 Binary phase diagram of Fe-Al. (Reprinted with permission from Haidara et al. [29], Elsevier).

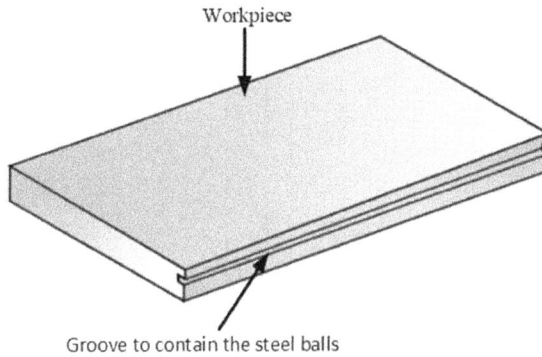

FIGURE 4.6   Schematic of the grove present to contain the tracer steel balls.

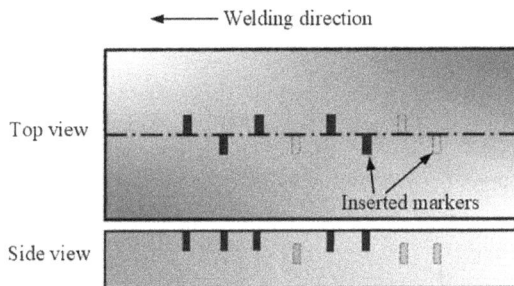

FIGURE 4.7   Arrangement of markers inside the base sheet material.

during interdiffusion, and flow of alloying elements. $FeAl_2$, Fe_2Al_5, and $FeAl_3$ are the common IMCs of aluminium and steel. This IMCs are considered to be the hardest (greater than 1000 VHN) IMCs present at the interface [28].

Understanding material flow in dissimilar FSW is complicated as it holds the key to an efficient joint. The material flow solely depends on the essential welding parameters like welding speed, the rotational speed of the tool and tool profile. If the combination of these welding parameters is not adequately taken, defects can be generated at the weld zone, making the component to be scrapped [30]. Colligan [31] reported the material flow behavior while joining two different aluminium alloys with the help of steel balls of 0.38 diameter. As shown in Figure 4.6, tracer steel balls are placed at the rectangular groove machined into the edge butt profiles of the plates. Seidal et al. [32] reported the material flow behavior of AA 2195 aluminium alloy with the Marker insertion technique (MIT).

The authors planted the AA 5454-H32 marker along the weld path. The markers are placed as shown in Figure 4.7, at the advancing side (AS) and retreating side (RS) of the weld line and different positions of the plate-like, at the top and middle. After the welding is done, the samples are sliced in the transverse direction to visualize the markers' movement to understand the material flow of the base material. There are many ways to monitor the movement of the marker in the base material. For example, Schmidth et al. [33] used copper as marker material and monitored the material flow with X-ray and computer tomography (CT). Guerra et al. [34] and Li et al. [35] use the differential etching technique to monitor the movement of the marker material.

Kumar and Kailas [36], without using external marker material, explained the material flow with the help of the movement of the rotating tool. They reported the pin and the shoulder (Figure 4.8) are responsible for the material flow. The pin moves the material layer by layer, whereas the shoulder moves the material in bulk.

Material flow in dissimilar welding is far more complicated than joining similar alloys. Above mentioned literature on material flow is for joining either similar alloys of aluminium or dissimilar

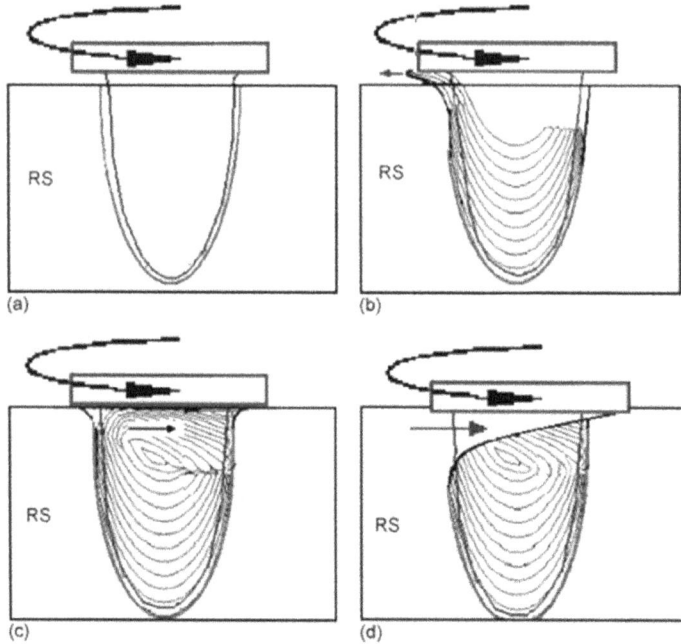

FIGURE 4.8 Schematic of material flow (a) formation cavity during plunging, (b) stacking of layers in pin driven flow, (c) interaction of pin driven and shoulder driven material, and (d) movement of base material in weld nugget. (Reprinted with permission from Kumar and Kailas [36], Elsevier).

aluminium alloys. Material flow in joining dissimilar alloys is challenging because the tool will easily deform the soft material but cannot deform the hard steel material and due to vast differences in metallurgical aspects. Yasui et al. [37] reported the influence of weld line shape on material flow. The authors used two different weld lines (straight and curved) with two different marker materials (aluminium oxide and pure aluminium). Morisada et al. [38] reported the material flow with the help of an X-ray transmission system. It has been reported that the uniform material flow is observed at 400 rpm in both the welding and transverse directions. Defects have been observed in the advancing side because of the inability of steel material to flow since the steel material is not adequately deformed under the tool's action.

The use of aluminium and steel dissimilar components in manufacturing industries is one way to reduce weight, reduce fuel consumption, and attain green sustainable manufacturing technology. However, the joining of these two materials is complicated because of vast differences in melting point temperatures, mechanical and chemical properties. Fusion welding has been used, but defects like the formation of hard and brittle IMCs, solidification, and embrittlement defects always hindered the construction of a sound weld. Over the years, many different types of joining methods have been used to achieve a sound joint. Joining processes like friction welding [39], friction stir knead welding [40], surface activated bonding [41], cold metal transfer [42] and laser penetration welding [43] are used to join aluminium to steel. But, the formation of IMCs during laser welding and cold metal transfer, limitations in using optimized setup and equipment constantly hampered the methods to flourish.

Regardless of recent growth in joining processes, solid-state processes like FSW and FSSW are considered practical joining techniques in modern industries. FSW and FSSW provide more advantages over other solid-state joining methods like diffusion welding, explosion welding, magnetic pulse welding, and ultrasonic welding [44]. Conventional FSW and FSSW are cost and time-productive. They can readily join various plate thicknesses. Moreover, due to the low heat generation in these processes, the thickness of the IMC layer is thinner [45].

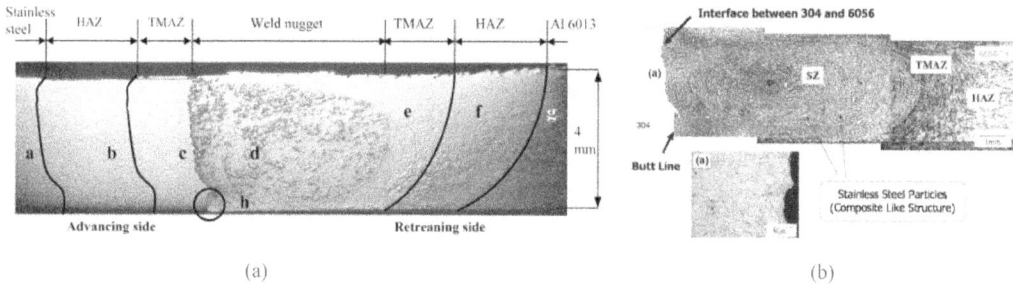

FIGURE 4.9 Macrograph of the joint configuration of (a) Al 6013-T4 and X5CRNi18-10 stainless steel and (b) Al 6056 and AISI 304 stainless steel. (Reprinted with permission from Uzun et al. [50] and Lee et al. [51] respectively, Elsevier).

The microstructure evolution of the dissimilar welding is different as it has TMAZ and zones from both the alloys along with the BM. The bonding of the Al and steel particles takes place at the stirred zone. Since the aluminium base materials are heavily deformed, recrystallized aluminium alloy has been found at the interface, whereas steel particles are present as fragments of different shapes and sizes. Therefore, the amount and size of the steel particles should be less. Otherwise, it will hinder the flow of the aluminium matrix, which leads to metallurgical defects. Large dispersed steel particles in the stirred zone can easily lead to voids as well as tunnel and wormhole defects. Steel particles in the stirred zone trigger solid brittle IMC layers [46–49]. Uzun et al. [50] conducted of FSW of Al 6013-T4 and X5CRNi18-10 stainless steel. As shown in the macrograph (Figure 4.9(a)), the authors have reported the accumulation of steel particles at the aluminium stir zone. About 0.5 μm of thin IMC layer is also reported at the interface. Lee et al. [51] joined stainless steel with an aluminium alloy. Authors have reported that some steel particles (Figure 4.9(b)), which are sheared off due to the action of the rotating tool get dispersed towards the aluminium alloy in a stirred zone. IMC layer of about 250 nm has been generated at the interface.

Dehghani et al. [52] carried out experiments to study the effects of parameters during FSW of an aluminium alloy and mild steel. As shown in Figure 4.10, at the interface, a tunneling defect along with the IMC layer is also formed. The authors reported that with the increase in welding speed, there is a considerable decrease in the thickness of the IMC layer. IMCs like Fe_2Al_5 and Al_6Fe formed at the stirred zone. Kundu et al. [53] investigated the formation of Al_3Fe during FSW of IF-steel and pure aluminium.

FSW of dissimilar aluminium and steel alloys also affects mechanical properties like tensile strength and microhardness. That depends on the amount of mixing of steel particles into the aluminium matrix in the stirred zone. Extensive heat input to the base materials deforms a large amount of Fe particles, hindering the material from flowing, leading to defects (voids and crack) [54–56]. This defect drastically lowers the tensile strength of the joint. Furthermore, substantial temperature cycles lead to the formation of IMC layers forcing the strength to drop. High heat input gathered harder and brittle IMCs at the stirred zone, leading to a brittle fracture of tensile specimens [57, 58]. The hardness values have been reported by many authors, as provided in Table 4.1.

FSSW is a variant of the FSW process in which the rotating tool is retracted without traversing along the joint. Like FSW, FSSW also provides fantastic joint properties in dissimilar welding. In automobile industries, dissimilar Al-steel components are fabricated with the help of the FSSW process. FSSW is considered to be the alternate spot joining process over resistance spot joining and riveting. The critical process parameters in FSSW of Al-steel materials are rotational speed of the tool, plunge speed, plunge depth, dwell time, and position of base materials. The material flow and joint characteristics is the same as mentioned in the FSW process. Dwell time and rotational speed of the tool dictate the material flow and the amount of heat input. Higher input to the base materials will lead to the formation of the IMC layer, the same as the FSW [64]. FSSW of similar alloys

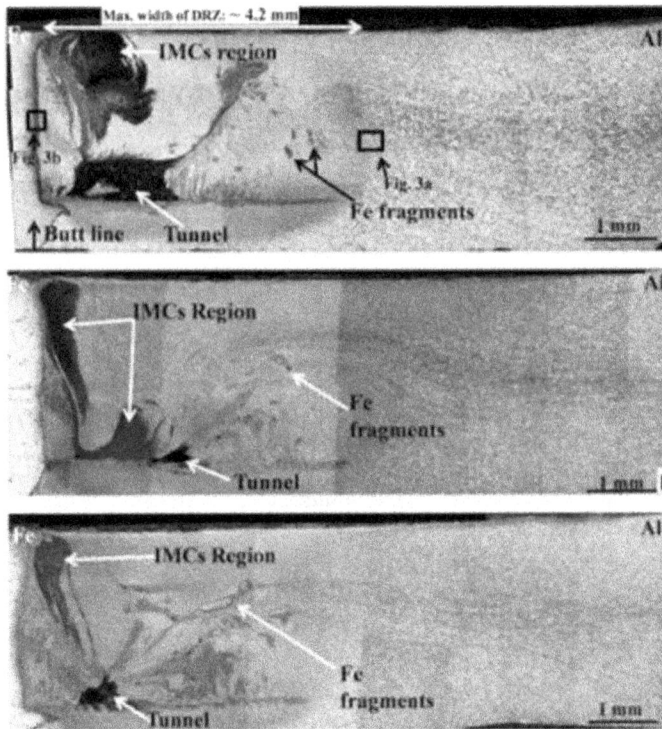

FIGURE 4.10 Macrograph showing the defects, IMC region and Fe particles in aluminium side. (Reprinted with permission from Dehghani et al. [52], Elsevier).

TABLE 4.1
Hardness for Aluminium–Steel Joints

Base Materials Used in the Study	Hardness	Reference
A1100H24 and Low carbon steel	376 HV	[59]
AA6061 and AISI 1081	320 HV	[60]
Al5083 and St-12	335 HV	[61]
Al5754 and DP600 steel	350 HV	[62]
Al 6061 and AISA 1081	400 HV	[63]

has been reported in the literature [65–71]. Chen et al. [72] investigated the joint characteristics and formation of the IMC layer during FSSW of AA 6111-T4 and DC04 alloy in lap fashion. Authors have used a different technique called abrasion circle to understand the effect of process parameters on FSSW. They reported a failure load of 3.5 kN at a lower dwell time of 1.1 sec. The authors did not report formation of IMC layer in the interface. Sun et al. [73] investigated the effects of process parameters in joining dissimilar AA 6061-T6 alloy and mild steel. They reported that 700 rpm and 2 sec dwell time is the optimized process parameters. The authors didn't report any formation of IMCs. Bozzi et al. [74] reported forming IMC layers in FSSW of Al-6061-T6 and IF-steel. The authors reported that the formation of IMC layers (thickness of 8 μm) had increased their failure load by 4.2 kN.

The majority of FSSW is conducted in lap fashion with always aluminium alloy at the top. Keeping the aluminium alloy at the top will allow the tool to get quickly plunged prevents tool wear, resulting in forming a recrystallized stirred zone with a small particle of steel. The plunge depth of the tool is responsible for the formation of defects. Less penetration of tool allows fewer steel particles in the stirred zone, and higher penetration will lead to excessive heat input leading to tool wear and the formation of IMCs. Moreover, no penetration into steel base material (means the tool plunged into aluminium alloy thickness) will lead to weaker joints [75]. In FSSW, joining is established by two means, i.e., by metallurgical bonding and mechanical interlocking. Metallurgical bonding allows the IMCs to form at the interface, whereas mechanical interlocking with hook defect makes the joint weaker. Keyhole features will always be there in case of the FSSW process. The keyhole feature at the end of the FSSW process degrades the aesthetic appearance of the joint and generates a considerable amount of stress concentration in the stirred joint allowing the joint to a lower failure load. In the literature survey, the authors have reported that the significant thin layer of the IMC layer can enhance the joint strength. Refill FSSW can be used to remove the keyhole feature in FSSW of dissimilar alloy [76].

4.3 JOINING OF ALUMINIUM ALLOY TO COPPER ALLOY

Over the years, dissimilar aluminium and copper joints are emerging as a boon to the manufacturing and electrical industries. The industries are trying to alter some of the copper parts with aluminium parts because aluminium is cheaper. It can carry an excellent electrical current per mass unit and poses good thermal and electrical conductivity. Due to these properties, aluminium and copper joints have found their purpose in fabricating bus bars, housing for heat sink and transformer conductors, and heat exchanger tubes [77]. Like other dissimilar alloys joining aluminium and copper is also very complicated due to huge differences in mechanical, metallurgical, and chemical properties. Fusion welding of aluminium and copper has an adverse effect on the joints. Fusion welding fuels the formation of hard and brittle IMC layers at the interface. Aluminum and copper are very irreconcilable to each other and become very reactive above 120°C. Apart from forming the IMCs, fusion welding also triggers solidification and hydrogen cracking defects [78]. The Al-Cu binary phase diagram is shown in Figure 4.11.

Apart from the fusion welding processes, friction welding [80], ultrasonic welding [81, 82], cold rolling, explosive welding [83], diffusion welding [84], FSW [85], and FSSW [86] are also used two join aluminium to copper alloys.

Firouzdor and Kou [87] joined the AA6061 and commercially available copper alloy. As shown in Figure 4.12, the authors reported a different approach to join the alloys in a lap fashion. They used

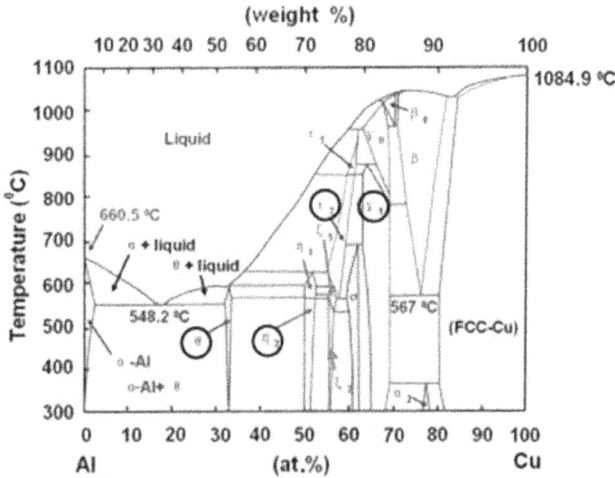

FIGURE 4.11 Al-Cu phase diagram. (Reprinted with permission from Tan et al. [79], Elsevier).

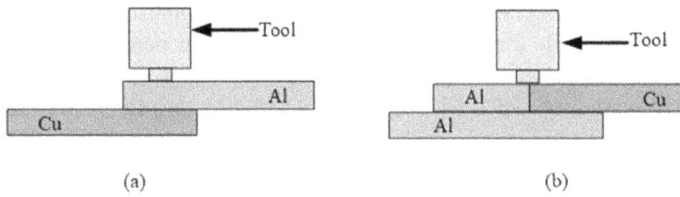

FIGURE 4.12   Schematic of (a) Conventional lap welding and (b) modified lap welding.

an extra aluminium sheet and placed it in a butt fashion. The authors revealed that a better mixing had been achieved between copper particles and aluminium matrix with this modified lap fashion method.

Like any other dissimilar welding, the position of aluminium and copper base materials during dissimilar welding is significant. The uniform material flow in the welding depends on the position of the base materials. In butt welding, the copper alloy is proposed to be placed on the advancing side to achieve a successful joint. During butt welding, the copper particles move from the advancing side to the retreating side in a whirlpool pattern provides better circulation of the copper particles in the stirred zone, leading to the defect-free joint. However, when the copper alloy is put on the retreating side, the tool finds it very difficult to deform the hard copper alloy from the retreating side leading to tunneling and void defects at the lower side of the stirred zone [88, 89]. Moreover, in lap welding, the aluminium alloy should always be kept at the top since the thermal conductivity of aluminium is less and allows a sufficient amount of heat to form at the nugget zone, which ensures a successful joint [90].

The rotational speed of the tool is a vital input process parameter that determines the plastic deformation of the base materials, generation of frictional heat by the shoulder and the amount of forging pressure needed to accomplished the joint. In dissimilar welding, the rotational speed the tool controls the material flow, formation of IMCs, stirred zone, and tool wear. Higher tool rotational speed produces excessive heat at the interface, which can easily trigger the appearance of the IMC layer. On the contrary, a higher rotational speed is required to deform the copper particles into the stirred zone. Higher tool rotational speed is also responsible for the tool wear due to continuous abrasion of tool on the copper alloy. Low rotational speed circulates low heat input, which hinders the material flow. This unsteady material flow due to low heat input leads to volumetric defects like wormhole, void, and tunnel. Moreover, due to less stirring action, the tool deforms fewer copper particles into the nugget, making the joint weaker to sustain heavy failure loads [91, 92].

The microstructure of any FSW consists of four distinct parts: nugget zone, TMAZ, HAZ, and BM. TMAZ zone is both ways, one at the copper side and another one at the aluminium side. The macrographs of all the zones and flow patterns are depicted in Figure 4.13. Esmaeili et al. [93] reported that a broad non-crystallized TMAZ is found in the aluminium side and slender at the joint

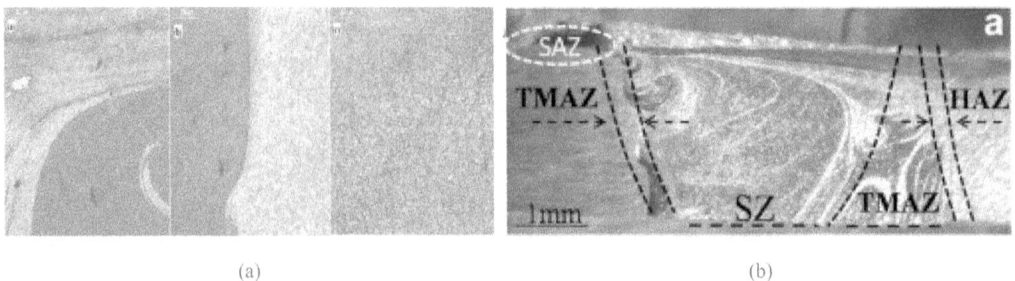

FIGURE 4.13   (a) Flow and grain patterns and (b) Macrograph of all the zones. (Reprinted with permission from Bakkiyaraj et al. [99] and Esmaeili et al. [93] respectively, Elsevier).

interface due to tool offset towards the aluminum alloy. On the other hand, HAZ occurs only in aluminium zone, which comprises coarse grains. A similar type of microstructural evolution has been discussed in the literature [94–98].

At the interface of the Al-Cu joint, three different patterns have been observed. They are lamellar intercalated structures, homogenous mixtures and composite type structures. The lamellar structure consists of swirl and vortex intercalated lamellae [100, 101]. This structure is present inside the stirred zone surrounding the pin. A similar lamellar formation is also reported by Liu et al. [102]. As shown in Figure 4.14 (a), Homogenous mixtures are reported by [103]. The authors reported this structure as a rich intermetallic structure in the shear zone. The possible reason for the formation of these structures is the heavy stirring with high heat input. The third structure is the composite structure [104]. Combining copper particles and intermetallic particles over the aluminium matrix shown in Figure 4.14 (b) reported these structures at the interface. Heavy offset towards the aluminium alloy leads to the formation of these composite structures. The pin sheared off some of the copper particles from the advancing side and deposit them on the aluminium side.

Mechanical properties like joint efficiency and hardness are the standards for an acceptable join. Joint efficiency is a mathematical ratio of weld tensile strength to the tensile strength of the base material. Process parameters of the FSW process heavily influenced the joint strength. The formation of brittle IMCs and other metallurgical defects can reduce the joint strength to a vast extent. Different types of intermetallic phases generated at the interface is presented in Table 4.2.

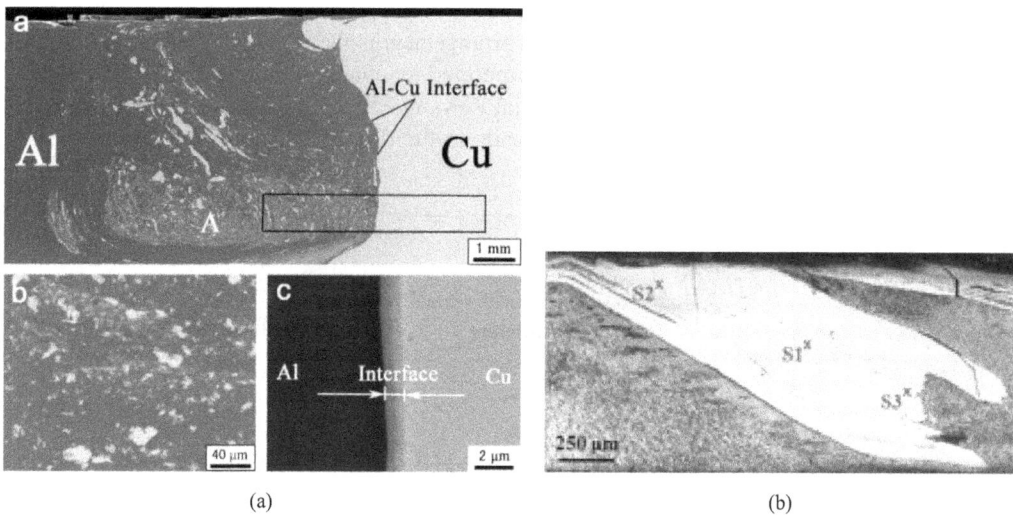

FIGURE 4.14   (a) Composite pattern and (b) Homogeneous pattern in the microstructure. (Reprinted with the permission from Xue et al. [105] and Galvão et al. [106] respectively, Elsevier).

TABLE 4.2
Types of Intermetallic Phases

Base Materials	Intermetallic Phases	Thickness	Reference
5A02 aluminium and pure copper	$CuAl_2$, Cu_9Al_4 and Cu_3Al_2	1 μm	[79]
AA 1060 and pure copper	$CuAl_2$ and Cu_9Al_4	1 μm	[105]
AA 1050 and pure copper	$CuAl_2$ and Cu_9Al_4	0.2 μm	[107]
AA 6082 and pure copper	$CuAl_2$ and Cu_9Al_4	3 μm	[108]
AA 1100 and pure copper	$CuAl_2$, $CuAl$ and Cu_9Al_4	1.7-3.4 μm	[109]

High rotational speed induces heavy stirring, allowing the copper particles to detach from the copper material and get fused in the aluminium matrix, leading to the formation of IMCs, which eventually weakens the joint. With proper and optimized heat input to the interface, refine the grains and increase the joint strength. Lower tool rotational speed and larger tool offset towards the aluminium alloy hinder the mixing of the copper particles leaking to weaker joints. According to the literature, TMAZ region is the failure-prone area. Brittle fracture and de-cohesive fracture are the two failure modes for determining the failure in the TMAZ. The brittle fracture occurs due to the presence of a sufficient amount of brittle intermetallic compounds. De-cohesive fracture occurs due to the presence of metallurgical defects because of unsteady mixing of the base materials [110, 111]. The hardness of a joint depends upon the different zones present in the microstructure. The area near the stirred zone possesses more hardness because of more recrystallized grain present there. The increase in hardness is more accumulation of copper particles and a more homogenous lamellar structure. Higher rotational speed and lower welding speed lead to a rise in heat input, allowing the IMC layer to increase the hardness [112]. HAZ and TMAZ also possess significant hardness as hardness in the HAZ are governed by the recovery and growth of the grains by the thermal cycle and the changes in the density of the precipitates [113]. The hardness of TMAZ in the nugget is higher due to the presence of intermetallic phases. Literature on TMAZ hardness is given in Table 4.3.

Manufacturing and electrical industries are trying to substitute seam welding techniques with a spot-welding process. Mechanical fasteners and riveting are used, but they impart weight to the fabricated components. Also, the high electrical conductivity required for the joint is cannot be achieved with mechanical fastening and riveting. FSSW is preferred over FSW mainly because of the time-saving nature of the process. The lap arrangement is the standard configuration to join aluminium and copper in FSSW. Su et al. [115] reported the material flow in FSSW. The authors reported two distinct ways through which material flows in FSSW: Inner flow and outer flow. As shown in Figure 4.15, inner flow zone corresponds to the downward movement of upper sheet

TABLE 4.3
Literature Related to the TMAZ Hardness

Base Materials	TMAZ Microstructure	Hardness (HV)	Reference
AA1050 and pure copper	Intercalated lamellae	195	[97]
AA 5086 and pure copper	Composite structure	130	[98]
AA 6061 and pure copper	Intermetallic phase (CuAl)	760	[112]
AA 1100 and pure copper	Intermetallic phase (Al/CuAl$_2$ eutectics)	350	[113]
AA 5083 and copper-DHP	Homogenous mixture (CuAl$_2$, Cu$_9$Al$_4$)	701 (Cu$_9$Al$_4$) 518 (CuAl$_2$)	[114]

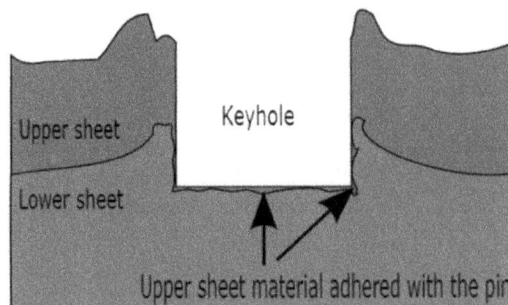

FIGURE 4.15   Patterns of material flow in FSSW.

material with the pin. It becomes dynamically recrystallized, settled around the periphery of the pin. On the other hand, in outer flow, the heavy stirring of the tool allows the material to displace upwards in a spiral fashion.

Yang et al. [116] explained the material flow with help of their zones: transition zone, stir zone and the transition zone (Figure 4.16). The transition zone consists of a mixture of upper sheet and lower sheet material and got adhered to the pin periphery. In the stir zone, the material around the pin got detached and formed a stirred zone around the pin. As the tool progresses, the stirred zone expands and the material present outside the stirred zone is forced towards stirred zone by the pin. Finally, the torsion region comprises material at the bottom of the pin pushed downward with torsion.

The entire FSSW joint is divided into four zones: nugget zone surrounded by keyhole TMAZ, HAZ and base material. The different zones have been shown in Figure 4.17. When the rotating tool is plunging to the sheets, the lower sheet material is squeezed upward by stirring and forming a hook. Heideman et al. [117] reported forming a copper hook into aluminium sheet as a copper ring, as shown in Figure 4.18. below along with the microstructure evolution. The authors also concluded that these copper rings could improve the interlocking of the two alloys. Mubiayi and Akinlabi [118]

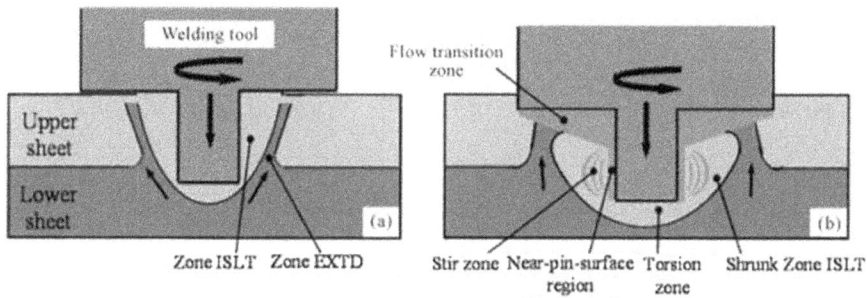

FIGURE 4.16 Schematic of the material flow zones, (a) before shoulder contacts, (b) after shoulder penetrates. (Reprinted with permission from Yang et al. [116], Elsevier).

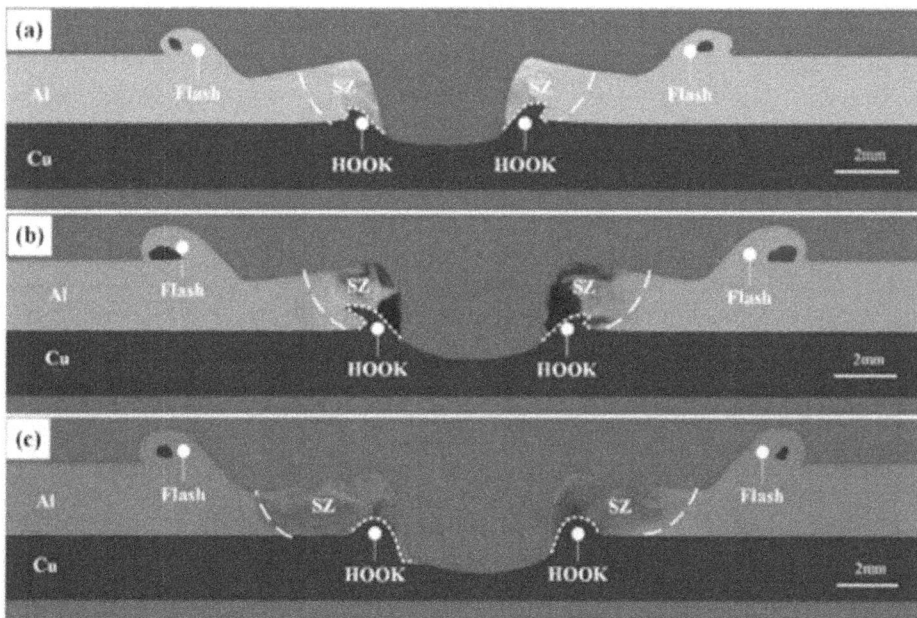

FIGURE 4.17 Macrograph of dissimilar Al-Cu joint at various dwell times ((a)-(c)). (Reprinted with the permission from Li et al. [119], Elsevier).

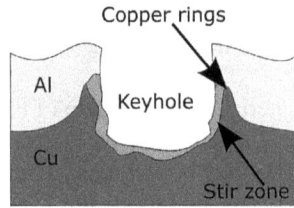

FIGURE 4.18 Schematic of the formation of Onion hook rings.

TABLE 4.4
Failure Loads of Aluminum–Copper Joints

Base Materials	Failure Load (kN)	Rotational Speed (rpm)	Tool Geometry	Reference
5083Al and C10100 Copper	2600	2600	Flat shoulder with cylindrical pin	[121]
AA1060 and pure copper	5225	800	Flat shoulder with flat pin	[126]
5052 Al and C27200 Cu	3908	1350	Flat shoulder with cylindrical pin	[127]
Pure Al and pure copper	4610	1400	Flat shoulder with tapered pin	[128]
AA 6061-T6 and pure copper	1728	2500	Pinless tool with flat shoulder	[129]

also reported this hook feature in the joints. In addition, the authors found that with the increase in plunge depth, the length of copper rings is also increasing.

Zhou et al. [120] reported that an uninterrupted constant laminar IMC layer was generated under high rotational speed. Lower rotational speed IMC like CuAl2 is distributed in the whole nugget zone. Siddharth and Senthilkumar [121] joined AA 5083 and C10100 by the FSSW process. Authors have reported that the material flow depends on variation in plunge depth. In addition, the material flow depends on the tool and the placement of the fringes outside the weld spot, forming rings at the vicinity of the stirred zone. FSSW of aluminium and copper experience plastic deformation and thermal cycle during welding. Solid copper particles detached from the base material got deposited in the liquidus state, which simultaneously forms the unstable, brittle IMCs at the interface.

The hardness of the friction stir spot welded aluminium and copper joint is depends on the influence of the temperature cycles, deformation and accumulation of copper particles in the aluminium matrix [122]. The hardness value of the stirred zone is the highest of all the zones because of the presence of copper particles at the interface. Key reason for the increase in hardness at the stirred zone surrounding the keyhole is hard and brittle IMC layers [123, 124]. The stirred zone consists of copper particles that deform by the stirring action of the tool and the high-temperature evolution makes the grains recrystallized and eventually increasing the hardness. Xiao et al. [125] reported the hardness value of 380 and 525 HV for Al_2Cu and Al_4Cu9, respectively.

Tensile failure load in dissimilar aluminium and copper joints depends upon the tool geometry and the rotational speed of the tool. Mubiayi and Akinlabi [126] has recorded a failure load of 5.225 kN and 4.844 kN for 1800 and 1200 rpm, respectively, when welded with the flat pin and flat shoulder. More literature on failure load is given in Table 4.4.

4.4   FSW OF ALUMINIUM AND POLYMER

Joining aluminium to non-metallic polymers is a standard way to reduce the weight of the components, make them cost-effective, and reduce the fuel consumption to get a greener environment. These polymers possess good specific strength, low thermal expansion, high fatigue and fracture strength and good corrosion resistance [130, 131]. Polymer is an excellent aspect of the automobile

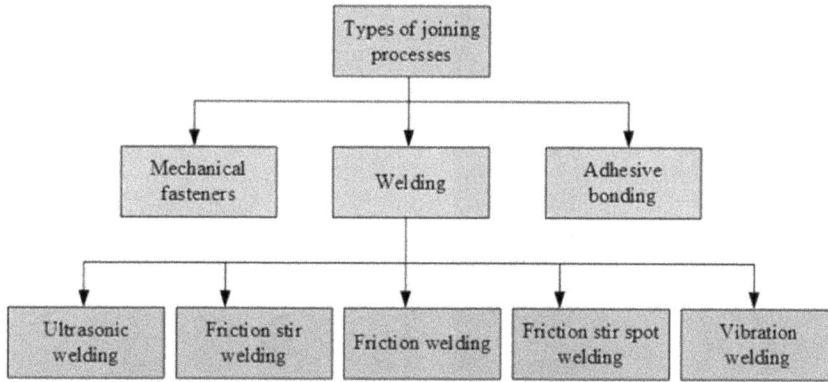

FIGURE 4.19 Flow chart showing different joining processes.

industry because components like vehicle panels and battery housing are fabricated using aluminium and polymers. Polyethylene is one of the promising polymers which is used by many manufacturing industries to fabricate lightweight components. Audi reports the first hybrid metallic and
non-metallic component. The front end of the Audi A6 is manufactured out of a steel sheet with an
elastomer modified polyamide PA6-GF30 [132, 133].

The fabrication of dissimilar materials requires a promising joining technique. For example, as
shown in Figure 4.19, aluminium and polymer dissimilar joining can be achieved by three joining
processes; adhesion bonding, mechanical fasteners, and welding [134, 135].

Adhesive bonding is not recommended because of its long curing and bonding cycles and its affinity towards the moisture content in the environment. The moisture can degrade the strength of the
joint to a large extent. Mechanical fasteners like rivets, bolts, and screws. The main limitation of using
mechanical fasteners is that the hole prepared to place the fasteners generates stress concentration.
Solid welding processes are found to be more efficient in joining aluminium and polymer. Ultrasonic
welding is used only to spot weld the aluminium and polymer. Ultrasonic spot-welding tooling and
machines are not that cost-effective, which hinders the use of this joining process. The limitation with
friction welding is that it always required a flat surface and high tooling and machine costs [136].

Friction stir welding has many advantages over other joining processes mentioned above, like
low tooling and machine cost, low processing temperature, low process time, and defect-free joints.
The mechanism of the FSW process is given in the literature [137]. FSW of a polymer is not considered a solid-state process because polymer contains molecules with varying chain lengths. The
shorter molecule chains melt early during FSW, leaving behind the longer molecule chains in the
solid-state. In 1997 for the first time, FSW had been used to join polymers. To date, many polymers,
such as Polyethylene [138], Acrylonitrile Butadiene styrene [139], and Polymethyl Methacrylate
(PMMA) [140], have been joined using FSW.

A variety of polymers have been joined with aluminium alloys. Khodabakshi et al. [141] joined
AA5059 and High-density Polyethylene (HDPE). The joint and the microstructural evolution have
been given in Figure 4.20. Authors have reported that the joining has been established by mechanical interlocking and chemical adhesion of the polymeric layers and aluminium fragments. Chemical
adhesion bonding occurs when the chemical charges of aluminium atoms move towards the polymer
molecules. Liu et al. [142] successfully joined AA 6061 and MC Nylon- 6 with FSW. The authors
have reported about 5–8 MPa joint shear strength. Rahmat et al. [143] joined AA 7075 and polycarbonate sheet. The authors reported joining has been achieved by interlocking polycarbonate molecules and aluminium alloy at the interface. A similar type of joining by interlocking is also reported
by other literature [144, 145].

Mechanical properties allow us to understand the strength and quality of joints. Khodabakhshi
et al. [141] reported a reduction in joint strength of AA5059 and HDPE dissimilar joint. The possible

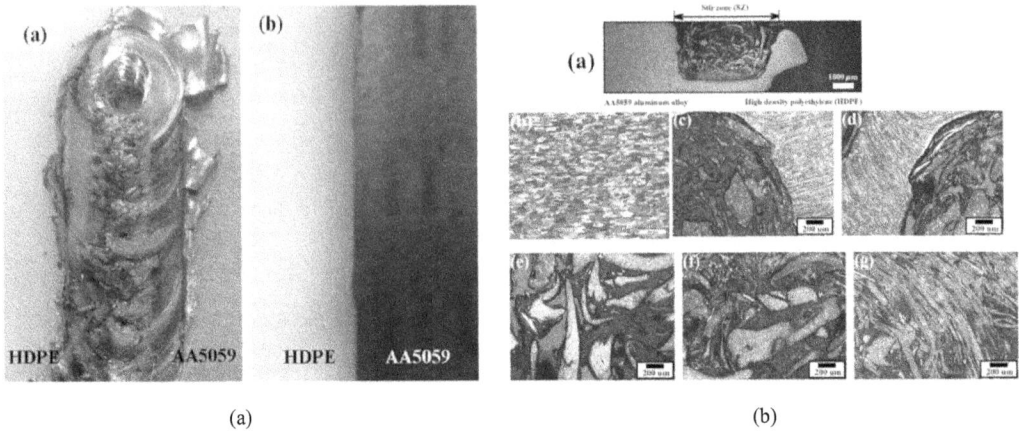

(a)

(b)

FIGURE 4.20   (a) The appearance of the butt joint and (b) microstructural evolution of the joints. (Reprinted with the permission from Khodabakshi et al. [141], Elsevier).

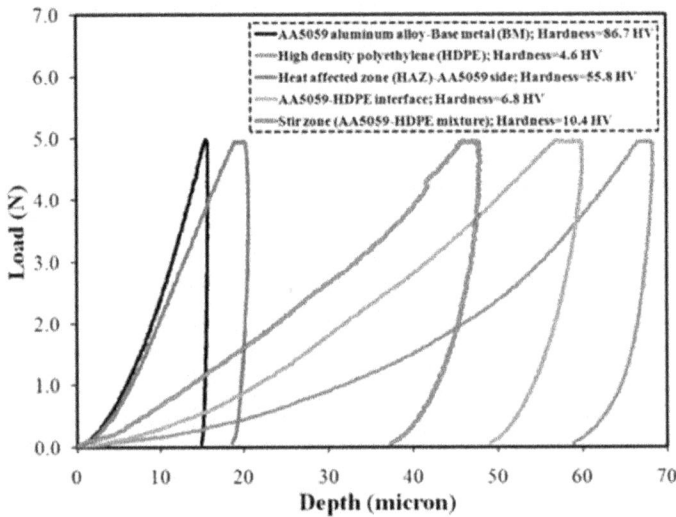

FIGURE 4.21   Load-depth curves during hardness measurement. (Reprinted with permission from Khodabakshi et al. [141], Elsevier).

reason is stress concentration at the edges of the aluminium particles deposited in the solidified HDPE. Derazkola et al. [146] investigated the strength of the joint and reported that with $w =$ 1250 rev/min and $\upsilon = 50$ mm/min joint quality of 45.5 MPa could be achieved. Moshwan et al. [147] studied the effects of process parameters in the joining of polycarbonate and AA 7075 aluminium alloy. Authors have reported a 10% decrease in joint strength compared to polycarbonate base material. Hardness is defined as the ability of the joints to withstand the applied loads. Sahu et al. [148] investigated the microhardness of AA 6063 alloy and polypropylene joint. The authors reported a decreasing trend from aluminium side to the weld centre and then some increasing trend from the weld centre to the interface, followed by a decrease in microhardness in the polymer side. Khodabakhshi et al. [141] use the load-displacement (see Figure 4.21) indentation method to evaluate the hardness of HDPE and AA5059 aluminium joint. The authors reported that there is a decrease in hardness in HAZ, TMAZ and stirred zone. Similar findings are also reported in the literature [149].

4.5 JOINING OF ALUMINIUM TO TITANIUM ALLOY

The fabrication of aluminium and titanium joints is one of the most discussed research topics in the 21st century. Aluminium and titanium dissimilar components have already found their application in the automotive, cryogenic, nuclear and manufacturing industries. The requirement of weight reduction, cost, and increase in efficiency of the components has allowed the researchers to focus on the dissimilar joints. However, joining aluminium and titanium alloys is not so simple because of the enormous differences in mechanical and metallurgical properties and the formation of brittle intermetallic phases at the joint interface. Thermal distortion and residual stresses are also generating in the joint due to the significant contrast in the coefficient of thermal expansion of this alloy. Fusion welding has been tried to join. Still, hydrogen embrittlement, solidification defects, inclusion defects, formation of Ti_xAl_y intermetallic phases prevented the process from flourishing [150, 151].

Aluminium's less dense, highly conductive, high corrosion resistance, and highly malleable properties make it suitable for dissimilar joining with titanium. Titanium also possesses some good properties like high stiffness, toughness, strength, and corrosion resistance. Friction stir welding technologies are the best way to join Al-Ti alloys because here joining is established with the help of coalescence of the base materials under plastic deformation and frictional heat. FSW technologies don't require filler material. The heat generated here is lower than the melting point temperature of the base materials and improved mechanical properties [152]. Some literature on Al-Ti joint is given in Table 4.5.

Li et al. [153] investigated the different microstructural zones in the joint shown in Figure 4.22. Deformed aluminium is found at the interface and hook structure on the titanium side. The authors reported that an increase in tool rotational speed could decrease the tensile strength and the average elongation. Aonuma & Nakata [155] reported that the refined recrystallized aluminium grains and

TABLE 4.5
Joint Characteristics of the Al-Ti Dissimilar Joint

Base Materials	Joining Arrangement	Joint Properties	Reference
Ti-6Al-4V and Al-6Mg	L shape joint	IMCs: TiAl, Ti(Al)$_3$, Ti$_3$Al UTS: 92% Al	[153]
Ti-6Al-4V and AA6061 -T6	Butt joint	UTS: 348 MPa Hardness: 135-140 HV	[154]
Ti-6Al-4V and 7075-T651/2024-T3	Butt joint	IMC: TiAl$_3$ UTS: 311 MPa	[155]
TC1 Ti and LF6 Al	Lap joint	IMCs: TiAl, Ti(Al)$_3$, Ti$_3$Al Max. Shear strength: 48 MPa	[156]
Ti-6Al-4V and AA1060	Lap joint	IMC: TiAl$_3$	[157]
Pure titanium and Al-5083	Butt joint	Hardness: 16% more than the Al and 60% more than Ti	[158]

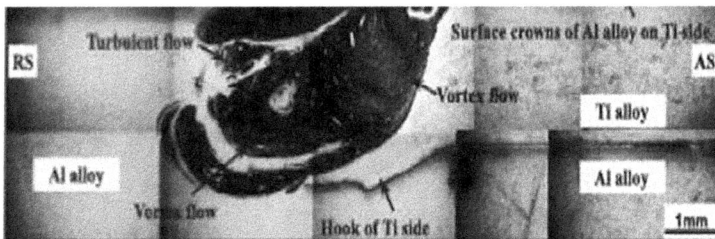

FIGURE 4.22 Microstructural evolution of the joint. (Reprinted with permission from Li et al. [153], Elsevier).

coarse titanium grains are present by the action of the pin. High rotational speed leads to the formation of the lamellae structure in the joint. Chen et al. [156] reported that the increase in rotational speed of the tool increases the frictional, allowing the IMCs to form at the interface, making the joints weak. Also, at lower rotational speed (~600 rpm), flash and groove were generated in the weld.

4.6 TOOL WEAR

In dissimilar welding, the tool wear occurs due to the load exerted by the tool during the rotation and translation. At elevated temperatures, the tool sometimes gets plastically deformed also. Sometimes at the plunging stage, the amount of stresses generated exceeds the load-bearing ability of the tool, which leads to the failure of the tool. It is understood from the literature that the tool plays an important role. The tool allows the material to flow from the advancing side to the retreating side, gives forging pressure to generate the right amount of frictional heat and contains the softened material to not come out of the cavity. Out of shoulder and pin, the pin is prone to failure as it faces plastic deformation. This is because, at the initial plunge of the tool, a huge of stresses are generated, which are easily responsible for the wear and tear of the tool. The rotating tool must have good strength, high toughness, ability to sustain high fracture load, and high load-bearing capacity retains and the tool material must keep its properties above 900°C [159–161]. Therefore, the selection of tool material is essential.

As it is understood from the literature that the rotating tool is considered to be the integral part of the FSW and FSSW. The tool allows the material to flow to obtain a join between the alloys. A well-known FSW and FSSW tool consists of a shoulder that is responsible for generating frictional heat. The forging pressure and pin allow the initial plastic deformation of the base material. Tool steel H-13 is used commercially to join soft materials like aluminium and magnesium because it can easily deform those materials and the temperature generated during is also less (around 550°C). But when it comes to joining two dissimilar alloys (Al-St, Al-Cu), tool steel is not that effective. The possible temperature generated here is about 600–1200°C. The tool steel can't withstand the temperature cycle that arises during welding leading to the deformation of the tool. It is also found that sometimes the pin got fractures during the initial plunging stage [162]. There are two basic categories of material used to fabricate the tools, super abrasive tool materials and refractory metal tools [163]. PCBN comes from the family of super abrasive tool materials. It is considered to second hardest material after diamond. Like the diamond, it has good strength and hardness at elevated temperatures besides excellent thermal and chemical properties [164]. As shown in Figure 4.23,

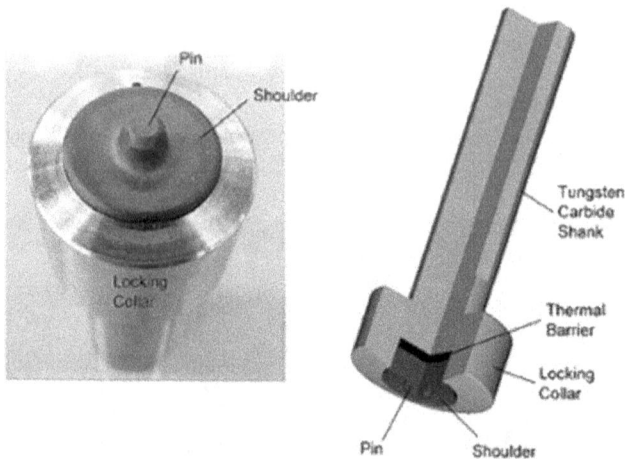

FIGURE 4.23 PCBN tool. (Reprinted with permission from Sato et al. [166], Elsevier).

PCBN tools are used to join many high melting point temperature steel alloys like Inconel alloy 600, SS 304L and SAF 2507 super duplex steel [165–167]. However, due to high stresses and bending fracture generation during plunging, the PCBN tools tend to fail [168]. Tungsten and molybdenum come under the refractory material family. Tungsten rhenium and tungsten carbide tools are the commonly used tools. Bozzi et al. [169] used the W-Re tool to perform FSSW of Al 6061 and IF-steel. Pourali et al. [170] used a conical tungsten carbide tool to join aluminium and steel in a lap fashion. Although they are hard enough to sustain deformation and wear at elevated temperatures, they suffer a fracture during the plunging because of the enormous plunging load and high ductility to the brittle transition temperature.

There exist some other processing techniques through which we can limit the tool wear some extent. They are preheating the harder alloy, friction surfacing, providing tool offset, and positioning the base materials according to the tool rotation direction. Hybrid FSW and FSSW processes are new processes through which we can reduce the tool wear. Out of the new hybrid processes, the thermal assisted processes are beneficial. This thermal-assisted FSW process deploys a preheat source that preheats the harder base material to allow easy stirring without any abrupt loads during welding. Preheating reduces the tool wear, allows a proper plastic flow, improves material flow velocity, and improves joint strength. In addition, preheating makes the materials more ductile and softer so that the tool experiences less wear [171]. Yaduwanshi et al. [172] conducted plasma arc-assisted FSW. The authors reported a decrease in plunging force by 22–28%. In addition, there is a substantial increase in tensile strength and hardness. There are many other ways through which a material can be preheated. They are electrically assisted FSW [173], GTAW assisted FSW [174], laser-assisted FSW [175] and induction coil assisted FSW [176]. An increase in intragranular corrosion during preheating of austenitic stainless steel and embrittlement developed during preheating of precipitation-hardened steels and ferrite steels are two limitations [177].

Friction surfacing is a solid-state process in which a consumable rod is plastically deformed over a substrate and a metallurgical bond develops. The applied pressure and frictional heat are established to accomplish the bond [178]. As proposed by Huang et al. [179], the Al6082-T6 coating layer produced by the friction surfacing process on TC4 titanium alloy prevented the tool from contacting titanium sheet surface during lap welding with another Aluminium plate. Hence, tool wear was not observed. To provide efficient coating, axial force, rotational speed, traverse speed, and other FSW parameters should be optimized. If not optimized, excessive heat affected zone formation, thinner substrate, the weak interface will result in interface failure [180–182].

Providing tool offset towards softer material is another way to reduce tool wear, as only a small portion of the tool will contact the harder base material. The tool cannot be plunged into the central weld line symmetrically. This will lead to excessive heat input into the interface and eventually lead to the aluminium alloy's over-melting. In dissimilar butt-welding of Al-St, steel is frequently kept at the advancing side while the welding tool is kept at a particular offset from the centre weld line towards the aluminium alloy. The offset towards the lower melting point alloy stops the overheating of the low melting point alloy [183]. The offset technique is depicted shown in Figure 4.24. Due to

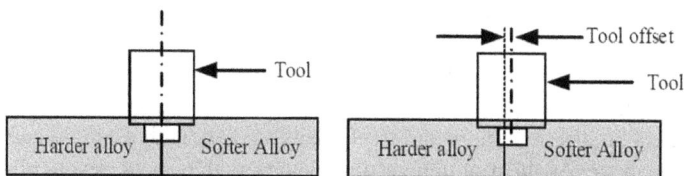

FIGURE 4.24   Schematic of tool offset arrangement.

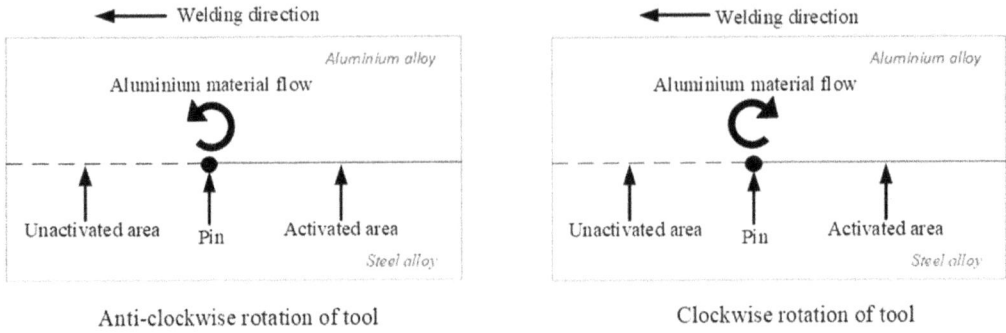

FIGURE 4.25 Schematic of the position of base materials.

the offset provided towards the aluminium alloy, aluminium alloy faces massive plastic deformation and on the other hand, steel alloy undergoes limited deformation. Due to the controlled deformation in steel, the small fragments of the steel get mixed with the aluminium alloy [184].

During the joint formation, the base material position is crucial because excessive heating will exhaust the aluminium alloy, eventually allowing defects to form at the stirred zone. So, in this regard, the soft aluminium alloy must be placed at the retreating side of the joint formation as shown in Figure 4.25, there are two ways to set the base sheet materials concerning the rotational direction of the tool. The activated area corresponds to the location from which the tool has passed. The area got activated with the heat generated by the tool. When the tool is rotating in a clockwise direction, then the steel sheet is on the retreating side. The deformed aluminium alloy moved with the tool from advancing to the retreating side, which the plasticized material come in contact with the oxide film of the steel faying surfaces. Furthermore, the material deposited will try to displace from the cavity due to the tangential velocity of the tool. Due to this join formed with this combination is very week. During the rotation of the tool in an anti-clockwise direction, the steel sheet is on the advancing side. The tool breaks the oxide layer situated on the steel sheet's faying surface, allowing the steel activated area to be exposed to the aluminium alloy. The plasticized aluminium alloy got deposited into the activated area behind the pin leading to a strong joint [185, 186].

Tool offset and base material position is far more cost-effective than using the expensive tool in dissimilar welding. But there are limitations that have to be taken care of in providing tool offset and arranging the base material. It is not recommended to give zero offsets because completely sifting the tool towards the softer alloy generates weak joints. With zero offset, the tool will not be able plastically to deform the harder alloy and due to which there will not be any interaction between harder material particles and liquid aluminium. So, there must be some part of the pin geometry on the hard material side. But this will allow the tool to have rubbing action against the harder alloy resulting in tool abrasion.

Das and Narayanan [187] have introduced a new novel FSSW method to join dissimilar alloys to avoid tool wear. The base materials used by the author are AA 6061-T6 aluminium alloy and CRCA/IS-513 steel. In this method, the rotating tool does not touch the faying surfaces to be joined. Instead, it only touches an aluminium consumable present at the top of the aluminium and steel base material. The consumable material used is the same as the aluminium base material. The authors used a cylindrical flat pin less tool. The pinless tool is used because it allows uniform heat generation. Moreover, with the pinless tool, the keyhole feature can be avoided at the end of the experiment. As shown in Figure 4.26, the aluminum and steel sheets are kept with a certain gap in between, and consumable is kept above. As the tool plunges into a consumable sheet, severe plastic deformation and frictional heating are generated in the consumable along the interface gap. Due to this, the consumable material is plasticized and softens by the frictional heat. With the help of the forging

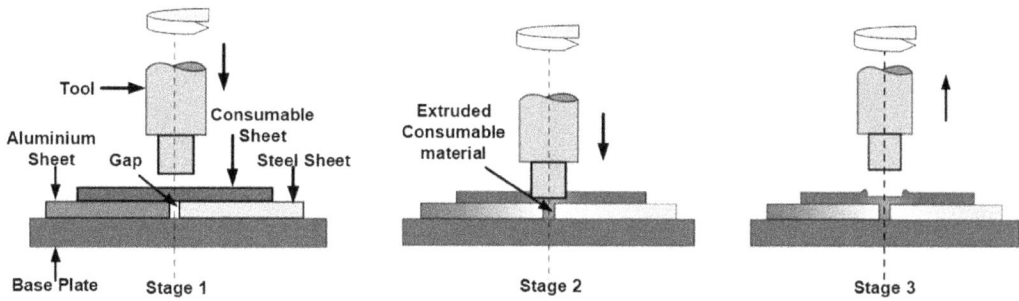

FIGURE 4.26 Schematic of FSSW of dissimilar alloys with a consumable sheet.

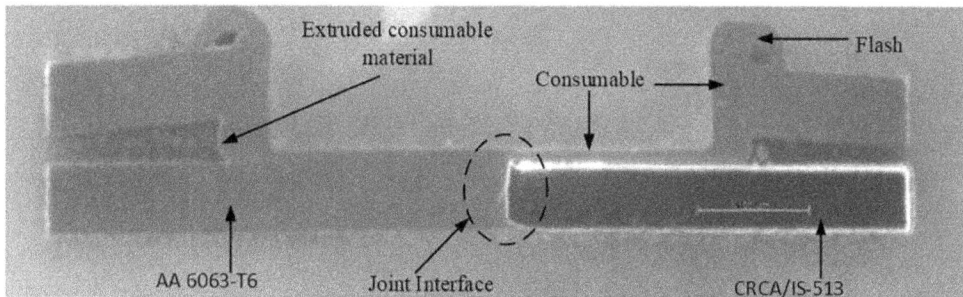

FIGURE 4.27 Macrograph of the joint made between Aluminium sheet and Steel sheet.

pressure provided by the rotating tool plasticized material is extruded through the gap get bonded with the aluminium and steel alloys. After the tool is dwelled for some time it is retracted. In the entire process, the tool will never come in contact with the sheets to be joined. Since the tool is not touching the interface, the formation IMCs is also neglected. The macrograph of the joint formed has been taken in a Nikon stereo microscope and shown below in Figure 4.27.

REFERENCES

[1] R. Jain, S. K. Pal, and S. B. Singh, "Numerical modeling methodologies for friction stir welding process," in *Computational Methods and Production Engineering*, Elsevier, 2017, pp. 125–169. doi:10.1016/B978-0-85709-481-0.00005-7.

[2] G. Wang, Y. Zhao, and Y. Hao, "Friction stir welding of high-strength aerospace aluminum alloy and application in rocket tank manufacturing," *J. Mater. Sci. Technol.*, vol. 34, no. 1, pp. 73–91, Jan. 2018, doi:10.1016/j.jmst.2017.11.041.

[3] P. Dabeer and G. Shinde, "Perspective of friction stir welding tools," *Mater. Today Proc.*, vol. 5, no. 5, pp. 13166–13176, 2018, doi:10.1016/j.matpr.2018.02.307.

[4] A. C. F. Silva, J. De Backer, and G. Bolmsjö, "Temperature measurements during friction stir welding," *Int. J. Adv. Manuf. Technol.*, vol. 88, no. 9–12, pp. 2899–2908, Feb. 2017, doi:10.1007/s00170-016-9007-4.

[5] Z. Harchouche, M. Zemri, and A. Lousdad, "Analytical modeling and analysis of the matter flow during friction stir welding," *Solid State Phenom.*, vol. 297, pp. 1–16, Sep. 2019, doi:10.4028/www.scientific.net/SSP.297.1.

[6] P. L. Threadgill, A. J. Leonard, H. R. Shercliff, and P. J. Withers, "Friction stir welding of aluminium alloys," *Int. Mater. Rev.*, vol. 54, no. 2, pp. 49–93, Mar. 2009, doi:10.1179/174328009X411136.

[7] G. M. Xie, Z. Y. Ma, and L. Geng, "Development of a fine-grained microstructure and the properties of a nugget zone in friction stir welded pure copper," *Scr. Mater.*, vol. 57, no. 2, pp. 73–76, Jul. 2007, doi:10.1016/j.scriptamat.2007.03.048.

[8] M. Fujimoto, M. Inuzuka, S. Koga, and Y. Seta, "Development of friction spot joining," *Weld. World*, vol. 49, no. 3–4, pp. 18–21, Mar. 2005, doi:10.1007/BF03266470.

[9] S. Venukumar, S. Yalagi, and S. Muthukumaran, "Comparison of microstructure and mechanical properties of conventional and refilled friction stir spot welds in AA 6061-T6 using filler plate," *Trans. Nonferrous Met. Soc. China*, vol. 23, no. 10, pp. 2833–2842, Oct. 2013, doi:10.1016/S1003-6326(13)62804-6.

[10] Z. Xu, Z. Li, S. Ji, and L. Zhang, "Refill friction stir spot welding of 5083-O aluminum alloy," *J. Mater. Sci. Technol.*, vol. 34, no. 5, pp. 878–885, May 2018, doi:10.1016/j.jmst.2017.02.011.

[11] X. Yang et al., "Microstructure and properties of probeless friction stir spot welding of AZ31 magnesium alloy joints," *Trans. Nonferrous Met. Soc. China*, vol. 29, no. 11, pp. 2300–2309, Nov. 2019, doi:10.1016/S1003-6326(19)65136-8.

[12] R. Sakano, "Development of spot FSW robot system for automobile body members," 2004.

[13] N. Bhardwaj, R. G. Narayanan, U. S. Dixit, and M. S. J. Hashmi, "Recent developments in friction stir welding and resulting industrial practices," *Adv. Mater. Process. Technol.*, vol. 5, no. 3, pp. 461–496, Jul. 2019, doi:10.1080/2374068X.2019.1631065.

[14] P. K. Rana, R. G. Narayanan, and S. V. Kailas, "Friction stir spot welding of AA5052-H32/HDPE/ AA5052-H32 sandwich sheets at varying plunge speeds," *Thin-Walled Struct.*, vol. 138, pp. 415–429, May 2019, doi:10.1016/j.tws.2019.02.016.

[15] N. Kumar, W. Yuan, and R. S. Mishra, "Challenges and opportunities for friction stir welding of dissimilar alloys and materials," in *Friction Stir Welding of Dissimilar Alloys and Materials*, Elsevier, 2015, pp. 123–126. doi:10.1016/B978-0-12-802418-8.00007-2.

[16] F. Khodabakhshi and A. P. Gerlich, "Potentials and strategies of solid-state additive friction-stir manufacturing technology: A critical review," *J. Manuf. Process.*, vol. 36, pp. 77–92, Dec. 2018, doi:10.1016/j.jmapro.2018.09.030.

[17] Z. Shen, Y. Ding, and A. P. Gerlich, "Advances in friction stir spot welding," *Crit. Rev. Solid State Mater. Sci.*, vol. 45, no. 6, pp. 457–534, Nov. 2020, doi:10.1080/10408436.2019.1671799.

[18] M. Schwartz, *Innovations in Materials Manufacturing, Fabrication, and Environmental Safety*. CRC Press, 2010. doi:10.1201/b10386.

[19] M. Tisza and I. Czinege, "Comparative study of the application of steels and aluminium in lightweight production of automotive parts," *Int. J. Lightweight Mater. Manuf.*, vol. 1, no. 4, pp. 229–238, Dec. 2018, doi:10.1016/j.ijlmm.2018.09.001.

[20] M. Ghosh, A. Kar, K. Kumar, and S. V. Kailas, "Structural characterisation of reaction zone for friction stir welded aluminium–stainless steel joint," *Mater. Technol.*, vol. 27, no. 2, pp. 169–172, Apr. 2012, doi :10.1179/175355509X12608916825994.

[21] M. P. Mubiayi, E. T. Akinlabi, and M. E. Makhatha, "Current state of friction stir spot welding between aluminium and copper," *Mater. Today Proc.*, vol. 5, no. 9, pp. 18633–18640, 2018, doi:10.1016/j.matpr.2018.06.208.

[22] H. K. D. H. Bhadeshia and T. DebRoy, "Critical assessment: Friction stir welding of steels," *Sci. Technol. Weld. Join.*, vol. 14, no. 3, pp. 193–196, Apr. 2009, doi:10.1179/136217109X421300.

[23] M. Bevilacqua, F. Ciarapica, A. Forcellese, and M. Simoncini, "Comparison among the environmental impact of solid state and fusion welding processes in joining an aluminium alloy," *Proc. Inst. Mech. Eng. Part B J. Eng. Manuf.*, vol. 234, no. 1–2, pp. 140–156, Jan. 2020, doi:10.1177/0954405419845572.

[24] P. Kah, R. Rajan, J. Martikainen, and R. Suoranta, "Investigation of weld defects in friction-stir welding and fusion welding of aluminium alloys," *Int. J. Mech. Mater. Eng.*, vol. 10, no. 1, p. 26, Dec. 2015, doi:10.1186/s40712-015-0053-8.

[25] S. D. Meshram, A. G. Paradkar, G. M. Reddy, and S. Pandey, "Friction stir welding: An alternative to fusion welding for better stress corrosion cracking resistance of maraging steel," *J. Manuf. Process.*, vol. 25, pp. 94–103, Jan. 2017, doi:10.1016/j.jmapro.2016.11.005.

[26] M. J. Torkamany, S. Tahamtan, and J. Sabbaghzadeh, "Dissimilar welding of carbon steel to 5754 aluminum alloy by Nd:YAG pulsed laser," *Mater. Des.*, vol. 31, no. 1, pp. 458–465, Jan. 2010, doi:10.1016/j.matdes.2009.05.046.

[27] B. Li, Z. Zhang, Y. Shen, W. Hu, and L. Luo, "Dissimilar friction stir welding of Ti-6Al–4V alloy and aluminum alloy employing a modified butt joint configuration: Influences of process variables on the weld interfaces and tensile properties," *Mater. Des.*, vol. 53, pp. 838–848, Jan. 2014, doi:10.1016/j.matdes.2013.07.019.

[28] K. P. Mehta, "A review on friction-based joining of dissimilar aluminum–steel joints," *J. Mater. Res.*, vol. 34, no. 1, pp. 78–96, Jan. 2019, doi:10.1557/jmr.2018.332.

[29] F. Haidara, M.-C. Record, B. Duployer, and D. Mangelinck, "Phase formation in Al–Fe thin film systems," *Intermetallics*, vol. 23, pp. 143–147, Apr. 2012, doi:10.1016/j.intermet.2011.11.017.

[30] A. P. Reynolds, "Flow visualization and simulation in FSW," *Scr. Mater.*, vol. 58, no. 5, pp. 338–342, Mar. 2008, doi:10.1016/j.scriptamat.2007.10.048.

[31] K. Colligan, "Material Flow Behavior during Friction Stir Welding of Aluminum," p. 9.

[32] T. U. Seidel and A. P. Reynolds, "Visualization of the material flow in AA2195 friction-stir welds using a marker insert technique," *Metall. Mater. Trans. A*, vol. 32, no. 11, pp. 2879–2884, Nov. 2001, doi:10.1007/s11661-001-1038-1.

[33] H. N. B. Schmidt, T. L. Dickerson, and J. H. Hattel, "Material flow in butt friction stir welds in AA2024-T3," *Acta Mater.*, vol. 54, no. 4, pp. 1199–1209, Feb. 2006, doi:10.1016/j.actamat.2005.10.052.

[34] M. Guerra, C. Schmidt, J. C. McClure, L. E. Murr, and A. C. Nunes, "Flow patterns during friction stir welding," *Mater. Charact.*, vol. 49, no. 2, pp. 95–101, Sep. 2002, doi:10.1016/S1044-5803(02)00362-5.

[35] Y. Li, L. E. Murr, and J. C. McClure, "Flow visualization and residual microstructures associated with the friction-stir welding of 2024 aluminum to 6061 aluminum," *Mater. Sci. Eng. A*, vol. 271, no. 1–2, pp. 213–223, Nov. 1999, doi:10.1016/S0921-5093(99)00204-X.

[36] K. Kumar and S. V. Kailas, "The role of friction stir welding tool on material flow and weld formation," *Mater. Sci. Eng. A*, vol. 485, no. 1–2, pp. 367–374, Jun. 2008, doi:10.1016/j.msea.2007.08.013.

[37] T. Yasui, N. Ando, S. Morinaka, H. Mizushima, and M. Fukumoto, "Effect of weld line shape on material flow during friction stir welding of aluminum and steel," *IOP Conf. Ser. Mater. Sci. Eng.*, vol. 61, p. 012009, Aug. 2014, doi:10.1088/1757-899X/61/1/012009.

[38] Y. Morisada, T. Imaizumi, and H. Fujii, "Clarification of material flow and defect formation during friction stir welding," *Sci. Technol. Weld. Join.*, vol. 20, no. 2, pp. 130–137, Feb. 2015, doi:10.1179/1362171 814Y.0000000266.

[39] M. Sahin, "Joining of stainless-steel and aluminium materials by friction welding," *Int. J. Adv. Manuf. Technol.*, vol. 41, no. 5–6, pp. 487–497, Mar. 2009, doi:10.1007/s00170-008-1492-7.

[40] M. Geiger et al., "Friction Stir Knead Welding of steel aluminium butt joints," *Int. J. Mach. Tools Manuf.*, vol. 48, no. 5, pp. 515–521, Apr. 2008, doi:10.1016/j.ijmachtools.2007.08.002.

[41] M. M. R. Howlader, T. Kaga, and T. Suga, "Investigation of bonding strength and sealing behavior of aluminum/stainless steel bonded at room temperature," *Vacuum*, vol. 84, no. 11, pp. 1334–1340, Jun. 2010, doi:10.1016/j.vacuum.2010.02.014.

[42] R. Cao, G. Yu, J. H. Chen, and P.-C. Wang, "Cold metal transfer joining aluminum alloys-to-galvanized mild steel," *J. Mater. Process. Technol.*, vol. 213, no. 10, pp. 1753–1763, Oct. 2013, doi:10.1016/j.jmatprotec.2013.04.004.

[43] S. Chen, J. Huang, K. Ma, X. Zhao, and A. Vivek, "Microstructures and mechanical properties of laser penetration welding joint with/without Ni-foil in an overlap steel-on-aluminum configuration," *Metall. Mater. Trans. A*, vol. 45, no. 7, pp. 3064–3073, Jun. 2014, doi:10.1007/s11661-014-2241-1.

[44] Y. Kusuda, "Honda develops robotized FSW technology to weld steel and aluminum and applied it to a mass-production vehicle," *Ind. Robot Int. J.*, vol. 40, no. 3, pp. 208–212, Apr. 2013, doi:10.1108/01439911311309889.

[45] V. Firouzdor and S. Kou, "Formation of liquid and intermetallics in Al-to-Mg friction stir welding," *Metall. Mater. Trans. A*, vol. 41, no. 12, pp. 3238–3251, Dec. 2010, doi:10.1007/s11661-010-0366-4.

[46] A. Yazdipour and A. Heidarzadeh, "Effect of friction stir welding on microstructure and mechanical properties of dissimilar Al 5083-H321 and 316L stainless steel alloy joints," *J. Alloys Compd.*, vol. 680, pp. 595–603, Sep. 2016, doi:10.1016/j.jallcom.2016.03.307.

[47] R. P. Mahto, R. Bhoje, S. K. Pal, H. S. Joshi, and S. Das, "A study on mechanical properties in friction stir lap welding of AA 6061-T6 and AISI 304," *Mater. Sci. Eng. A*, vol. 652, pp. 136–144, Jan. 2016, doi:10.1016/j.msea.2015.11.064.

[48] K. K. Ramachandran, N. Murugan, and S. Shashi Kumar, "Effect of tool axis offset and geometry of tool pin profile on the characteristics of friction stir welded dissimilar joints of aluminum alloy AA5052 and HSLA steel," *Mater. Sci. Eng. A*, vol. 639, pp. 219–233, Jul. 2015, doi:10.1016/j.msea.2015.04.089.

[49] M. Elyasi, H. Aghajani Derazkola, and M. Hosseinzadeh, "Investigations of tool tilt angle on properties friction stir welding of A441 AISI to AA1100 aluminium," *Proc. Inst. Mech. Eng. Part B J. Eng. Manuf.*, vol. 230, no. 7, pp. 1234–1241, Jul. 2016, doi:10.1177/0954405416645986.

[50] H. Uzun, C. Dalle Donne, A. Argagnotto, T. Ghidini, and C. Gambaro, "Friction stir welding of dissimilar Al 6013-T4 To X5CrNi18-10 stainless steel," *Mater. Des.*, vol. 26, no. 1, pp. 41–46, Feb. 2005, doi:10.1016/j.matdes.2004.04.002.

[51] W.-B. Lee, M. Schmuecker, U. A. Mercardo, G. Biallas, and S.-B. Jung, "Interfacial reaction in steel–aluminum joints made by friction stir welding," *Scr. Mater.*, vol. 55, no. 4, pp. 355–358, Aug. 2006, doi:10.1016/j.scriptamat.2006.04.028.

[52] M. Dehghani, A. Amadeh, and S. A. A. Akbari Mousavi, "Investigations on the effects of friction stir welding parameters on intermetallic and defect formation in joining aluminum alloy to mild steel," *Mater. Des.*, vol. 49, pp. 433–441, Aug. 2013, doi:10.1016/j.matdes.2013.01.013.

[53] S. Kundu, D. Roy, R. Bhola, D. Bhattacharjee, B. Mishra, and S. Chatterjee, "Microstructure and tensile strength of friction stir welded joints between interstitial free steel and commercially pure aluminium," *Mater. Des.*, vol. 50, pp. 370–375, Sep. 2013, doi:10.1016/j.matdes.2013.02.017.

[54] H. K. D. H. Bhadeshia and T. DebRoy, "Critical assessment: Friction stir welding of steels," *Sci. Technol. Weld. Join.*, vol. 14, no. 3, pp. 193–196, Apr. 2009, doi:10.1179/136217109X421300.

[55] C. A. Maltin, L. J. Nolton, J. L. Scott, A. I. Toumpis, and A. M. Galloway, "The potential adaptation of stationary shoulder friction stir welding technology to steel," *Mater. Des.*, vol. 64, pp. 614–624, Dec. 2014, doi:10.1016/j.matdes.2014.08.017.

[56] R. A. Prado, L. E. Murr, K. F. Soto, and J. C. McClure, "Self-optimization in tool wear for friction-stir welding of Al 6061+20% Al2O3 MMC," *Mater. Sci. Eng. A*, vol. 349, no. 1–2, pp. 156–165, May 2003, doi:10.1016/S0921-5093(02)00750-5.

[57] L. H. Shah and M. Ishak, "Review of research progress on aluminum–steel dissimilar welding," *Mater. Manuf. Process.*, vol. 29, no. 8, pp. 928–933, Aug. 2014, doi:10.1080/10426914.2014.880461.

[58] H. Das, A. Kumar, K. V. Rajkumar, T. Saravanan, T. Jayakumar, and T. K. Pal, "Nondestructive evaluation of friction stir-welded aluminum alloy to coated steel sheet lap joint," *J. Mater. Eng. Perform.*, vol. 24, no. 11, pp. 4192–4199, Nov. 2015, doi:10.1007/s11665-015-1713-9.

[59] A. Elrefaey, M. Gouda, M. Takahashi, and K. Ikeuchi, "Characterization of aluminum/steel lap joint by friction stir welding," *J. Mater. Eng. Perform.*, vol. 14, no. 1, pp. 10–17, Feb. 2005, doi:10.1361/10599490522310.

[60] W. H. Jiang and R. Kovacevic, "Feasibility study of friction stir welding of 6061-T6 aluminium alloy with AISI 1018 steel," *Proc. Inst. Mech. Eng. Part B J. Eng. Manuf.*, vol. 218, no. 10, pp. 1323–1331, Oct. 2004, doi:10.1243/0954405042323612.

[61] M. Movahedi, A. H. Kokabi, S. M. Seyed Reihani, W. J. Cheng, and C. J. Wang, "Effect of annealing treatment on joint strength of aluminum/steel friction stir lap weld," *Mater. Des.*, vol. 44, pp. 487–492, Feb. 2013, doi:10.1016/j.matdes.2012.08.028.

[62] Z. Shen, Y. Chen, M. Haghshenas, and A. P. Gerlich, "Role of welding parameters on interfacial bonding in dissimilar steel/aluminum friction stir welds," *Eng. Sci. Technol. Int. J.*, vol. 18, no. 2, pp. 270–277, Jun. 2015, doi:10.1016/j.jestch.2014.12.008.

[63] C. M. Chen and R. Kovacevic, "Joining of Al 6061 alloy to AISI 1018 steel by combined effects of fusion and solid state welding," *Int. J. Mach. Tools Manuf.*, vol. 44, no. 11, pp. 1205–1214, Sep. 2004, doi:10.1016/j.ijmachtools.2004.03.011.

[64] E. Fereiduni, M. Movahedi, and A. H. Kokabi, "Aluminum/steel joints made by an alternative friction stir spot welding process," *J. Mater. Process. Technol.*, vol. 224, pp. 1–10, Oct. 2015, doi:10.1016/j.jmatprotec.2015.04.028.

[65] S. Mehrez, M. Paidar, K. Cooke, R. V. Vignesh, O. O. Ojo, and B. Babaei, "A comparative study on weld characteristics of AA5083-H112 to AA6061-T6 sheets produced by MFSC and FSSW processes," *Vacuum*, vol. 190, p. 110298, Aug. 2021, doi:10.1016/j.vacuum.2021.110298.

[66] D. Labus Zlatanovic et al., "An experimental study on lap joining of multiple sheets of aluminium alloy (AA 5754) using friction stir spot welding," *Int. J. Adv. Manuf. Technol.*, vol. 107, no. 7–8, pp. 3093–3107, Apr. 2020, doi:10.1007/s00170-020-05214-z.

[67] A. Kumar Pandey and S. S. Mahapatra, "Investigation of weld zone obtained by friction stir spot welding (FSSW) of aluminium-6061 alloy," *Mater. Today Proc.*, vol. 18, pp. 4491–4500, 2019, doi:10.1016/j.matpr.2019.07.419.

[68] N. Pathak, K. Bandyopadhyay, M. Sarangi, and S. K. Panda, "Microstructure and mechanical performance of friction stir spot-welded aluminum-5754 sheets," *J. Mater. Eng. Perform.*, vol. 22, no. 1, pp. 131–144, Jan. 2013, doi:10.1007/s11665-012-0244-x.

[69] M. Tutar, H. Aydin, C. Yuce, N. Yavuz, and A. Bayram, "The optimisation of process parameters for friction stir spot-welded AA3003-H12 aluminium alloy using a Taguchi orthogonal array," *Mater. Des.*, vol. 63, pp. 789–797, Nov. 2014, doi:10.1016/j.matdes.2014.07.003.

[70] F. Yusof, Y. Miyashita, N. Seo, Y. Mutoh, and R. Moshwan, "Utilising friction spot joining for dissimilar joint between aluminium alloy (A5052) and polyethylene terephthalate," *Sci. Technol. Weld. Join.*, vol. 17, no. 7, pp. 544–549, Oct. 2012, doi:10.1179/136217112x13408696326530.

[71] M. Merzoug, M. Mazari, L. Berrahal, and A. Imad, "Parametric studies of the process of friction spot stir welding of aluminium 6060-T5 alloys," *Mater. Des.*, vol. 31, no. 6, pp. 3023–3028, Jun. 2010, doi:10.1016/j.matdes.2009.12.029.

[72] Y. C. Chen, A. Gholinia, and P. B. Prangnell, "Interface structure and bonding in abrasion circle friction stir spot welding: A novel approach for rapid welding aluminium alloy to steel automotive sheet," *Mater. Chem. Phys.*, vol. 134, no. 1, pp. 459–463, May 2012, doi:10.1016/j.matchemphys.2012.03.017.

[73] Y. F. Sun, H. Fujii, N. Takaki, and Y. Okitsu, "Microstructure and mechanical properties of dissimilar Al alloy/steel joints prepared by a flat spot friction stir welding technique," *Mater. Des.*, vol. 47, pp. 350–357, May 2013, doi:10.1016/j.matdes.2012.12.007.

[74] S. Bozzi, A. L. Helbert-Etter, T. Baudin, B. Criqui, and J. G. Kerbiguet, "Intermetallic compounds in Al 6016/IF-steel friction stir spot welds," *Mater. Sci. Eng. A*, vol. 527, no. 16–17, pp. 4505–4509, Jun. 2010, doi:10.1016/j.msea.2010.03.097.

[75] K. Chen, X. Liu, and J. Ni, "Effects of process parameters on friction stir spot welding of aluminum alloy to advanced high-strength steel," *J. Manuf. Sci. Eng.*, vol. 139, no. 8, p. 081016, Aug. 2017, doi:10.1115/1.4036225.

[76] O. M. Ikumapayi and E. T. Akinlabi, "Recent advances in keyhole defects repairs via refilling friction stir spot welding," *Mater. Today Proc.*, vol. 18, pp. 2201–2208, 2019, doi:10.1016/j.matpr.2019.06.663.

[77] R. S. Mishra and Z. Y. Ma, "Friction stir welding and processing," *Mater. Sci. Eng. R Rep.*, vol. 50, no. 1–2, pp. 1–78, Aug. 2005, doi:10.1016/j.mser.2005.07.001.

[78] F. Ji, S. Xue, J. Lou, Y. Lou, and S. Wang, "Microstructure and properties of Cu/Al joints brazed with Zn–Al filler metals," *Trans. Nonferrous Met. Soc. China*, vol. 22, pp. 281–287, Feb. 2012, doi:10.1016/S1003-6326(11)61172-2.

[79] C. W. Tan, Z. G. Jiang, L. Q. Li, Y. B. Chen, and X. Y. Chen, "Microstructural evolution and mechanical properties of dissimilar Al–Cu joints produced by friction stir welding," *Mater. Des.*, vol. 51, pp. 466–473, Oct. 2013, doi:10.1016/j.matdes.2013.04.056.

[80] G. B. Sathishkumar et al., "Friction welding of similar and dissimilar materials: A review," *Mater. Today Proc.*, p. S2214785321021088, Mar. 2021, doi:10.1016/j.matpr.2021.03.089.

[81] M. P. Matheny and K. F. Graff, "11 - Ultrasonic welding of metals," in *Power Ultrasonics*, J. A. Gallego-Juárez and K. F. Graff, Eds. Oxford: Woodhead Publishing, 2015, pp. 259–293. doi:10.1016/B978-1-78242-028-6.00011-9.

[82] S. Matsuoka and H. Imai, "Direct welding of different metals used ultrasonic vibration," *J. Mater. Process. Technol.*, vol. 209, no. 2, pp. 954–960, Jan. 2009, doi:10.1016/j.jmatprotec.2008.03.006.

[83] H. Okamura and K. Aota, "Joining of dissimilar materials with friction stir welding," *Weld. Int.*, vol. 18, no. 11, pp. 852–860, Nov. 2004, doi:10.1533/wint.2004.3344.

[84] J. P. Bergmann, F. Petzoldt, R. Schürer, and S. Schneider, "Solid-state welding of aluminum to copper—case studies," *Weld. World*, vol. 57, no. 4, pp. 541–550, Jul. 2013, doi:10.1007/s40194-013-0049-z.

[85] S. A. Hussein, A. S. M. Tahir, and A. B. Hadzley, "Characteristics of aluminum-to-steel joint made by friction stir welding: A review," *Mater. Today Commun.*, vol. 5, pp. 32–49, Dec. 2015, doi:10.1016/j.mtcomm.2015.09.004.

[86] K. P. Mehta and V. J. Badheka, "A review on dissimilar friction stir welding of copper to aluminum: Process, properties, and variants," *Mater. Manuf. Process.*, vol. 31, no. 3, pp. 233–254, Feb. 2016, doi:10.1080/10426914.2015.1025971.

[87] V. Firouzdor and S. Kou, "Al-to-Cu friction stir lap welding," *Metall. Mater. Trans. A*, vol. 43, no. 1, pp. 303–315, Jan. 2012, doi:10.1007/s11661-011-0822-9.

[88] P. Xue, B. L. Xiao, D. R. Ni, and Z. Y. Ma, "Enhanced mechanical properties of friction stir welded dissimilar Al–Cu joint by intermetallic compounds," *Mater. Sci. Eng. A*, vol. 527, no. 21–22, pp. 5723–5727, Aug. 2010, doi:10.1016/j.msea.2010.05.061.

[89] P. Xue, D. R. Ni, D. Wang, B. L. Xiao, and Z. Y. Ma, "Effect of friction stir welding parameters on the microstructure and mechanical properties of the dissimilar Al–Cu joints," *Mater. Sci. Eng. A*, vol. 528, no. 13–14, pp. 4683–4689, May 2011, doi:10.1016/j.msea.2011.02.067.

[90] M. Akbari, R. Abdi Behnagh, and A. Dadvand, "Effect of materials position on friction stir lap welding of Al to Cu," *Sci. Technol. Weld. Join.*, vol. 17, no. 7, pp. 581–588, Oct. 2012, doi:10.1179/1362171812Y.0000000049.

[91] E. T. Akinlabi, "Effect of shoulder size on weld properties of dissimilar metal friction stir welds," *J. Mater. Eng. Perform.*, vol. 21, no. 7, pp. 1514–1519, Jul. 2012, doi:10.1007/s11665-011-0046-6.

[92] L. Zhou et al., "Microstructure evolution and mechanical properties of friction stir spot welded dissimilar aluminum-copper joint," *J. Alloys Compd.*, vol. 775, pp. 372–382, Feb. 2019, doi:10.1016/j.jallcom.2018.10.045.

[93] A. Esmaeili, H. R. Zareie Rajani, M. Sharbati, M. K. B. Givi, and M. Shamanian, "The role of rotation speed on intermetallic compounds formation and mechanical behavior of friction stir welded brass/aluminum 1050 couple," *Intermetallics*, vol. 19, no. 11, pp. 1711–1719, Nov. 2011, doi:10.1016/j.intermet.2011.07.006.

[94] A. Abdollah-Zadeh, T. Saeid, and B. Sazgari, "Microstructural and mechanical properties of friction stir welded aluminum/copper lap joints," *J. Alloys Compd.*, vol. 460, no. 1–2, pp. 535–538, Jul. 2008, doi:10.1016/j.jallcom.2007.06.009.

[95] T. Saeid, A. Abdollah-Zadeh, and B. Sazgari, "Weldability and mechanical properties of dissimilar aluminum–copper lap joints made by friction stir welding," *J. Alloys Compd.*, vol. 490, no. 1–2, pp. 652–655, Feb. 2010, doi:10.1016/j.jallcom.2009.10.127.

[96] H. Bisadi, A. Tavakoli, M. Tour Sangsaraki, and K. Tour Sangsaraki, "The influences of rotational and welding speeds on microstructures and mechanical properties of friction stir welded Al5083 and commercially pure copper sheets lap joints," *Mater. Des.*, vol. 43, pp. 80–88, Jan. 2013, doi:10.1016/j.matdes.2012.06.029.

[97] H. Barekatain, M. Kazeminezhad, and A. H. Kokabi, "Microstructure and mechanical properties in dissimilar butt friction stir welding of severely plastic deformed aluminum AA 1050 and commercially pure copper sheets," *J. Mater. Sci. Technol.*, vol. 30, no. 8, pp. 826–834, Aug. 2014, doi:10.1016/j.jmst.2013.11.007.

[98] A. O. Al-Roubaiy, S. M. Nabat, and A. D. L. Batako, "Experimental and theoretical analysis of friction stir welding of Al–Cu joints," *Int. J. Adv. Manuf. Technol.*, vol. 71, no. 9–12, pp. 1631–1642, Apr. 2014, doi:10.1007/s00170-013-5563-z.

[99] M. Bakkiyaraj, S. S. Bernard, G. Saikrishnan, S. Guruyogesh, T. G. Guruprasanna, and K. Dineshkumar, "Effect of tool offset condition on mechanical and metallurgical properties of FSW dissimilar Al-Cu joint," *Mater. Today Proc.*, vol. 43, pp. 824–827, 2021, doi:10.1016/j.matpr.2020.06.529.

[100] L. E. Murr, R. D. Flores, O. V. Flores, J. C. McClure, G. Liu, and D. Brown, "Friction-stir welding: Microstructural characterization," *Mater. Res. Innov.*, vol. 1, no. 4, pp. 211–223, Mar. 1998, doi:10.1007/s100190050043.

[101] L. E. Murr, Y. Li, R. D. Flores, E. A. Trillo, and J. C. McClure, "Intercalation vortices and related microstructural features in the friction-stir welding of dissimilar metals," *Mater. Res. Innov.*, vol. 2, no. 3, pp. 150–163, Nov. 1998, doi:10.1007/s100190050078.

[102] H. J. Liu et al., "Weld appearance and microstructural characteristics of friction stir butt barrier welded joints of aluminium alloy to copper," *Sci. Technol. Weld. Join.*, vol. 17, no. 2, pp. 104–110, Feb. 2012, doi:10.1179/1362171811Y.0000000086.

[103] I. Galvão, J. Oliveira, A. Loureiro, and D. Rodrigues, "Formation and distribution of brittle structures in friction stir welding of aluminium and copper: Influence of process parameters," *Sci. Technol. Weld. Join.*, vol. 16, no. 8, pp. 681–689, Nov. 2011, doi:10.1179/1362171811Y.0000000057.

[104] C. W. Tan, Z. G. Jiang, L. Q. Li, Y. B. Chen, and X. Y. Chen, "Microstructural evolution and mechanical properties of dissimilar Al–Cu joints produced by friction stir welding," *Mater. Des.*, vol. 51, pp. 466–473, Oct. 2013, doi:10.1016/j.matdes.2013.04.056.

[105] P. Xue, B. L. Xiao, D. R. Ni, and Z. Y. Ma, "Enhanced mechanical properties of friction stir welded dissimilar Al–Cu joint by intermetallic compounds," *Mater. Sci. Eng. A*, vol. 527, no. 21–22, pp. 5723–5727, Aug. 2010, doi:10.1016/j.msea.2010.05.061.

[106] I. Galvão, J. C. Oliveira, A. Loureiro, and D. M. Rodrigues, "Formation and distribution of brittle structures in friction stir welding of aluminium and copper: Influence of shoulder geometry," *Intermetallics*, vol. 22, pp. 122–128, Mar. 2012, doi:10.1016/j.intermet.2011.10.014.

[107] C. Genevois, M. Girard, B. Huneau, X. Sauvage, and G. Racineux, "Interfacial reaction during friction stir welding of Al and Cu," *Metall. Mater. Trans. A*, vol. 42, no. 8, pp. 2290–2295, Aug. 2011, doi:10.1007/s11661-011-0660-9.

[108] M. N. Avettand-Fenoël, R. Taillard, G. Ji, and D. Goran, "Multiscale study of interfacial intermetallic compounds in a dissimilar Al 6082-T6/Cu friction-stir weld," *Metall. Mater. Trans. A*, vol. 43, no. 12, pp. 4655–4666, Dec. 2012, doi:10.1007/s11661-012-1277-3.

[109] M. F. X. Muthu and V. Jayabalan, "Tool travel speed effects on the microstructure of friction stir welded aluminum–copper joints," *J. Mater. Process. Technol.*, vol. 217, pp. 105–113, Mar. 2015, doi:10.1016/j.jmatprotec.2014.11.007.

[110] S. Shankar, P. Vilaça, P. Dash, S. Chattopadhyaya, and S. Hloch, "Joint strength evaluation of friction stir welded Al-Cu dissimilar alloys," *Measurement*, vol. 146, pp. 892–902, Nov. 2019, doi:10.1016/j.measurement.2019.07.019.

[111] P. Xue, B. L. Xiao, D. Wang, and Z. Y. Ma, "Achieving high property friction stir welded aluminium/copper lap joint at low heat input," *Sci. Technol. Weld. Join.*, vol. 16, no. 8, pp. 657–661, Nov. 2011, doi:10.1179/1362171811Y.0000000018.

[112] J. Ouyang, E. Yarrapareddy, and R. Kovacevic, "Microstructural evolution in the friction stir welded 6061 aluminum alloy (T6-temper condition) to copper," *J. Mater. Process. Technol.*, vol. 172, no. 1, pp. 110–122, Feb. 2006, doi:10.1016/j.jmatprotec.2005.09.013.

[113] M. Sarvghad-Moghaddam, R. Parvizi, A. Davoodi, M. Haddad-Sabzevar, and A. Imani, "Establishing a correlation between interfacial microstructures and corrosion initiation sites in Al/Cu joints by SEM–EDS and AFM–SKPFM," *Corros. Sci.*, vol. 79, pp. 148–158, Feb. 2014, doi:10.1016/j.corsci.2013.10.039.

[114] I. Galvão, R. M. Leal, A. Loureiro, and D. M. Rodrigues, "Material flow in heterogeneous friction stir welding of aluminium and copper thin sheets," *Sci. Technol. Weld. Join.*, vol. 15, no. 8, pp. 654–660, Nov. 2010, doi:10.1179/136217110X12785889550109.

[115] P. Su, A. Gerlich, T. H. North, and G. J. Bendzsak, "Material flow during friction stir spot welding," *Sci. Technol. Weld. Join.*, vol. 11, no. 1, pp. 61–71, Feb. 2006, doi:10.1179/174329306X77056.

[116] Q. Yang, S. Mironov, Y. S. Sato, and K. Okamoto, "Material flow during friction stir spot welding," *Mater. Sci. Eng. A*, vol. 527, no. 16–17, pp. 4389–4398, Jun. 2010, doi:10.1016/j.msea.2010.03.082.

[117] R. Heideman, C. Johnson, and S. Kou, "Metallurgical analysis of Al/Cu friction stir spot welding," *Sci. Technol. Weld. Join.*, vol. 15, no. 7, pp. 597–604, Oct. 2010, doi:10.1179/136217110X12785889549985.

[118] M. P. Mubiayi and E. T. Akinlabi, "Friction stir spot welding between copper and aluminium: Microstructural evolution," *Hong Kong*, p. 6, 2015.

[119] G. Li, L. Zhou, W. Zhou, X. Song, and Y. Huang, "Influence of dwell time on microstructure evolution and mechanical properties of dissimilar friction stir spot welded aluminum–copper metals," *J. Mater. Res. Technol.*, vol. 8, no. 3, pp. 2613–2624, May 2019, doi:10.1016/j.jmrt.2019.02.015.

[120] L. Zhou et al., "Microstructure evolution and mechanical properties of friction stir spot welded dissimilar aluminum-copper joint," *J. Alloys Compd.*, vol. 775, pp. 372–382, Feb. 2019, doi:10.1016/j.jallcom.2018.10.045.

[121] S. Siddharth and T. Senthilkumar, "Optimization of friction stir spot welding process parameters of dissimilar Al 5083 and C 10100 joints using response surface methodology," *Russ. J. Non-Ferr. Met.*, vol. 57, no. 5, pp. 456–466, Aug. 2016, doi:10.3103/S1067821216050151.

[122] H. R. Zareie Rajani, A. Esmaeili, M. Mohammadi, M. Sharbati, and M. K. B. Givi, "The role of metal-matrix composite development during friction stir welding of aluminum to brass in weld characteristics," *J. Mater. Eng. Perform.*, vol. 21, no. 11, pp. 2429–2437, Nov. 2012, doi:10.1007/s11665-012-0178-3.

[123] N. Panaskar and R. Terkar, "A review on recent advances in friction stir lap welding of aluminium and copper," *Mater. Today Proc.*, vol. 4, no. 8, pp. 8387–8393, 2017, doi:10.1016/j.matpr.2017.07.182.

[124] S. Shankar and S. Chattopadhyaya, "Friction stir welding of commercially pure copper and 1050 aluminum alloys," *Mater. Today Proc.*, vol. 25, pp. 664–667, 2020, doi:10.1016/j.matpr.2019.07.719.

[125] Y. Xiao, H. Ji, M. Li, and J. Kim, "Ultrasound-assisted brazing of Cu/Al dissimilar metals using a Zn–3Al filler metal," *Mater. Des. 1980-2015*, vol. 52, pp. 740–747, Dec. 2013, doi:10.1016/j.matdes.2013.06.016.

[126] M. P. Mubiayi and E. T. Akinlabi, "Evolving properties of friction stir spot welds between AA1060 and commercially pure copper C11000," *Trans. Nonferrous Met. Soc. China*, vol. 26, no. 7, pp. 1852–1862, Jul. 2016, doi:10.1016/S1003-6326(16)64296-6.

[127] S. Siddharth, T. Senthilkumar, and M. Chandrasekar, "Development of processing windows for friction stir spot welding of aluminium Al5052/copper C27200 dissimilar materials," *Trans. Nonferrous Met. Soc. China*, vol. 27, no. 6, pp. 1273–1284, Jun. 2017, doi:10.1016/S1003-6326(17)60148-1.

[128] A. Boucherit, M.-N. Avettand-Fènoël, and R. Taillard, "Effect of a Zn interlayer on dissimilar FSSW of Al and Cu," *Mater. Des.*, vol. 124, pp. 87–99, Jun. 2017, doi:10.1016/j.matdes.2017.03.063.

[129] A. Garg and A. Bhattacharya, "Similar and dissimilar joining of AA6061-T6 and copper by single and multi-spot friction stirring," *J. Mater. Process. Technol.*, vol. 250, pp. 330–344, Dec. 2017, doi:10.1016/j.jmatprotec.2017.07.029.

[130] M. Wahba, "Laser direct joining of AZ91D thixomolded Mg alloy and amorphous polyethylene terephthalate," *J. Mater. Process. Technol.*, p. 9, 2011.

[131] G. S. Cole and A. M. Sherman, "Light weight materials for automotive applications," *Mater. Charact.*, vol. 35, no. 1, pp. 3–9, Jul. 1995, doi:10.1016/1044-5803(95)00063-1.

[132] M. Grujicic et al., "An overview of the polymer-to-metal direct-adhesion hybrid technologies for load-bearing automotive components," *J. Mater. Process. Technol.*, vol. 197, no. 1–3, pp. 363–373, Feb. 2008, doi:10.1016/j.jmatprotec.2007.06.058.

[133] M. Haghshenas and F. Khodabakhshi, "Dissimilar friction-stir welding of aluminum and polymer: A review," *Int J Adv Manuf Technol*, p. 26, 2019.

[134] A. Pramanik et al., "Joining of carbon fibre reinforced polymer (CFRP) composites and aluminium alloys – A review," *Compos. Part Appl. Sci. Manuf.*, vol. 101, pp. 1–29, Oct. 2017, doi:10.1016/j.compositesa.2017.06.007.

[135] S. Shankar, A. Kaushal, S. Chattopadhyaya, P. Vilaça, and F. Bennis, "Joining of aluminium to polymer by friction stir welding: An overview," *IOP Conf. Ser. Mater. Sci. Eng.*, vol. 1104, no. 1, p. 012005, Mar. 2021, doi:10.1088/1757-899X/1104/1/012005.

[136] Y. Huang et al., "Friction stir welding/processing of polymers and polymer matrix composites," *Compos. Part Appl. Sci. Manuf.*, vol. 105, pp. 235–257, Feb. 2018, doi:10.1016/j.compositesa.2017.12.005.

[137] K. M. Venkatesh, M. Arivarsu, M. Manikandan, and N. Arivazhagan, "Review on friction stir welding of steels," *Mater. Today Proc.*, vol. 5, no. 5, pp. 13227–13235, 2018, doi:10.1016/j.matpr.2018.02.313.

[138] S. Saeedy and M. K. B. Givi, "Investigation of the effects of critical process parameters of friction stir welding of polyethylene," *Proc. Inst. Mech. Eng. Part B J. Eng. Manuf.*, vol. 225, no. 8, pp. 1305–1310, Aug. 2011, doi:10.1243/09544054JEM1989.

[139] N. Sadeghian and M. K. Besharati Givi, "Experimental optimization of the mechanical properties of friction stir welded Acrylonitrile Butadiene Styrene sheets," *Mater. Des.*, vol. 67, pp. 145–153, Feb. 2015, doi:10.1016/j.matdes.2014.11.032.

[140] F. Simões and D. M. Rodrigues, "Material flow and thermo-mechanical conditions during Friction Stir Welding of polymers: Literature review, experimental results and empirical analysis," *Mater. Des.*, vol. 59, pp. 344–351, Jul. 2014, doi:10.1016/j.matdes.2013.12.038.

[141] F. Khodabakhshi et al., "Microstructure-property characterization of a friction-stir welded joint between AA5059 aluminum alloy and high density polyethylene," *Mater. Charact.*, vol. 98, pp. 73–82, Dec. 2014, doi:10.1016/j.matchar.2014.10.013.

[142] F. C. Liu, J. Liao, and K. Nakata, "Joining of metal to plastic using friction lap welding," *Mater. Des. 1980-2015*, vol. 54, pp. 236–244, Feb. 2014, doi:10.1016/j.matdes.2013.08.056.

[143] S. M. Rahmat, M. Hamdi, F. Yusof, and R. Moshwan, "Preliminary study on the feasibility of friction stir welding in 7075 aluminium alloy and polycarbonate sheet," *Mater. Res. Innov.*, vol. 18, no. sup6, pp. S6-515–S6-519, Dec. 2014, doi:10.1179/1432891714Z.0000000001035.

[144] H. A. Derazkola, R. Kashiry Fard, and F. Khodabakhshi, "Effects of processing parameters on the characteristics of dissimilar friction-stir-welded joints between AA5058 aluminum alloy and PMMA polymer," *Weld. World*, vol. 62, no. 1, pp. 117–130, Jan. 2018, doi:10.1007/s40194-017-0517-y.

[145] H. A. Derazkola and M. Elyasi, "The influence of process parameters in friction stir welding of Al-Mg alloy and polycarbonate," *J. Manuf. Process.*, vol. 35, pp. 88–98, Oct. 2018, doi:10.1016/j.jmapro.2018.07.021.

[146] H. A. Derazkola, F. Khodabakhshi, and A. Simchi, "Friction-stir lap-joining of aluminium-magnesium/poly-methyl-methacrylate hybrid structures: Thermo-mechanical modelling and experimental feasibility study," *Sci. Technol. Weld. Join.*, vol. 23, no. 1, pp. 35–49, Jan. 2018, doi:10.1080/13621718.2017.1323441.

[147] R. Moshwan, S. M. Rahmat, F. Yusof, M. A. Hassan, M. Hamdi, and M. Fadzil, "Dissimilar friction stir welding between polycarbonate and AA 7075 aluminum alloy," *Int. J. Mater. Res.*, vol. 106, no. 3, pp. 258–266, Mar. 2015, doi:10.3139/146.111172.

[148] S. K. Sahu, K. Pal, R. P. Mahto, and P. Dash, "Monitoring of friction stir welding for dissimilar Al 6063 alloy to polypropylene using sensor signals," *Int. J. Adv. Manuf. Technol.*, vol. 104, no. 1–4, pp. 159–177, Sep. 2019, doi:10.1007/s00170-019-03855-3.

[149] G. Çam, S. Güçlüer, A. Çakan, and H. T. Serindag, "Mechanical properties of friction stir butt-welded Al-5086 H32 plate," *Mater. Werkst.*, vol. 40, no. 8, pp. 638–642, Aug. 2009, doi:10.1002/mawe.200800455.

[150] S. Mironov, Y. S. Sato, and H. Kokawa, "Friction-stir welding and processing of Ti-6Al-4V titanium alloy: A review," *J. Mater. Sci. Technol.*, vol. 34, no. 1, pp. 58–72, Jan. 2018, doi:10.1016/j.jmst.2017.10.018.

[151] A. Wu, Z. Song, K. Nakata, J. Liao, and L. Zhou, "Interface and properties of the friction stir welded joints of titanium alloy Ti6Al4V with aluminum alloy 6061," *Mater. Des.*, vol. 71, pp. 85–92, Apr. 2015, doi:10.1016/j.matdes.2014.12.015.

[152] M. Kreimeyer, F. Wagner, and F. Vollertsen, "Laser processing of aluminum–titanium-tailored blanks," *Opt. Lasers Eng.*, vol. 43, no. 9, pp. 1021–1035, Sep. 2005, doi:10.1016/j.optlaseng.2004.07.005.

[153] B. Li, Z. Zhang, Y. Shen, W. Hu, and L. Luo, "Dissimilar friction stir welding of Ti–6Al–4V alloy and aluminum alloy employing a modified butt joint configuration: Influences of process variables on the weld interfaces and tensile properties," *Mater. Des.*, vol. 53, pp. 838–848, Jan. 2014, doi:10.1016/j.matdes.2013.07.019.

[154] K.-S. Bang, K. Lee, H. Bang, and H.-S. Bang, "Interfacial microstructure and mechanical properties of dissimilar friction stir welds between 6061-T6 aluminum and Ti-6%Al-4%V Alloys *," *Mater. Trans.*, vol. 52, pp. 974–978, 2011.

[155] M. Aonuma and K. Nakata, "Dissimilar metal joining of 2024 and 7075 aluminium alloys to titanium alloys by friction stir welding," *Mater. Trans.*, vol. 52, pp. 948–952, 2011.

[156] Y. Chen, Q. Ni, and L. Ke, "Interface characteristic of friction stir welding lap joints of Ti/Al dissimilar alloys," *Trans. Nonferrous Met. Soc. China*, vol. 22, no. 2, pp. 299–304, Feb. 2012, doi:10.1016/S1003-6326(11)61174-6.

[157] Y. Wei, J. Li, J. Xiong, F. Huang, F. Zhang, and S. H. Raza, "Joining aluminum to titanium alloy by friction stir lap welding with cutting pin," *Mater. Charact.*, vol. 71, pp. 1–5, Sep. 2012, doi:10.1016/j.matchar.2012.05.013.

[158] M. Sadeghi-Ghogheri, M. Kasiri-Asgarani, and K. Amini, "Friction stir welding of dissimilar joint of aluminum alloy 5083 and commercially pure titanium," *Met. Mater.*, vol. 54, no. 01, pp. 71–75, 2016, doi:10.4149/km_2016_1_71.

[159] A. Bist, J. S. Saini, and B. Sharma, "A review of tool wear prediction during friction stir welding of aluminium matrix composite," *Trans. Nonferrous Met. Soc. China*, vol. 26, no. 8, pp. 2003–2018, Aug. 2016, doi:10.1016/S1003-6326(16)64318-2.

[160] S. H. C. Park, Y. S. Sato, H. Kokawa, K. Okamoto, S. Hirano, and M. Inagaki, "Boride formation induced by pcBN tool wear in friction-stir-welded stainless steels," *Metall. Mater. Trans. A*, vol. 40, no. 3, pp. 625–636, Mar. 2009, doi:10.1007/s11661-008-9709-9.

[161] P. Sahlot, K. Jha, G. K. Dey, and A. Arora, "Wear-induced changes in FSW tool pin profile: Effect of process parameters," *Metall. Mater. Trans. A*, vol. 49, no. 6, pp. 2139–2150, Jun. 2018, doi:10.1007/s11661-018-4580-9.

[162] J. Pratap Kumar, A. Raj, K. Arul, and V. Mohanavel, "A literature review on friction stir welding of dissimilar materials," *Mater. Today Proc.*, p. S2214785321033745, May 2021, doi:10.1016/j.matpr.2021.04.449.

[163] F. C. Liu, Y. Hovanski, M. P. Miles, C. D. Sorensen, and T. W. Nelson, "A review of friction stir welding of steels: Tool, material flow, microstructure, and properties," *J. Mater. Sci. Technol.*, vol. 34, no. 1, pp. 39–57, Jan. 2018, doi:10.1016/j.jmst.2017.10.024.

[164] C. D. Sorensen, T. W. Nelson, and S. Packer, "Third International Symposium of Friction Stir Welding," 2001.

[165] Y. S. Sato, P. Arkom, H. Kokawa, T. W. Nelson, and R. J. Steel, "Effect of microstructure on properties of friction stir welded Inconel Alloy 600," *Mater. Sci. Eng. A*, vol. 477, no. 1–2, pp. 250–258, Mar. 2008, doi:10.1016/j.msea.2007.07.002.

[166] Y. S. Sato, T. W. Nelson, and C. J. Sterling, "Recrystallization in type 304L stainless steel during friction stirring," *Acta Mater.*, vol. 53, no. 3, pp. 637–645, Feb. 2005, doi:10.1016/j.actamat.2004.10.017.

[167] Y. S. Sato, T. W. Nelson, C. J. Sterling, R. J. Steel, and C.-O. Pettersson, "Microstructure and mechanical properties of friction stir welded SAF 2507 super duplex stainless steel," *Mater. Sci. Eng. A*, vol. 397, no. 1–2, pp. 376–384, Apr. 2005, doi:10.1016/j.msea.2005.02.054.

[168] M. Collier, R. Steel, T. W. Nelson, C. Sorensen, and S. Packer, "Grade development of polycrystalline cubic boron nitride for friction stir processing of ferrous alloys," *Mater. Sci. Forum*, vol. 426–432, pp. 3011–3016, Aug. 2003, doi:10.4028/www.scientific.net/MSF.426-432.3011.

[169] S. Bozzi, A. L. Helbert-Etter, T. Baudin, V. Klosek, J. G. Kerbiguet, and B. Criqui, "Influence of FSSW parameters on fracture mechanisms of 5182 aluminium welds," *J. Mater. Process. Technol.*, vol. 210, no. 11, pp. 1429–1435, Aug. 2010, doi:10.1016/j.jmatprotec.2010.03.022.

[170] M. Pourali, A. Abdollah-Zadeh, T. Saeid, and F. Kargar, "Influence of welding parameters on intermetallic compounds formation in dissimilar steel/aluminum friction stir welds," *J. Alloys Compd.*, vol. 715, pp. 1–8, Aug. 2017, doi:10.1016/j.jallcom.2017.04.272.

[171] J. Tang and Y. Shen, "Effects of preheating treatment on temperature distribution and material flow of aluminum alloy and steel friction stir welds," *J. Manuf. Process.*, vol. 29, pp. 29–40, Oct. 2017, doi:10.1016/j.jmapro.2017.07.005.

[172] D. K. Yaduwanshi, S. Bag, and S. Pal, "Effect of preheating in hybrid friction stir welding of aluminum alloy," *J. Mater. Eng. Perform.*, vol. 23, no. 10, pp. 3794–3803, Oct. 2014, doi:10.1007/s11665-014-1170-x.

[173] I. Das Chowdhury et al., "Study of mechanical properties of mild steel joint made by electrically assisted friction stir welding using DC and AC," *Mater. Today Proc.*, vol. 44, pp. 3959–3966, 2021, doi:10.1016/j.matpr.2020.10.011.

[174] H. Bang, H. Bang, G. Jeon, I. Oh, and C. Ro, "Gas tungsten arc welding assisted hybrid friction stir welding of dissimilar materials Al6061-T6 aluminum alloy and STS304 stainless steel," *Mater. Des.*, vol. 37, pp. 48–55, May 2012, doi:10.1016/j.matdes.2011.12.018.

[175] W.-S. Chang, S. R. Rajesh, C.-K. Chun, and H.-J. Kim, "Microstructure and mechanical properties of hybrid laser-friction stir welding between AA6061-T6 Al alloy and AZ31 Mg alloy," *J. Mater. Sci. Technol.*, vol. 27, no. 3, pp. 199–204, Jan. 2011, doi:10.1016/S1005-0302(11)60049-2.

[176] X. Long and S. K. Khanna, "Modelling of electrically enhanced friction stir welding process using finite element method," *Sci. Technol. Weld. Join.*, vol. 10, no. 4, pp. 482–487, Jul. 2005, doi:10.1179/174329305X46664.

[177] W. M. Thomas, P. L. Threadgill, and E. D. Nicholas, "Feasibility of friction stir welding steel," *Sci. Technol. Weld. Join.*, vol. 4, no. 6, pp. 365–372, Dec. 1999, doi:10.1179/136217199101538012.

[178] S. M. Bararpour, H. Jamshidi Aval, and R. Jamaati, "Modeling and experimental investigation on friction surfacing of aluminum alloys," *J. Alloys Compd.*, vol. 805, pp. 57–68, Oct. 2019, doi:10.1016/j.jallcom.2019.07.010.

[179] Y. Huang, Z. Lv, L. Wan, J. Shen, and J. F. dos Santos, "A new method of hybrid friction stir welding assisted by friction surfacing for joining dissimilar Ti/Al alloy," *Mater. Lett.*, vol. 207, pp. 172–175, Nov. 2017, doi:10.1016/j.matlet.2017.07.081.

[180] J. Gandra, R. M. Miranda, and P. Vilaça, "Performance analysis of friction surfacing," *J. Mater. Process. Technol.*, vol. 212, no. 8, pp. 1676–1686, Aug. 2012, doi:10.1016/j.jmatprotec.2012.03.013.

[181] H. K. Rafi, G. D. J. Ram, G. Phanikumar, and K. P. Rao, "Friction surfaced tool steel (H13) coatings on low carbon steel: A study on the effects of process parameters on coating characteristics and integrity," *Surf. Coat. Technol.*, vol. 205, no. 1, pp. 232–242, Sep. 2010, doi:10.1016/j.surfcoat.2010.06.052.

[182] T. Shinoda, J. Q. Li, Y. Katoh, and T. Yashiro, "Effect of process parameters during friction coating on properties of non-dilution coating layers," *Surf. Eng.*, vol. 14, no. 3, pp. 211–216, Jan. 1998, doi:10.1179/sur.1998.14.3.211.

[183] A. Kar, S. Suwas, and S. V. Kailas, "Significance of tool offset and copper interlayer during friction stir welding of aluminum to titanium," *Int. J. Adv. Manuf. Technol.*, vol. 100, no. 1–4, pp. 435–443, Jan. 2019, doi:10.1007/s00170-018-2682-6.

[184] B. Sadeghian, A. Taherizadeh, and M. Atapour, "Simulation of weld morphology during friction stir welding of aluminum- stainless steel joint," *J. Mater. Process. Technol.*, vol. 259, pp. 96–108, Sep. 2018, doi:10.1016/j.jmatprotec.2018.04.012.

[185] P. K. Sahu, S. Pal, S. K. Pal, and R. Jain, "Influence of plate position, tool offset and tool rotational speed on mechanical properties and microstructures of dissimilar Al/Cu friction stir welding joints," *J. Mater. Process. Technol.*, vol. 235, pp. 55–67, Sep. 2016, doi:10.1016/j.jmatprotec.2016.04.014.

[186] M. Raturi, A. Garg, and A. Bhattacharya, "Joint strength and failure studies of dissimilar AA6061-AA7075 friction stir welds: Effects of tool pin, process parameters and preheating," *Eng. Fail. Anal.*, vol. 96, pp. 570–588, Feb. 2019, doi:10.1016/j.engfailanal.2018.12.003.

[187] S. Das and R. G. Narayanan, "Case studies on friction stir welding of aluminum and steel sheets with a consumable," *International Conference on Precision, Meso, Micro and Nano Engineering (COPEN-11)*, IIT Indore, India, December 12–14, 2019, Paper id: COPEN11_003, 2019.

5 Cryorolling of Aluminum Alloy Sheets and Their Characterization: A Review

Kandarp Changela
Saint-Gobain Research India, Indian Institute of Technology Madras Research Park, Chennai, India

K. Hariharan
Indian Institute of Technology Madras, Chennai, India

D. Ravi Kumar
Indian Institute of Technology Delhi, New Delhi, India

CONTENTS

5.1   INTRODUCTION

One of the key requirements of aerospace and automobile industries is lightweight materials for greater fuel efficiency and reduced emissions. The use of aluminum alloys as lightweight structural materials has become inevitable due to their higher strength to weight ratio when compared to steel. The increasing need for complex parts with minimum material wastage and maximum energy saving has also created a huge interest in these materials. The strength of heat-treatable Al alloys is mainly achieved by precipitation (age) hardening at high temperatures [1, 2]. These alloys are extensively used in automobile, aerospace, and marine industries because of their properties such as high fatigue strength and high corrosion resistance [3, 4]. Non heat-treatable Al alloys are usually strengthened by strain hardening at room temperature. Despite the conventional strengthening methods, the tensile strength of aluminum alloys is usually lower and it is limited to 200–300 MPa. Further improvement in strength can be attained using various thermo-mechanical processes leading to diverse microstructural evolution and property modification.

DOI: 10.1201/9781003226703-5

Grain refinement in metallic materials to a very fine size of the order of a few hundred nanometers increases strength and toughness simultaneously [5]. The effect of grain size on the yield strength of polycrystalline materials at low temperatures (typically up to 0.5 Tm, where Tm is the melting temperature) is given by Hall [6] and greatly extended by Petch [7]. The well-established Hall-Petch relation is given by

$$\sigma_y = \sigma_o + K_{hp}d^{-1/2} \tag{5.1}$$

where σ_y is the yield strength, σ_o is the lattice friction, K_{hp} is the Hall Petch constant, and d is the grain size.

Smaller grain size results in a larger area of grain boundary network, and hence the mobility of dislocations becomes more difficult. This leads to an increase in stress required for plastic deformation (yield stress). In addition to that, reducing grain size up to nanometer scale can also alter the physical, chemical, and mechanical behavior of materials [8].

Grain refinement in metallic materials to a very fine size of the order of a few hundred nanometers increases the potential demand of such materials for the structural and functional applications [5]. Ultrafine-grained (UFG) materials are defined as materials with grain size in the nanoscale (in the range of 100 nm–1000 nm) with a homogeneously distributed microstructure. The mechanical properties such as strength, hardness, toughness, and fatigue life are usually higher in UFG materials, but ductility and formability are lower when compared to coarse-grained materials [9]. Ultrafine-grained materials, owing to their excellent strength and toughness, are considered as the next generation materials. Recently it has been shown that UFG materials can be used for bio-implant applications also [10].

In the last decade, severe plastic deformation (SPD) has been established as an effective method for the production of bulk ultrafine-grained (UFG) metallic materials [11–14]. Severe plastic deformation is a method of processing metals with large plastic deformation (plastic strain > 1.5) to create bulk UFG materials with an average grain size in the sub-micrometer or the nanometer range. In SPD, the metal undergoes intense plastic strain, leading to large shear deformation. The original coarse grains become smaller and smaller with an increasing strain that results in nanoscale grains with high crystal misorientation. These highly misoriented sub-micrometer sized grains lead to a bulk material with improved mechanical properties [15]. In order to fabricate UFG materials, many different SPD processes have been proposed, developed, and evaluated. These techniques include equal channel angular pressing (ECAP) [16], high pressure torsion (HPT) [17], accumulative roll bonding (ARB) [18], multi-axial forging [19], asymmetric rolling (ASR) [20], cryorolling (CYR) [21], and constrained groove pressing (CGP) [22]. These SPD methods are schematically shown in Figure 5.1 [17, 20, 23, 24].

Among these techniques, accumulative roll bonding, asymmetric rolling, cryorolling, and constrained groove pressing are capable of producing bulk UFG materials in the form of thin sheets. This chapter presents a detailed review of the research on cryorolling of aluminium alloys that has been reported in the literature. The effect of subsequent heat treatment of the processed sheets on microstructure, mechanical properties, and formability is also reviewed.

5.2 CRYOROLLING

Significant research has been conducted on different materials to improve the strength through various deformation techniques at a cryogenic temperature such as cryorolling (CYR) [25], asymmetric cryorolling [26], Cryo-equal channel angular pressing [27] and cryoforging [28]. Among these, cryorolling constitutes an interesting thermo-mechanical processing route for sheet metal applications and it causes significant microstructural refinement and enhances the mechanical strength of a sheet by athermal (work) hardening [25]. Figure 5.1 (f) illustrates the schematic of the cryorolling process.

One of the serious limitations of some SPD methods is that the rate of grain refinement saturates at a certain strain level and a steady state grain structure is obtained [29]. For example, grain

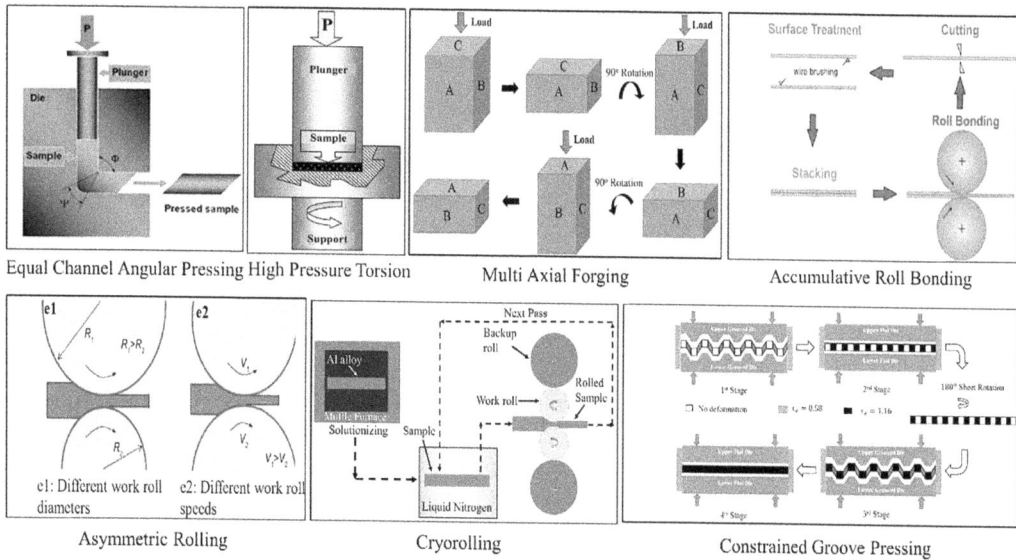

FIGURE 5.1 Schematic of different SPD processes [17, 20, 23, 24].

refinement becomes limited during ECAP of an Al alloy as high angle grain boundaries typically saturate at 70% deformation [30]. Prangnell and Huang [31] performed cryogenic plane strain deformation in an Al-0.13%Mg square billet and subsequently rolled at room temperature. They observed that steady state grain size decreased due to extremely low temperature of deformation. There are two possible inter-related phenomena to limit the grain refinement in Al alloys. The first one is high rate of dynamic recovery due to plastic strain accumulation and the second is high stacking fault energy (SFE) due to narrow stacking fault. Dynamic recovery is a thermally activated process associated with dislocation climb and cross slip, which predominantly occurs at a fixed deformation condition in high SFE materials. This limitation of grain refinement during some SPD processes is thought to be related to fixed deformation conditions such as strain rate, strain path and deformation temperature or Zener–Holloman parameter (Z) [31]. Therefore, it is expected that the grain refinement could be enhanced by changing the deformation condition. One of the methods to improve the strength and augment the dislocation multiplication is to suppress the dynamic recovery by lowering the deformation temperature to cryogenic levels [23].

Wang et al. [21] initially performed cryorolling on pure copper up to a true strain of 2.3 and obtained ultrafine-grained copper with moderate ductility. In the cryorolling of Al and Cu alloys, it is possible to produce UFG sheets by deformation at extremely low temperatures, which suppresses the dynamic recovery and enhances the dislocation density. It is well known that high dislocation density plays a major role in refining the microstructure during cryorolling [32]. However, the as-cryorolled alloys have limited ductility at room temperature, which limits their industrial applications. The ductility can, however, be improved by subsequent low-temperature annealing treatment without compromising on the maximum strength obtained. The high dislocation density acts as a driving force for the initiation of large number of nucleation sites during the short annealing resulting in nanocrystalline materials with high angle boundaries [33].

Advantages of Cryorolling

- A less plastic strain is required to produce sub-microcrystalline structural features in some alloys when compared to other SPD processes.
- Cryorolling can be carried out using a conventional rolling mill, and hence, it is inexpensive.
- No interface bonding problems as in the case of accumulative roll bonding.

Limitations of Cryorolling

- Extremely low temperature is required during cryorolling, which is very hazardous for human skin, and hence careful handling is necessary.
- It is highly dependent on the crystal structure of the deforming material. For example, materials with a body-centered cubic (BCC) crystal structure are not feasible to rolling at liquid nitrogen temperature. It is suitable only for materials, which do not undergo ductile to brittle transition temperature, for example, aluminum alloys, copper alloys, etc.
- Scalability of the process to produce thin sheets in large quantity is difficult.

5.3 MICROSTRUCTURAL EVOLUTION DURING CRYOROLLING

Many researchers [34–37] studied the effect of cryorolling on microstructure of different materials and found that ultrafine-grained materials can be produced due to deformation at liquid nitrogen temperature. The deformation at extremely low temperature suppresses the dynamic recovery and enhances the dislocation density by athermal (work) hardening [25]. The structural features of cryorolled materials are quite complex, and they are characterized not only by high density of extrinsic dislocations, but also by the presence of non-equilibrium grain boundaries, twin boundaries (in case of low stacking fault energy materials), high internal stress, high lattice distortion, and changes in the local phase composition. The external factors that influence cryorolling process include applied total strain, strain rate, and soaking time in liquid nitrogen. For example, recently, Changela et al. [38] estimated the time required for samples to reach the liquid nitrogen temperature before cryorolling. Heat transfer coefficient was calculated for different immersion times using lumped capacitance method and found out the time required for the 3 mm thick Al alloy sheet samples used in their work to reach the liquid nitrogen temperature.

Stacking fault energy plays a major role in the formation of substructure features in face-centered cubic (FCC) materials at low-temperature deformation [39]. Recently, Dhal et al. [40] compared cryorolling behaviour of pure Al with Al alloys (AA 5083 and AA 2014) in terms of stacking fault energy. Pure Al being a high SFE material (166 mJ/m^2) is expected to recover faster during cryorolling when compared to AA 5083 (33 mJ/m^2) and AA 2014 (98 mJ/m^2) alloys which have lower SFE. Presence of alloying elements in Al alloys reduces the SFE and hence reduces the recovery kinetics. In this study, the authors also discussed the various mechanisms of grain refinement during cryorolling of pure Al and Al alloys through dislocation accumulation and restoration processes. Figure 5.2 [40] shows stage by stage progression of dynamic recovery during cryorolling of pure Al and Al alloys. In the initial stage of cryorolling at low plastic strains, various defects (dislocations, vacancies etc.) are generated due to disintegration of atomic bonds and occupy spaces within the grains as well as at the grain boundaries. Dislocations get interlinked with further deformation leading to tangled microstructure, as shown in Figure 5.2 (A2 and B2). When more plastic strain is applied, two opposite phenomena act simultaneously: dislocation multiplication and annihilation. Due to high stacking fault energy and absence of solute atoms, pure Al undergoes rapid dislocation interactions followed by polyslip deformation, which eventually leads to significant cell formation within the parent grains. An individual cell block consists of disordered dislocation structure with low energy configuration. In the case of cryorolling of Al alloys (A3 and A4 in Figure 5.2), the rate of strain hardening is much higher and the tangled microstructure leads to relatively ordered structure during the cell formation. The cell boundaries consist of dense dislocations, and subsequent annihilation of dislocations within the cells results in fine sub-grains with slightly higher misorientation [40].

Panigrahi and Jayaganthan [41] studied the microstructural evolution in AA 6063 alloy samples deformed by cold rolling and cryogenic rolling at different true strain levels – 0.4, 2.3, and 3.8 – as shown in Figure 5.3 [41]. In both the conditions, the grains are elongated along the rolling direction at a strain of 0.4 (Figures 5.3 (a) and (b)). The fragmentation and elongation of sub-grains along the rolling direction were observed at a strain of 2.3 (Figure 5.3 (c) and (d)). At a total true strain of around 3.8 (Figure 5.3 (e)), most of the dislocations inside the elongated grains in the cryorolled

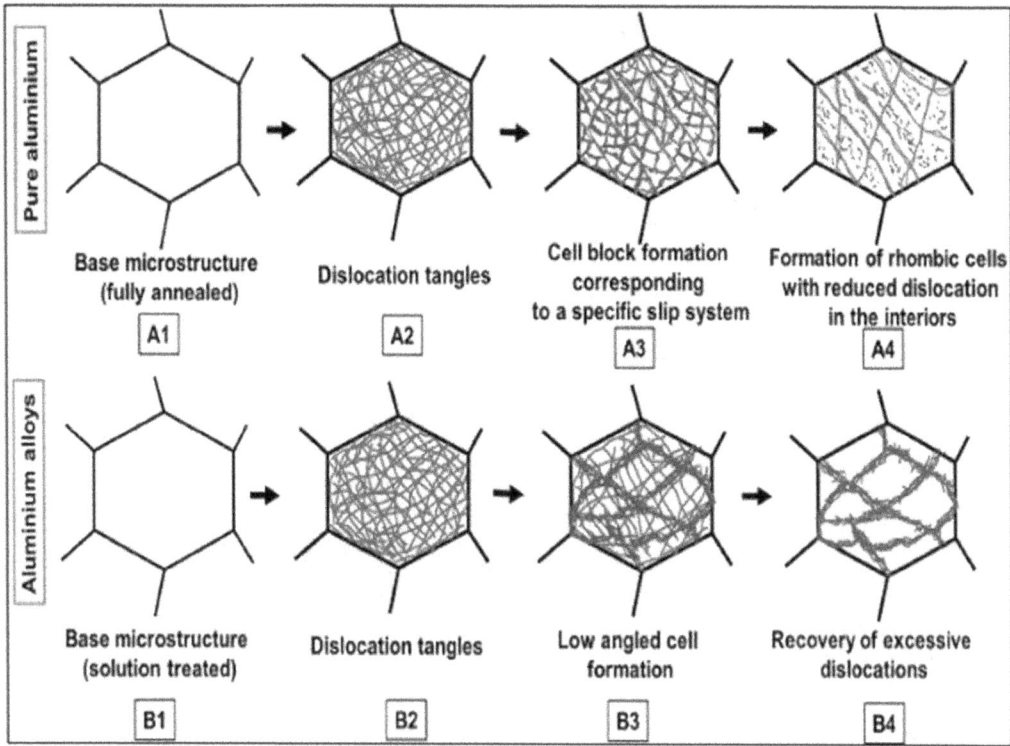

FIGURE 5.2   Development of sub-grains during cryorolling of pure Al and Al alloys [40].

sample were transformed into equiaxed sub-grains. However, such transformation was not observed in the cold-rolled samples at similar strain (Figure 5.3 (f)).

Feyissa et al. [42] investigated the microstructure evolution of cryorolled and cold rolled samples and observed that optical microscopy was not able to distinguish grain boundaries clearly in both the samples and grain size could not be measured from the optical microstructures (Figure 5.4 (a) and (b)). TEM analysis was performed to measure the grain size and investigate the microstructure evolution of cold rolled and cryorolled samples, as shown in Figure 5.4 (c) and (d). The cryorolled sample (Figure 5.4 (c)) showed highly dense dislocations with cellular structure and the average grain size was found to be 200 nm. These observations are consistent with the results of Singh et al. [43]. The high density of dislocations in cryorolled materials was mainly due to absence of adiabatic heating during cryorolling leading to suppression of dynamic recovery. Higher grain refinement was achieved in the cryorolled samples due to the presence of large number of sub-grains which could not be found in the cold rolled samples (Figure 5.4 (d)).

The dislocation structure and local misorientation are the important factors influencing the cryorolled microstructure. The local misorientation represents the misorientation between neighbouring grains or sub-grains. EBSD coupled with kernel average misorientation (KAM) is one of the tools used to analyse dislocation structures and local misorinetation. Recently, Changela et al. [23, 44] used KAM analysis to identify local misorientation and KAM distributions for cryorolled AA 5083 and AA 6061 alloys. Figure 5.5 [23] shows the KAM maps and distributions of solutionized and cryorolled AA 5083 and AA 6061 samples. KAM value in CYR 5083 is almost three times of the value in solutionized condition (Figure 5.5 (b) and (d)) while it is almost double in the case of 6061 alloy (Figure 5.5 (f) and (h)). Higher KAM value represents higher dislocation density [45]. Additionally, KAM peak in cryorolled samples shifted towards higher angles when compared to solutionized samples indicating severe strain field around dislocations.

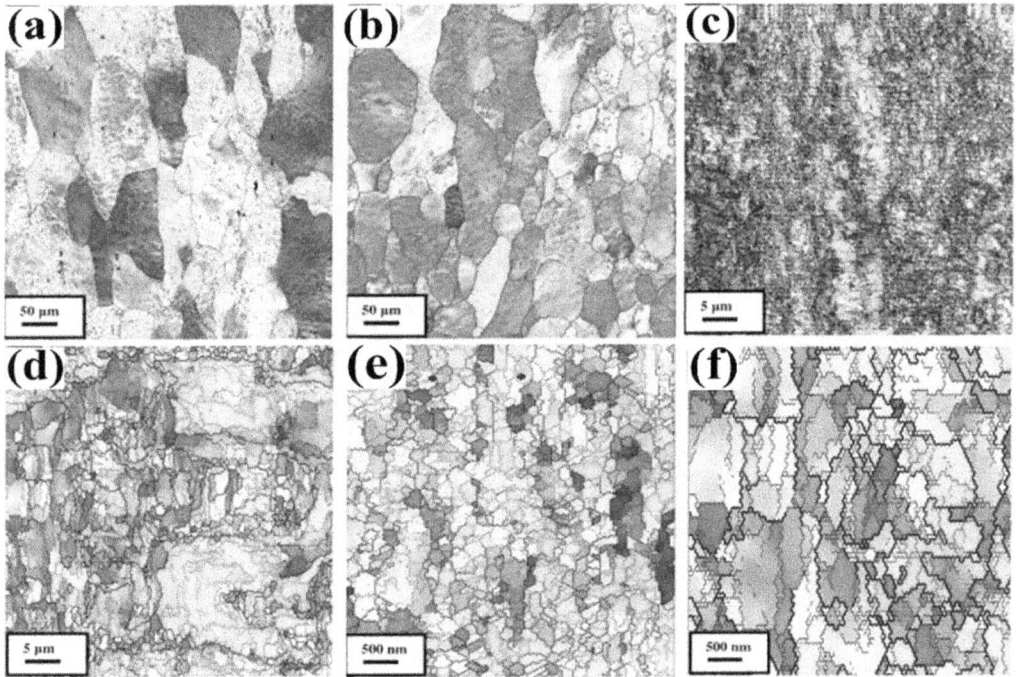

FIGURE 5.3 EBSD micrographs of AA 6063 alloy in cryorolled (CYR) and cold-rolled conditions at different strains: (a) CYR 0.4, (b) CR 0.4, (c) CYR 2.3, (d) CR 2.3, (e) CYR 3.8, and (f) CR 3.8 [41].

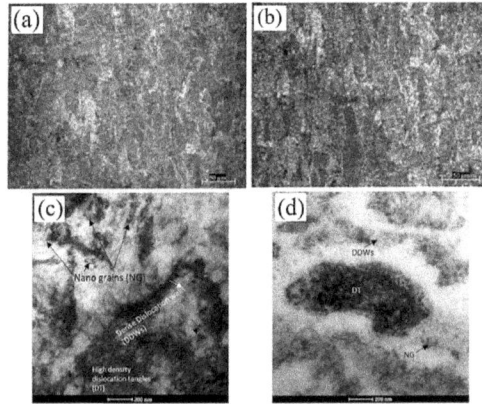

FIGURE 5.4 Optical and TEM microstructures of AA 5083 alloys in cryorolled (a) and (c), and cold rolled (b) and (d) conditions [42].

Jayaganthan et al. [46] investigated the microstructure of solutionized and cryorolled AA 7075 alloy using electron backscattered diffraction (EBSD) and transmission electron microscopy (TEM). Figure 5.6 (a) and (b) show the misorientation distribution obtained from EBSD. The solutionized sample shows higher fraction of high angle grain boundaries (HAGBs) when compared to cryorolled sample. The TEM image of cryorolled sample with 90% thickness reduction (Figure 5.6 (c)) shows the elongated grains in submicron range of 300–400 nm. A large number of dislocations are present within the nanometer size grain. Panigrahi and Jayaganthan [47] found a combination of unrecrystallized and equiaxed recrystallized grains in cryorolled AA 7075 alloy at a higher true strain of 3.4 whereas a large fraction of unrecrystalized grains are found in room temperature rolled

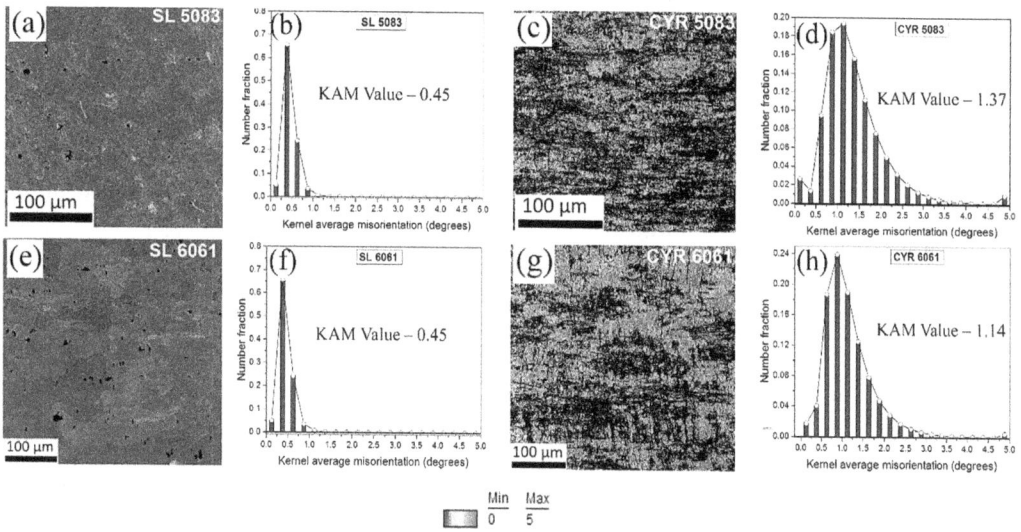

FIGURE 5.5 KAM maps and distribtions of (a) and (b) SL 5083, (c) and (d) CYR 5083, (e) and (f) SL 6061, and (g) and (h) CYR 6061 [23].

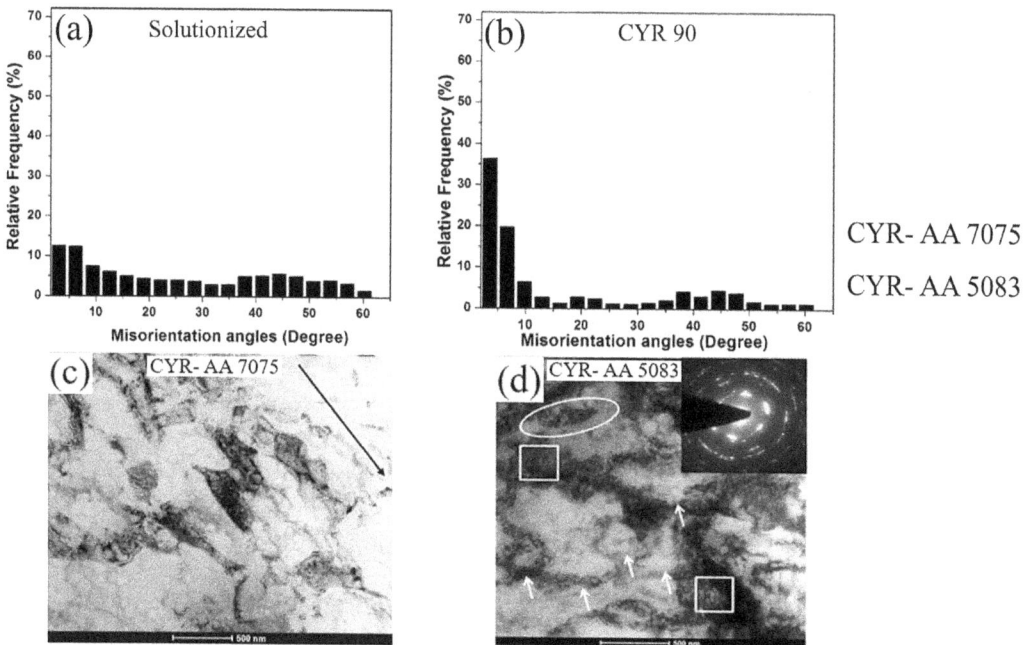

FIGURE 5.6 Misorientation distribution of AA 7075 alloy for the condition of (a) solutionized and (b) Cryorolled with 90% thickness reduction, and (c) TEM microstructure of cryorolled Al 7075 alloy sample at 90% of thickness reduction [46]. (d) TEM image of cryorolled AA 5083 alloy at 90% of thickness reduction [35].

sample. The TEM image of cryorolled AA 5083 alloy at 90% thickness reduction and corresponding SAED pattern (inside the image) are shown in Figure 5.6 (d). The diffused rings in SAED pattern indicates a large fraction of low angle grain boundaries in CYR AA 5083 alloy. Large number of sub-grains (indicated by oval shape) along with dislocation tangles (indicated by squares) and fine dislocation cells (indicated by arrows) are also observed.

5.4 EFFECT OF CRYOROLLING ON MECHANICAL PROPERTIES

Due to the microstructural evolution during cryorolling which was discussed earlier, the mechanical properties of aluminium alloy sheets are significantly affected. This section describes the effect of cryorolling on hardness and tensile properties of various aluminium alloys.

Panigrahi et al. [48] studied the hardness of AA 6063 alloy subjected to cryorolling and cold rolling. It has been found that the hardness of the cryorolled sheet were significantly higher than that of the cold-rolled sheet, and it was attributed to suppression of dynamic recovery and a higher dislocation density. Figure 5.7 shows the variation of hardness of the cryorolled sheet with respect to the cumulative true strain (after each pass of thickness reduction). The difference in hardness between cold and cryorolled increased with the amount of deformation (Figure 5.7 (a)). The similar behaviour of hardness has been observed by Feyissa et al. [42] in AA 5083 alloy, as shown in Figure 5.7 (b). Cryorolled sheets shows higher hardness than cold rolled sheets in any deformation stage which is attributed to suppression of dynamic recovery, high dislocation density, and grain refinement.

Figure 5.8 shows the variation of hardness with respect to plastic strain of cryorolled AA 6061 [32, 33, 49–52] and AA 5083 alloys [42, 53–56]. It shows that hardness increases with percentage

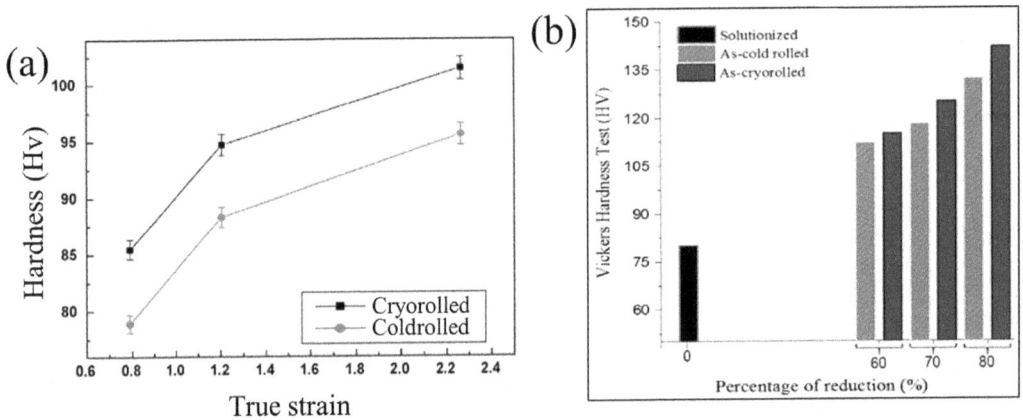

FIGURE 5.7 Variation of hardness with respect of amount of deformation in cold and cryorolled sample conditions (a) Al–Mg–Si alloy [32] and (b) AA 5083 alloy [42].

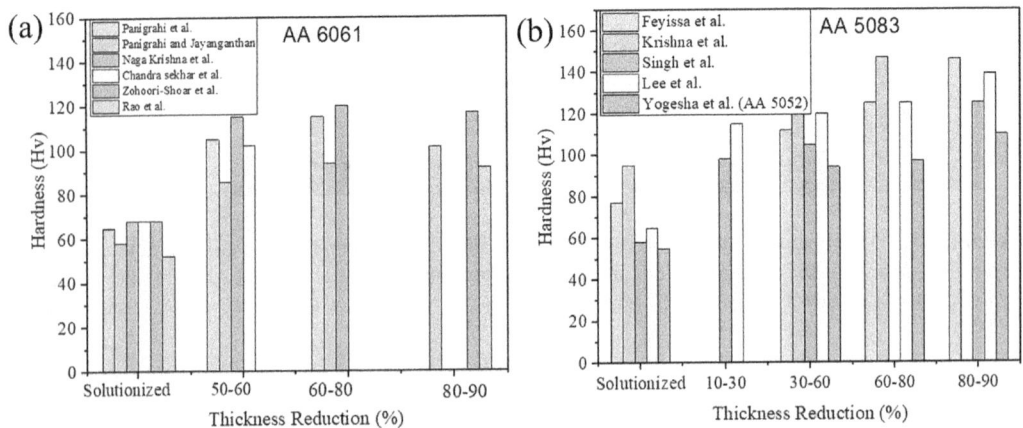

FIGURE 5.8 Variation of hardness with respect to percentage of thickness reduction of cryorolled aluminium alloys (a) AA 6061 [32, 33, 49–52] and (b) AA 5083 [42, 53–56].

of thickness reduction during cryorolling and it is mainly due to suppression of dynamic recovery and higher dislocation density. However, it has been observed that improvement in hardness is significant in the initial plastic deformation and then becomes stable at higher plastic strains value for both the aluminium alloys.

5.4.1 Tensile Properties

Panigrahi and Jayaganthan [57] studied the tensile properties of AA 6063 alloy subjected to cryorolling (CR) and room temperature rolling (RTR). It has been found that the tensile strength of the cryorolled sheets was significantly higher than that of the cold-rolled sheets. Figure 5.9 [57] shows the variation of yield strength (YS) and ultimate tensile strength (UTS) of CR and RTR AA 6063 alloy at different thickness reductions. It has been observed that the strength increment in RTR is comparatively slower than in CR samples. This is consistent with the published literature by some other researchers [21, 58] also and the improvement in strength of cryorolled samples is attributed to the effective suppression of dynamic recovery leading to higher dislocation density compared to cold-rolled sheets. Similar observation was made by Feyissa et al. [42] by comparing tensile properties of cold-rolled and cryorolled AA 5083 alloy. But it has also been observed that with improvement in strength of cryorolled sheets, ductility (total elongation) reduced to less than 5%. However, slightly contradicting results were found by Niranjani et al. [59]. They achieved almost similar yield strength (380–390 MPa) and ductility (4–5%) in the case of both cold-rolled and cryorolled sheets of AA 6061 alloy. The aging treatment at 125°C for 18h was also carried after both the rolling processes and found that cold rolled + aged samples showed better strength and ductility when compared to cryorolled + aged samples.

Sarkar et al. [60] investigated the tensile properties of cryorolled AA 2219 alloy at different strain levels in two conditions, with and without solutionizing treatment prior to cryorolling. Solutionizing was carried out by heating the sample at 535°C for one hour followed by water quenching. Figure 5.10 [60] shows the variation of mechanical properties of cryorolled AA 2219 alloy with respect to percentage thickness reduction in both the conditions. Sample without solutionizing showed higher strength than the solutionized one but the ductility of solutionized sample was higher. It was recommended that solutionizing followed by cryorolling is an effective route for Al alloys to get an optimum combination of mechanical properties. A similar study was conducted by Panigrahi and Jayaganthan [41]. In this study, AA 6063 alloy was solution treated at various temperatures between 400°C to 580°C

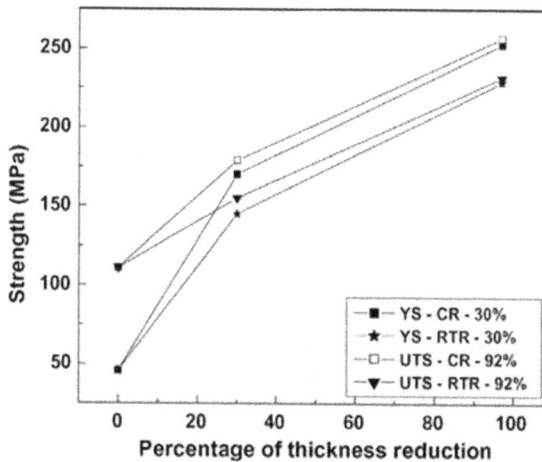

FIGURE 5.9 Comparison of room temperature rolled (RTR) and cryorolled (CR) AA 6063 alloy in yield strength (YS) and ultimate tensile strength (UTS) at different thickness reduction [57].

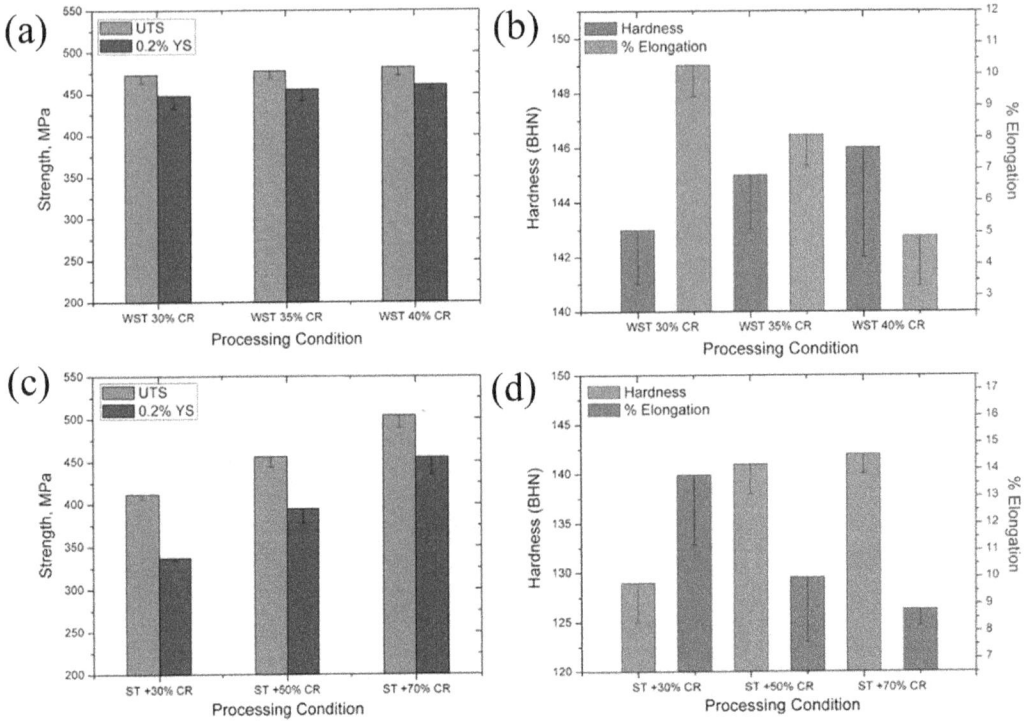

FIGURE 5.10 Variation of mechanical properties with respect to thickness reduction of cryorolled AA 2219 alloy in different conditions (a) YS and UTS for CR, (b) hardness and elongation for CR, (c) YS and UTS for ST+CR, and (d) hardness and elongation for ST+CR [60].

at different intervals for one hour and quenched in water. It has been concluded in this study that 520°C was the optimum solutionizing temperature before cryorolling to achieve optimum mechanical properties.

5.5   EFFECT OF HEAT TREATMENT AND POST PROCESSING ON CRYOROLLED SHEETS

As mentioned earlier, as-cryorolled sheets exhibit poor ductility and low strain hardening ability at room temperature, which limits their industrial applications. It is important to study the post-heat treatment parameters of cryorolled material to obtain sufficient ductility for sheet metal forming applications without losing strength significantly. Heat treatment or post processing after cryorolling proved to be an efficient route to achieve better mechanical properties with controlled microstructure. Liu et al. [61] showed that structural parameters of heavily cold worked metals could be estimated quantitatively. Modeling of severely deformed and subsequently annealed microstructure can be performed by using these quantitative parameters. The mechanism of substructure formation and grain refinement in severely deformed (cryorolled) and annealed sheets is quite complex. Initially, at low plastic strain, grains are elongated along the rolling direction with an increase in the dislocation density. The dislocation density continuously increases due to suppression of dynamic recovery associated with cross slip and climb of dislocations and gets accumulated with further deformation resulting in the formation of disordered cellular structures.

A detailed microstructural investigation and comparison of pure Al and Al alloys (AA 5083 and AA 2014) during cryorolling and post-annealing treatment has been done by Dhal et al. [40]. They observed that the variation in recovery, recrystallization, and grain growth was mainly due to the difference in strengthening mechanism and stacking fault energy among these alloys. The microstructure evolution during annealing of pure Al, AA 5083 and AA 2024 are shown in Figure 5.11. It was highlighted that there is a remarkable difference in the microstructure of aluminum alloys compared to pure aluminum after annealing of cryorolled samples. The suggested mechanisms in pure Al are rapid recovery and sub-grain growth due to cell block structure. The grain boundary migration was reduced in the case of pure Al due to low concentration of dislocations, presence of high angle grain boundaries in cryorolled samples, and the development of rhombic grain structure, as shown in Figure 5.11 (a). In contrast, the annealed microstructure of Al alloys (Figure 5.11 (b) and (c)) shows rapid transformation of low angle grain boundaries to stable high angle grain boundaries via grain boundary migration, high rate of static recovery during annealing treatment, and faster recrystallization kinetics.

Effect of annealing temperature on mechanical properties of cryorolled Al alloys was discussed by many researchers in the past [44, 55, 64, 68]. The hardness variation with respect to annealing temperature in cryorolled AA 5083 and AA 6061 alloys has been systematically studied in earlier work [44, 69]. Figure 5.12 [44] shows that hardness has been increased by almost 100% after cryorolling when compared to solutionized condition in both the aluminium alloys. A short annealing of cryorolled samples has been done at different temperatures with a fixed time of 15 minutes. The expected decreasing trend of hardness has been observed with increasing annealing temperature in both the aluminium alloys. The optimum annealing temperature has been found to be of 275°C in AA5083 alloy because hardness suddenly drops at 300°C while in the case of AA 6061 alloy, hardness gradually decreases up to 300°C and then a sudden drop at 325°C. After 325°C, hardness is almost constant which is almost equal to hardness in solutionized condition.

The effect of post annealing temperature on tensile properties has also been studied. It can be observed from Figure 5.13 [44, 55, 64, 68] that yield strength decreased, and elongation increased with increasing annealing temperature in the case of AA 7075 (Figure 5.13 (a)) and AA 5083

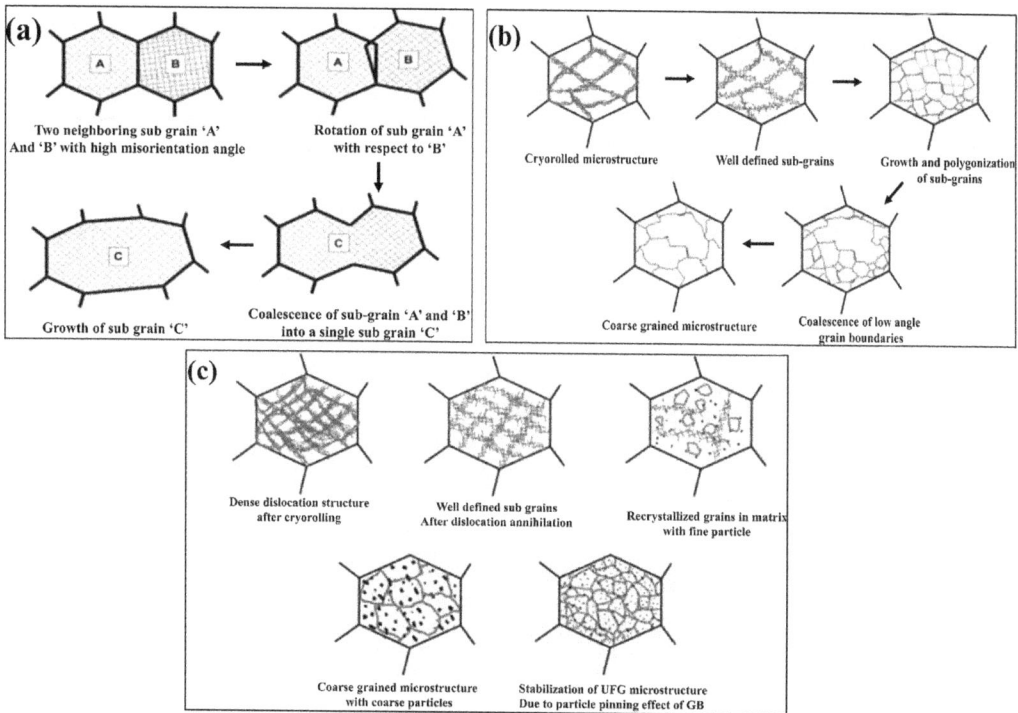

FIGURE 5.11 Variation in microstructure during annealing of (a) pure Al, (b) AA 5083, and (c) AA 2024 [40].

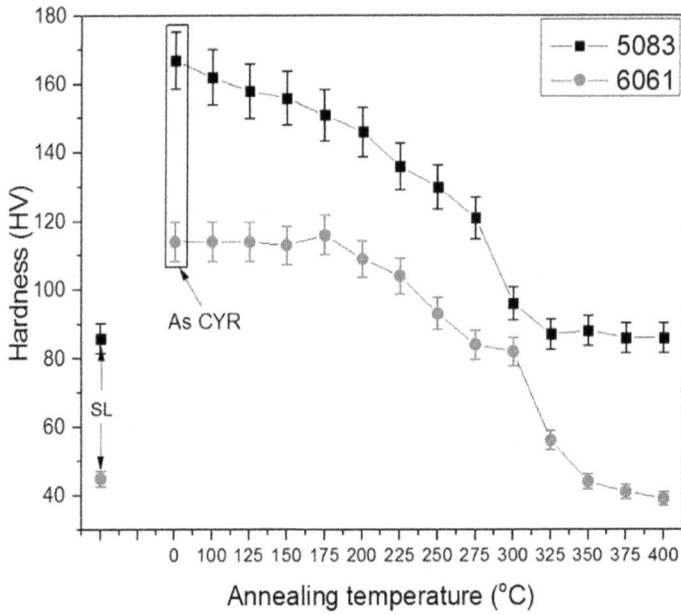

FIGURE 5.12   Hardness variation with annealing temperature in cryorolled AA5083 and AA6061 alloys [44].

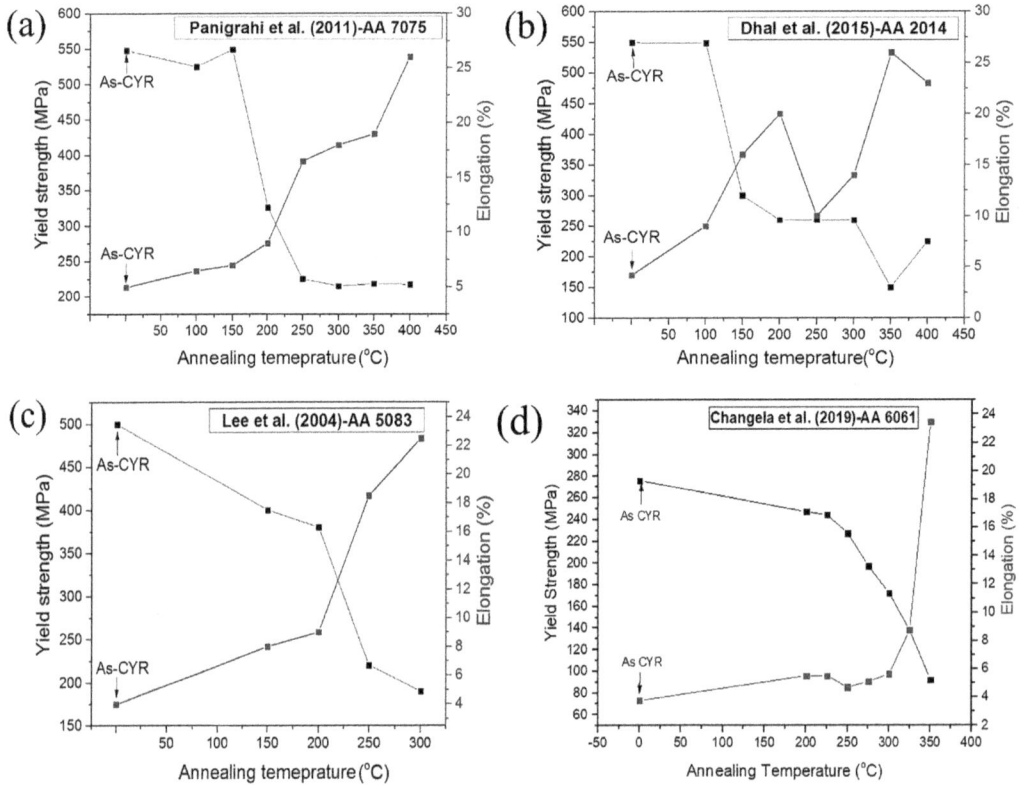

FIGURE 5.13   Effect of annealing temperature on tensile properties of cryorolled Al alloys for (a) AA 7075 [68], (b) AA 2014 [64], (c) AA 5083 [55], and (d) AA 6061 [44].

(Figure 5.13 (c)) alloys. This was attributed to the recovery and recrystallization of sub-grains. In the case of AA 2014 alloy (Figure 5.13 (b)), and AA 6061 alloy (Figure 5.13 (d)) a slightly different trend was observed in which total elongation increased up to 200°C, suddenly dropped at 250°C, and then increased again up to 350°C. The possible reason for different trends in AA 2014 and AA 6061 is the formation of precipitates. Coarse and brittle precipitate particles nucleated at the grain boundary at 250°C. Precipitates effectively act as barriers for the dislocation movement which resulted in decrease in ductility and strain hardening ability [64].

Cheng et al. [70] developed an effective approach to achieve both high strength and high ductility after cryorolling of AA 2014 alloy. They found that T phase particles developed by solutionizing treatment could effectively accumulate the dislocations during cryorolling. High dislocation density inside the grain acts as a driving force for the nucleation of sub-grains during annealing. Many other studies have been done in the past to simultaneously improve the strength and ductility by optimizing the heat treatment parameters [57, 71]. Panigrahi and Jayaganthan [72] proposed that cryorolling followed by aging was an effective method to achieve both strength and ductility simultaneously in AA 7075 alloy. Nano-sized precipitates were found to be uniformly distributed in cryorolled and cold rolled samples after reaching the peak aged condition. It was also observed that the density of nano-sized precipitates in the cryorolled + peak aged (100°C for 45 h) samples was more than that of the cold-rolled + peak aged samples. Dislocation annihilation during aging gives more space for effective accumulation of dislocations around the nano-sized precipitates, and hence substantial improvement in ductility was also observed in cryorolled + peak aged samples compared to as-cryorolled samples. Similarly, Changela et al. [44] also observed the different size of precipitates after annealing of cryorolled AA 6061 samples for 15 minutes. Figure 5.14 (a) and (b) [44] show the SEM microstructure of cryorolled AA 6061 alloys annealed at 200°C and 300°C for 15 minutes, respectively. The microstructure shows the spherical (β), rod (β') and needles (β'') types of precipitates formed after annealing of cryorolled AA6061 alloy. It has been reported that the maximum strengthening achieved in CYR 6061 alloy is mainly due to β' and β'' types of precipitates with the size of 430nm and 250nm, respectively. Shanmugasundaram et al. [73] observed that cold rolling was not sufficient to develop nanocrystalline/ultrafine-grained matrix structure in Al-Cu alloy due to low driving force available for recrystallization during subsequent annealing.

Dhal et al. [74] studied the mechanical properties and microstructural behavior of commercially pure Al subjected to cryorolling and subsequent annealing at 150°C for 30 minutes. It has been

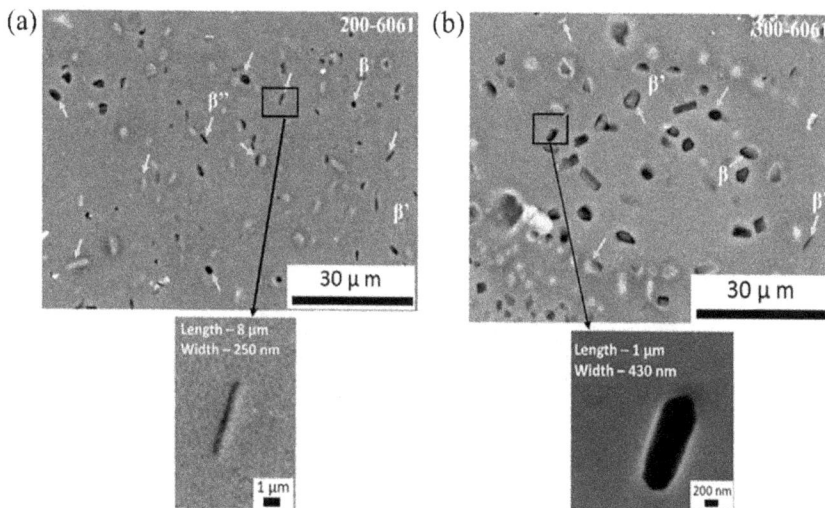

FIGURE 5.14 SEM images showing the precipitate distribution in cryorolled AA6061 alloy annealed at (a) 200°C and (b) 300°C [44].

found that a good combination of strength and ductility could be achieved by controlled annealing of cryorolled samples leading to a significant improvement in strain hardening capability. This has been attributed to the slight increase in grain size and a significant decrease in dislocation density. Variation of grain size was very minimal even after annealing of cryorolled sheets at 150°C for 30 minutes, and it proved that the annealing temperature was sufficient to activate the static recovery but not enough for recrystallization of new grains. Krishna et al. [53] observed that the UFG AA 5083 alloy developed by cryorolling exhibited reasonable peak shift as well as peak broadening in X-ray diffraction analysis, which was attributed to the formation of fine grains and a higher density of defects than in the cold rolling process.

Lee at al. [55] investigated the effect of annealing temperature on the mechanical properties and microstructure of cryorolled AA 5083 alloy. TEM micrographs of cryorolled AA 5083 annealed at different temperatures are shown in Figure 5.15. Annealing of cryorolled samples at 150°C (Figure 5.15 (a)) showed the recovered microstructure, including the formation of sub-grains and annihilation of dislocations. The rearrangement of dislocations at this temperature caused a rapid decrease in tensile strength and increased ductility. The study indicated that annealing of cryorolled samples at 200°C could yield high strength to ductility ratio and equiaxed grains in the order of < 200 nm with elongated sub-grains (Figure 5.15 (b)). Figure 5.15 (c) shows the microstructure of a cryorolled sample annealed at 250°C in which recrystallized and coarsened grains with a size of 1.5–2 μm have been found. The bimodal grain structure was also observed by Changela et al. [44] in cryorolling followed by short annealing at 275°C for 15 minutes. The bimodal microstructure consists of the presence of ultrafine grains randomly surrounded by recrystallized grains.

Huang et al. [75] studied Cu-Al alloys with three levels of Al content (2.3 %, 7.2 %, and 11.6 %). They reported a substantial improvement in the strength and ductility of nano-grained Cu-Al alloys with increasing solute content and decreasing stacking fault energy. Similarly, in Al-Mg alloys, distribution of Mg in Al matrix would directly affect the mechanical properties to a great extent. Yu et al. [76] observed higher suppression of precipitation in cryorolled Al-Mg alloy than in the case of cold rolling, resulting in a higher solute content of Mg in the Al matrix. A new mechanism was proposed for the enhancement of the ductility of cryorolled sheets during tensile deformation. Besides the common features of the ductile fracture such as uniform elongation, void nucleation and growth, and local necking before the final damage, a higher grain refinement near the fracture was observed in cryorolled sheets.

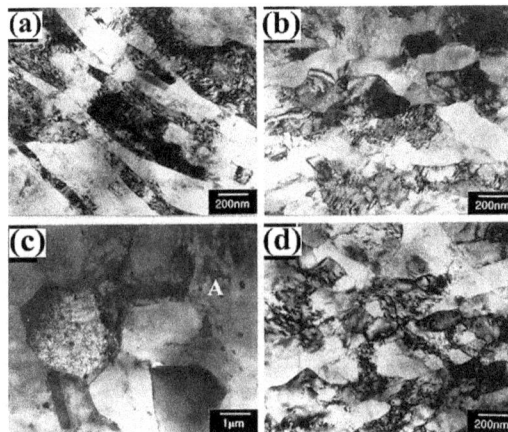

FIGURE 5.15 TEM micrographs of the cryorolled AA 5083 alloy annealed at various temperatures for 1 h: (a) 150°C, (b) 200°C, (c) 250°C, and (d) elongated sub-grains and equiaxed grains in the region marked A in (c) [55].

Post processing on cryorolled sheets such as warm rolling (WR) has also been found to be one of the potential routes to increase the yield strength and tensile strength of aluminum alloys without a significant decrease in ductility. Kang et al. [77] showed that warm rolling at 175°C after cryorolling of AA 5052 alloy was more effective than a single cryorolling or cold rolling to achieve a good combination of strength and ductility. The improvement in strength and ductility of cryorolled followed by warm rolling was attributed to the formation of fine dislocation free subcell structure during warm rolling [78]. Nageswara Rao et al. [52] compared cryorolling followed by warm rolling at 200°C (CR+WR) with cryorolling followed by short annealing (CR+SA) at 200°C for 10 minutes of AA 6061 alloy. They observed that CR+WR showed higher improvement in strength than CR+SA. Further improvement in strength and ductility of CR+WR sheets could be achieved by low temperature aging at 125°C for 27 h and it was attributed to the development of fine precipitates during artificial aging. Similar processing route (cryorolling followed by warm rolling) has been studied by Singh et al. [35] for AA 5083 alloy. They observed that 175°C was the optimum warm rolling temperature to get significant improvement in strength and ductility simultaneously. Warm rolling of cryorolled sheets at more than 175°C leads to reduction in strength due to significant effect of dynamic recovery during warm rolling. Mei et al. [79] studied the effect of warm rolling on the precipitation in cryorolled Al-Zn-Mg-Cu sheet. The hardness of CR+WR sheets was measured at different temperatures and aging time. It has been observed that maximum hardness was achieved at 100°C and 24 hrs which was considered to be peak aged (PA) condition. The CR+WR+PA sample condition showed better mechanical properties when compared to CR+PA sample condition. Recently, Bembalge and Panigrahi [80] developed processing maps by hot deformation of cryorolled AA 6063 alloy at different temperatures and strain rates. Compression and tensile tests of cryorolled samples have been conducted at temperatures of 300, 350, 400 and 450°C and strain rates of 0.001, 0.01, 0.1, 1, and 10 s^{-1}. The processing maps of cryorolled samples by compression and tensile modes consist of overlap of stability and instability regions. It has been found that dynamic recovery and recrystallization were the main mechanisms in stability region while instability was mainly due to stress induced cracking. It has also been found that maximum stability of cryorolled samples in tensile and compression modes was achieved at deformation temperatures of 350°C and 430°C, respectively.

5.6 FORMABILITY OF CRYOROLLED SHEETS AND EVALUATION

The forming behavior of cryorolled Al alloy sheets has been studied only by a few researchers. In these studies, formability of ultrafine-grained Al alloys sheets processed through cryorolling has been reported.

Taylor et al. [81] achieved high formability in biaxial stretching of high purity Al alloy sheet, which was cryorolled and annealed at 275°C for two minutes. The FLDs of the coarse-grained and the ultrafine-grained materials revealed that formability of cryorolled aluminum increased in the biaxial stretching region and the difference in forming limits between ultrafine-grained and coarse-grained sheets was observed to be less than 45% in biaxial tensile region. Weiss et al. [63] evaluated the stretch formability of cryorolled and annealed AA 2024 alloy sheets by the Erichsen cupping test. In this study, two conditions were chosen prior to cryorolling: solutionizing and natural aging. The study revealed that naturally aged samples showed lower formability compared to solutionized samples before cryorolling process due to the solute clusters developed during the aging process, which increases the dislocation density during cryorolling process.

Feyissa et al. [42] compared formability of cryorolled AA 5083 sheets with cold rolled sheets by limiting dome height (LDH) tests after short annealing. FLDs were also determined. Figure 5.16 (a) shows the effect of annealing time on LDH of cryorolled and cold rolled samples, which shows that cryorolled samples exhibit slightly higher LDH than cold-rolled samples for each annealing time indicating better stretchability. The FLDs (Figure 5.16 (b)) indicated that the forming limit strains of cryorolled sheets are almost equal to those of cold-rolled sheets. This was attributed to

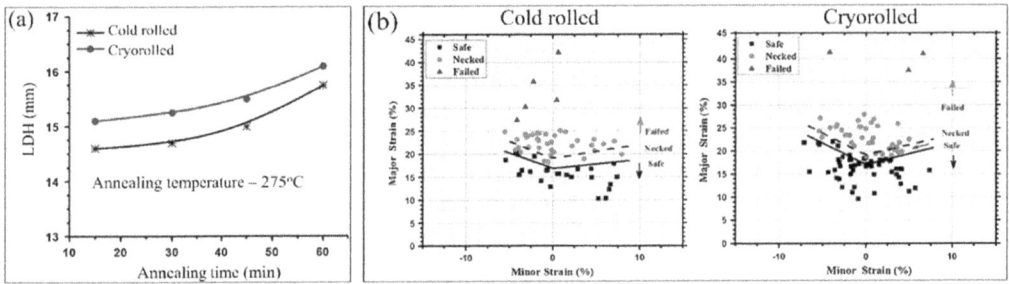

FIGURE 5.16   (a) Effect of annealing time on LDH and (b) FLDs of cold-rolled and cryorolled AA 5083 sheets [42].

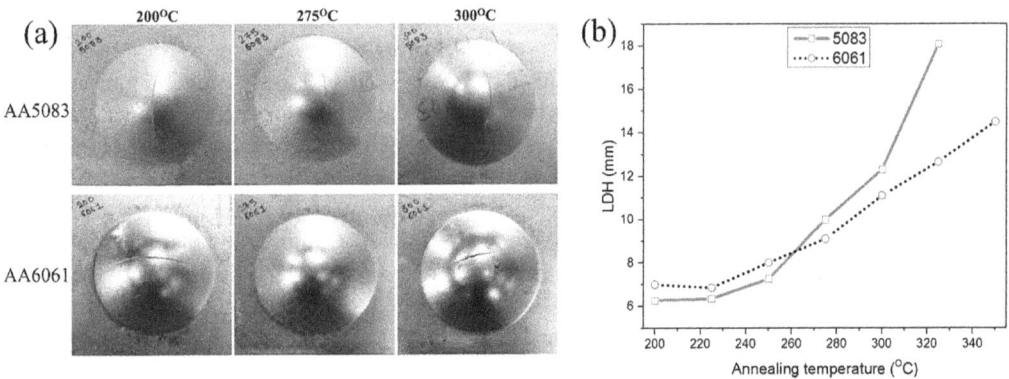

FIGURE 5.17   (a) LDH tested samples of AA5083 and AA6061 alloys (b) Variation of LDH with respect to annealing temperature [44].

the bimodal microstructure and higher strain hardening exponent of cryorolled and annealed sheet. Overall, this study concluded that cryorolling followed by a short annealing could be an effective route to produce high strength sheet metal parts without losing formability compared to conventional route.

Changela et al. [44] compared the formability of AA5083 and AA6061 alloy sheets in cryorolled and annealed condition by LDH tests. Figures 5.17 (a) and (b) [44] show the tested LDH samples and the variation of LDH with respect to annealing temperatures, respectively. As expected, the LDH increased with annealing temperature for both the Al alloys. However, the trend of LDH variation is different in both the alloys, particularly at higher annealing temperatures (275°C to 325°C), as shown in Figure 5.17 (b). Above 300°C, significant difference has been observed in LDH between the two alloys. In the case of AA6061 alloy, LDH values are lower than AA5083 alloy at higher annealing temperature. The important point observed by the authors that only "n" value is not sufficient to explain the formability due to significant effect of other microstructural parameters. The difference in formability between these two alloys has been correlated with the dislocation density based strain hardening model.

5.6.1   IMPROVEMENT IN FORMABILITY OF CRYOROLLED SHEETS

Different approaches, such as warm forming, hydroforming, etc. have been investigated by a few researchers [66, 82] to improve the formability of ultrafine-grained Al alloy sheet metals.

5.6.1.1 Warm Forming

Warm forming is referred to as plastic deformation of the sheet metals at elevated temperatures, but below the recrystallization temperature. Usually, the temperature interval for warm forming is 0.3–0.5 times the melting temperature of the material. Recently Satish et al. [66, 83] studied the formability of cryorolled AA 6061 sheets at different warm working temperatures and developed a new potential route to achieve both high strength and high formability without any post-heat treatment. Formability was characterized in terms of FLD, LDH, strain distribution and thinning. Stretch forming of cryorolled sheets was conducted in 200°C, 250°C, and 300°C temperatures and FLDs was plotted as shown in Figure 5.18. The higher limit strains in all modes of deformation indicated that formability of CYR sheets increased with warm working temperature. CYR sheets tested at 250°C and 300°C showed enhanced formability when compared to cold rolled and annealed condition (CLRA). They found that, for cryorolled AA 6061 sheets, 250°C was the optimum warm working temperature to get better formability without losing strength much. In this study, limiting dome height (LDH) test was also conducted to estimate formability of cold rolled and cryorolled AA 6061 sheets.

As shown in Figure 5.19 (a), it was observed that LDH of CYR-250°C and CYR-300°C tested samples was higher than CYR-200°C and CLRA tested samples. The load-displacement data was captured during LDH test of cryorolled samples, as shown in Figure 5.19 (b). It was found that peak load gradually decreased with increase in warm working temperatures. Hence, it was concluded that in addition to better formability, peak load was also lower when compared to CLRA samples.

5.6.1.2 Hydroforming

Hydroforming is one of the promising advanced sheet metal forming techniques that can be used in automobile and aerospace industries. In this technique, high pressure fluid is used to form sheet metal parts within die cavity. There are many advantages of hydroforming compared to conventional forming such as easy to form complex parts, uniform strain distribution and good surface quality. Recently, Feyissa and Ravi Kumar [82] investigated hydroforming of cryorolled AA 5083 sheets for the first time. They studied deep drawing of flat bottom square cup-shaped parts by hydroforming of cryorolled AA 5083 sheets annealed at 275°C for 15 minutes and compared with the conventional deep drawing process (Figure 5.20 (a)). It was demonstrated that the formability of cryorolled

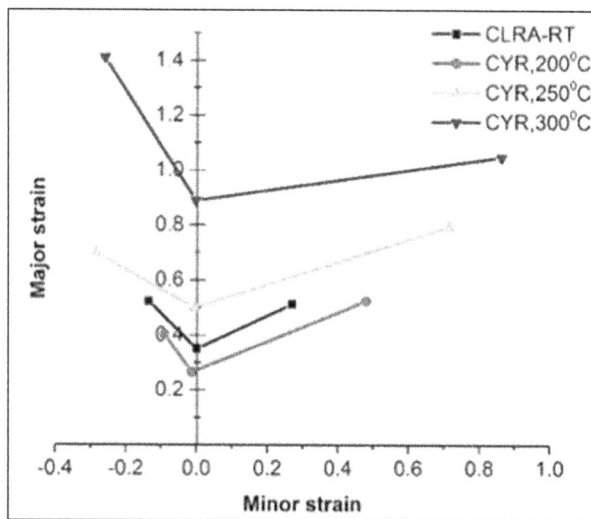

FIGURE 5.18 Comparison of FLDs of cryorolled AA 6061 sheets determined at different temperatures with that of cold rolled and annealed sheet [66].

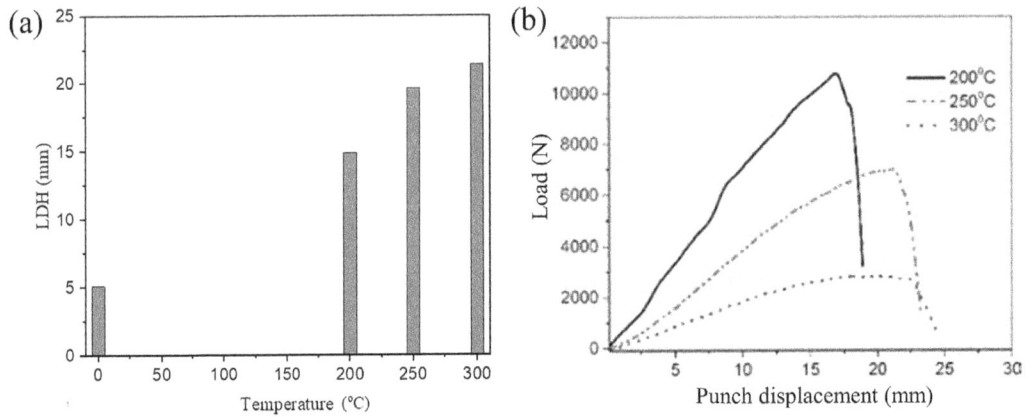

FIGURE 5.19 (a) Variation of LDH of cryorolled 6061 samples with respect to forming temperature and (b) load-displacement curves at different temperatures [66].

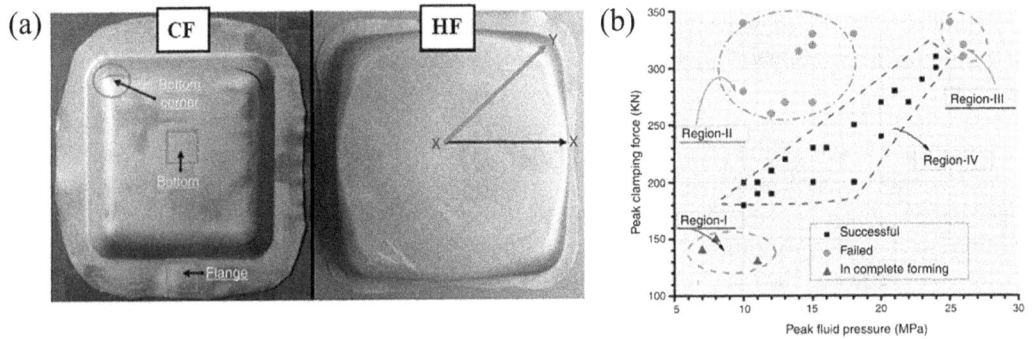

FIGURE 5.20 (a) Hydrofomed (HF) and conventionally formed (CF) square cups. (b) Process window of hydroforming square cups for cryorolled AA5083 sheets [82].

Al alloy sheets can be enhanced by hydroforming due to lower thinning and more uniform strain distribution compared to the conventional deep drawing process. The study demonstrated that combining two unconventional forming techniques, cryorolling with hydroforming, is a potential route to produce complex sheet metal parts from high strength Al alloy sheets. In this study, process window was also determined to successful drawing of square cup by hydroforming from cryorolled aluminium alloy sheet blanks. Using different combination of process parameters and large experimental data from hydroforming process, process window was divided into four regions, as shown in Figure 5.20 (b).

5.6.2 MICRO FORMING OF ULTRAFINE-GRAINED SHEETS

In recent years, micro forming emerged as an effective manufacturing process to fabricate micro components with high production capabilities and material utilization. Despite its potential applications, formability of sheet materials at the micro level is affected by the "size effect" [84]. Dhal et al. [85] studied the micro forming behavior of coarse-grained (CG) and ultrafine-grained (UFG) pure Al sheets. They observed that UFG sheets developed by cryorolling demonstrated improved micro formability with better surface quality than coarse-grained sheets. The drawn cups using a micro deep drawing process (both CG and UFG) are shown in Figure 5.21. It was observed that the formed

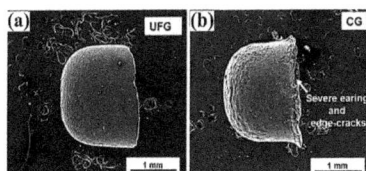

FIGURE 5.21 SEM images of micro deep-drawn cups for Al sheets with (a) ultrafine-grains and (b) coarse grains [85].

micro cups from UFG sheets appeared to have better surface quality. The walls of the cup were free from wrinkles and cracks, while CG sheets exhibited poor quality of micro cup with severe wrinkles, cracks, and earing. The improved micro formability with better surface quality in UFG sheets could be mainly due to the fact that grain migration and rotation were activated in equiaxed high angle grain boundary microstructure at microscale.

5.7 SUMMARY

A detailed review on the investigations carried out by many researchers on the effect of cryorolling and post processing treatments on microstructural, mechanical properties, and formability of various aluminium alloys is presented. It can be summarized that, in general, cryorolling is an effective SPD technique to achieve UFG structure in Al alloy sheets with superior mechanical properties when compared to conventional cold rolled conditions. Microstructural evolution during cryorolling process revealed the mechanisms by which UFG structure has been formed in cryorolled samples. Improvement in strength and hardness has been observed to be higher in cryorolled Al alloys than in cold rolling due to effective suppression of dynamic recovery leading to higher dislocation density. However, due to low ductility of the sheets in as-cryorolled condition, post cryorolling treatments such as low temperature annealing and warm rolling have been attempted to produce sheets with an optimum combination of strength and ductility. From limiting dome height (LDH) tests and forming limit diagrams, it has been found that formability of cryorolled and annealed Al alloy samples is slightly better than cold rolled and annealed sheets and it is attributed to significant recovery and partial recrystallization of sub-grains in the UFG sheets obtained after processing. Hydroforming and warm forming are the potential sheet metal forming techniques that can be used to improve the formability of cryorolled aluminium alloy sheets to make them suitable for a wide range of applications in the future.

REFERENCES

[1] U. Chakkingal, Aluminium and its alloys, in Z.X. Guo, *The deformation and processing of structural materials*, Woodhead Publishing Limited Abington, Cambridge, England, 2005.

[2] A.K. Vasudevan, R.D. Doherty, *Aluminum alloys - Contemporary research and applications*, Academic Press, Inc, London, 1989.

[3] W.S. Miller, L. Zhuang, J. Bottema, A.J. Wittebrood, P. De Smet, A. Haszler, A. Vieregge, Recent development in aluminium alloys for the automotive industry, *Mater. Sci. Eng. A.* 280 (2000) 37–49. DOI:10.1016/S0921-5093(99)00653-X.

[4] E.A. Starke, J.T. Staleyt, Application of modern aluminium alloys to aircraft, *Prog. Aerosp. Sci.* 32 (1996) 131–172. DOI:10.1016/0376-0421(95)00004-6.

[5] Y. Zhao, *Advanced mechanical properties and deformation mechanisms of bulk nanostructured materials*, Trans Tech Publications Ltd. Switzerland, 2011.

[6] E.O. Hall, The deformation and ageing of mild steel, *Proc. Phys. Soc.* 64 (1951) 747–752. DOI:10.1088/0370-1301/64/9/303.

[7] N.J. Petch, The cleavage strength of polycrystals, *J. Iron Steel Inst.* 174 (1953) 24.

[8] I. Sabirov, N.A. Enikeev, M.Y. Murashkin, R.Z. Valiev, *Bulk nanostructured materials with multifunctional properties*, Springer, New York, 2015.

[9] C.C. Koch, Ductility in nanostructured and ultra fine-grained materials: Recent evidence for optimism, *J. Metastable Nanocryst. Mater.* 18 (2003) 9–20. DOI:10.4028/www.scientific.net/JMNM.18.9.

[10] S.V. Muley, A.N. Vidvans, G.P. Chaudhari, S. Udainiya, An assessment of ultra-fine grained 316L stainless steel for implant applications, *Acta Biomater.* 30 (2016) 408–419. DOI:10.1016/j.actbio.2015.10.043.

[11] R.Z. Valiev, V. Korznikov, R.R. Mulyukov, Structure and properties of ultrafine-grained materials produced by severe plastic deformation, *Mater. Sci. Eng. A.* 168 (1993) 141–148. DOI:10.1016/0921-5093(93)90717-S.

[12] R.Z. Valiev, R.K. Islamgaliev, I.V. Alexandrov, Bulk nanostructured materials from severe plastic deformation, *Prog. Mater. Sci.* 45 (2000) 103–189. DOI:10.1016/S0079-6425(99)00007-9.

[13] R. Valiev, Nanostructuring of metals by severe plastic deformation for advanced properties, *Nat. Mater.* 3 (2004) 511–516. DOI:10.1038/nmat1180.

[14] K. Changela, S. Kumar, K. Hariharan, D. Ravi Kumar, Aging behavior of ultra-fine grained AA 6061 alloy subjected to constrained groove pressing followed by cold rolling, *IOP Conf. Ser. Mater. Sci. Eng.* 651 (2019) 012069. DOI:10.1088/1757-899X/651/1/012069.

[15] M. Zehetbauer, R.Z. Valiev, *Nanomaterials by severe plastic deformation*, WIlley-Vch Verlag GmbH & Co. KGaA, Weinheim, 2004.

[16] R.Z. Valiev, T.G. Langdon, Principles of equal-channel angular pressing as a processing tool for grain refinement, *Prog. Mater. Sci.* 51 (2006) 881–981. DOI:10.1016/j.pmatsci.2006.02.003.

[17] K. Edalati, Z. Horita, A review on high-pressure torsion (HPT) from 1935 to 1988, *Mater. Sci. Eng. A.* 652 (2016) 325–352. DOI:10.1016/j.msea.2015.11.074.

[18] M.R. Toroghinejad, F. Ashrafizadeh, R. Jamaati, On the use of accumulative roll bonding process to develop nanostructured aluminum alloy 5083, *Mater. Sci. Eng. A* 561 (2013) 145–151. DOI:10.1016/j.msea.2012.11.010.

[19] V.M. Imayev, G.A. Salishchev, M.R. Shagiev, A. V. Kuznetsov, R.M. Imayev, O.N. Senkov, F.H. Froes, Low-temperature superplasticity of submicrocrystalline Ti-48Al-2Nb-2Cr alloy produced by multiple forging, *Scr. Mater.* 40 (1999) 183–190. Doi:10.1016/S1359-6462(98)00419-9.

[20] H.L. Yu, C. Lu, A.K. Tieu, H.J. Li, A. Godbole, S.H. Zhang, Special rolling techniques for improvement of mechanical properties of ultrafine-grained metal sheets: A review, *Adv. Eng. Mater.* 18 (2016) 754–769. DOI:10.1002/adem.201500369.

[21] Y. Wang, M. Chen, F. Zhou, E. Ma, High tensile ductility in a nanostructured metal, *Nature* 419 (2002) 912–915. DOI:10.1038/nature01133.

[22] D.H. Shin, J.J. Park, Y.S. Kim, K.T. Park, Constrained groove pressing and its application to grain refinement of aluminum, *Mater. Sci. Eng. A* 328 (2002) 98–103. DOI:10.1016/S0921-5093(01)01665-3.

[23] K. Changela, H. Krishnaswamy, R.K. Digavalli, Development of combined groove pressing and rolling to produce ultra-fine grained Al alloys and comparison with cryorolling, *Mater. Sci. Eng. A.* 760 (2019) 7–18. DOI:10.1016/j.msea.2019.05.088.

[24] S. Kumar, K. Hariharan, R. Kumar, S. Kumar, Accounting bauschinger effect in the numerical simulation of constrained groove pressing process, *J. Manuf. Process.* 38 (2019) 49–62. DOI:10.1016/j.jmapro.2018.12.013.

[25] V. Parmar, K. Changela, B. Srinivas, M.M. Sankar, S. Mohanty, S.K. Panigrahi, K. Hariharan, D. Kalyanasundaram, Relationship between dislocation density and antibacterial activity of cryo-rolled and cold-rolled copper, *Mater.* 12 (2019) 200, 1–11. DOI:10.3390/ma12020200.

[26] H.L. Yu, A.K. Tieu, C. Lu, X.H. Liu, A. Godbole, C. Kong, Mechanical properties of Al-Mg-Si alloy sheets produced using asymmetric cryorolling and ageing treatment, *Mater. Sci. Eng. A* 568 (2013) 212–218. DOI:10.1016/j.msea.2013.01.048.

[27] A. Chatterjee, G. Sharma, A. Sarkar, J.B. Singh, J.K. Chakravartty, A study on cryogenic temperature ECAP on the microstructure and mechanical properties of Al-Mg alloy, *Mater. Sci. Eng. A* 556 (2012) 653–657. DOI:10.1016/j.msea.2012.07.043.

[28] A. Joshi, N. Kumar, K.K. Yogesha, R. Jayaganthan, S.K. Nath, Mechanical properties and microstructural evolution in Al 2014 alloy processed through multidirectional cryoforging, *J. Mater. Eng. Perform.* 25 (2016) 3031–3045. DOI:10.1007/s11665-016-2126-0.

[29] Y. Huang, P.B. Prangnell, The effect of cryogenic temperature and change in deformation mode on the limiting grain size in a severely deformed dilute aluminium alloy, *Acta Mater.* 56 (2008) 1619–1632. DOI:10.1016/j.actamat.2007.12.017.

[30] P.J. Apps, M. Berta, P.B. Prangnell, The effect of dispersoids on the grain refinement mechanisms during deformation of aluminium alloys to ultra-high strains, *Acta Mater.* 53 (2005) 499–511. DOI:10.1016/j.actamat.2004.09.042.

[31] P.B. Prangnell, Y. Huang, The effect of cryogenic deformation on the limiting grain size in an SMG Al-alloy, *J. Mater. Sci.* 43 (2008) 7280–7285. DOI:10.1007/s10853-008-2673-3.

[32] S.K. Panigrahi, R. Jayaganthan, A study on the mechanical properties of cryorolled Al-Mg-Si alloy, *Mater. Sci. Eng. A.* 480 (2008) 299–305. DOI:10.1016/j.msea.2007.07.024.

[33] S.K. Panigrahi, D. Devanand, R. Jayaganthan, A comparative study on mechanical properties of ultra-fine-grained Al 6061 and Al 6063 alloys processed by cryorolling, *Trans. Indian Inst. Met.* 61 (2008) 159–163. DOI:10.1007/s12666-008-0028-z.

[34] S.K. Panigrahi, R. Jayaganthan, V. Chawla, Effect of cryorolling on microstructure of Al–Mg–Si alloy, *Mater. Lett.* 62 (2008) 2626–2629. DOI:10.1016/j.matlet.2008.01.003.

[35] D. Singh, P.N. Rao, R. Jayaganthan, Effect of deformation temperature on mechanical properties of ultrafine grained Al–Mg alloys processed by rolling, *Mater. Des.* 50 (2013) 646–655. DOI:10.1016/j.matdes.2013.02.068.

[36] S. Goel, N. Keskar, R. Jayaganthan, I.V. Singh, D. Srivastava, G.K. Dey, N. Saibaba, Mechanical behaviour and microstructural characterizations of ultrafine grained Zircaloy-2 processed by cryorolling, *Mater. Sci. Eng. A* 603 (2014) 23–29. DOI:10.1016/j.msea.2014.02.025.

[37] Y. Xiong, Y. Yue, Y. Lu, T. He, M. Fan, F. Ren, W. Cao, Cryorolling impacts on microstructure and mechanical properties of AISI 316 LN austenitic stainless steel, *Mater. Sci. Eng. A* 709 (2018) 270–276. DOI:10.1016/j.msea.2017.10.067.

[38] K. Changela, H.B. Naik, K.P. Desai, H.K. Raval, Effect of rolling temperatures on mechanical and fracture behavior of AA 3003 alloy and pure Cu, *SN Appl. Sci.* 2 (2020) 1109. DOI:10.1007/s42452-020-2903-0.

[39] B. Srinivas, A. Dhal, S.K. Panigrahi, A mathematical prediction model to establish the role of stacking fault energy on the cryo-deformation behavior of FCC materials at different strain levels, *Int. J. Plast.* 97 (2017) 159–177. DOI:10.1016/j.ijplas.2017.05.014.

[40] A. Dhal, S.K. Panigrahi, M.S. Shunmugam, Insight into the microstructural evolution during cryo-severe plastic deformation and post-deformation annealing of aluminum and its alloys, *J. Alloys Compd.* 726 (2017) 1205–1219. DOI:10.1016/j.jallcom.2017.08.062.

[41] S.K. Panigrahi, R. Jayaganthan, Development of ultrafine-grained Al 6063 alloy by cryorolling with the optimized initial heat treatment conditions, *Mater. Des.* 32 (2011) 2172–2180. DOI:10.1016/j.matdes.2010.11.027.

[42] F. Feyissa, D.R. Kumar, P.N. Rao, Characterization of microstructure, mechanical properties and formability of cryorolled AA5083 alloy, *J. Mater. Eng. Perform.* 27 (2018) 1614–1627. DOI:10.1007/s11665-018-3243-8.

[43] D. Singh, P. Nageswararao, R. Jayaganthan, Microstructural studies of Al 5083 alloy deformed through cryorolling, *Adv. Mater. Res.* 585 (2012) 376–380. DOI:10.4028/www.scientific.net/AMR.585.376.

[44] K. Changela, H. Krishnaswamy, R.K. Digavalli, Mechanical behavior and deformation kinetics of aluminum alloys processed through cryorolling and subsequent annealing, *Metall. Mater. Trans. A.* 51 (2020) 648–666. DOI:10.1007/s11661-019-05532-2.

[45] Y. Zhong, F. Yin, T. Sakaguchi, K. Nagai, K. Yang, Dislocation structure evolution and characterization in the compression deformed Mn-Cu alloy, *Acta Mater.* 55 (2007) 2747–2756. DOI:10.1016/j.actamat.2006.12.012.

[46] R. Jayaganthan, H.G. Brokmeier, B. Schwebke, S.K. Panigrahi, Microstructure and texture evolution in cryorolled Al 7075 alloy, *J. Alloys Compd.* 496 (2010) 183–188. DOI:10.1016/j.jallcom.2010.02.111.

[47] S.K. Panigrahi, R. Jayaganthan, A comparative study on mechanical properties of Al 7075 alloy processed by rolling at cryogenic temperature and room temperature, *Mater. Sci. Forum,* 584–586 (2008) 734–740. DOI:10.4028/www.scientific.net/MSF.584-586.734.

[48] S.K. Panigrahi, R. Jayaganthan, V. Pancholi, Effect of plastic deformation conditions on microstructural characteristics and mechanical properties of Al 6063 alloy, *Mater. Des.* 30 (2009) 1894–1901. DOI:10.1016/j.matdes.2008.09.022.

[49] N.N. Krishna, M. Ashfaq, P. Susila, K. Sivaprasad, K. Venkateswarlu, Mechanical anisotropy and microstructural changes during cryorolling of Al–Mg–Si alloy, *Mater. Charac.* 107 (2015) 302–308. DOI:10.1016/j.matchar.2015.07.033.

[50] K. Chandra Sekhar, R. Narayanasamy, K. Venkateswarlu, Formability, fracture and void coalescence analysis of a cryorolled Al-Mg-Si alloy, *Mater. Des.* 57 (2014) 351–359. DOI:10.1016/j.matdes.2013.12.077.

[51] V. Zohoori-Shoar, A. Eslami, F. Karimzadeh, M. Abbasi-Baharanchi, Resistance spot welding of ultra-fine grained/nanostructured Al 6061 alloy produced by cryorolling process and evaluation of weldment properties, *J. Manuf. Process.* 26 (2017) 84–93. DOI:10.1016/j.jmapro.2017.02.003.

[52] P.N. Rao, D. Singh, R. Jayaganthan, Effect of post cryorolling treatments on microstructural and mechanical behaviour of ultra fine grained Al-Mg-Si alloy, *J. Mater. Sci. Technol.* 30 (2014) 998–1005. DOI:10.1016/j.jmst.2014.03.009.

[53] K.S.V.B.R. Krishna, K.C. Sekhar, R. Tejas, N.N. Krishna, K. Sivaprasad, R. Narayanasamy, K. Venkateswarlu, Effect of cryorolling on the mechanical properties of AA5083 alloy and the Portevin–Le Chatelier phenomenon, *Mater. Des.* 67 (2015) 107–117. DOI:10.1016/j.matdes.2014.11.022.

[54] D. Singh, P. Nageswara Rao, R. Jayaganthan, Microstructures and impact toughness behavior of Al 5083 alloy processed by cryorolling and afterwards annealing, *Int. J. Miner. Metall. Mater.* 20 (2013) 759–769. DOI:10.1007/s12613-013-0794-4.

[55] Y.B. Lee, D.H. Shin, K.T. Park, W.J. Nam, Effect of annealing temperature on microstructures and mechanical properties of a 5083 Al alloy deformed at cryogenic temperature, *Scri. Mater.* 51 (2004) 355–359. DOI:10.1016/j.scriptamat.2004.02.037.

[56] K.K. Yogesha, N. Kumar, A. Joshi, R. Jayaganthan, S.K. Nath, A comparative study on tensile and fracture behavior of Al-Mg alloy processed through cryorolling and cryo groove rolling, *Metallogr. Microstruct. Anal.* 5 (2016) 251–263. DOI:10.1007/s13632-016-0282-0.

[57] S.K. Panigrahi, R. Jayaganthan, Effect of rolling temperature on microstructure and mechanical properties of 6063 Al alloy, *Mater. Sci. Eng. A.* 492 (2008) 300–305. DOI:10.1016/j.msea.2008.03.029.

[58] J. Shi, L. Hou, J. Zuo, L. Zhuang, J. Zhang, A Cryogenic rolling-enhanced mechanical properties and microstructural evolution of 5052 Al-Mg alloy, *Mater. Sci. Eng. A* 701 (2017) 274–284. DOI:10.1016/j.msea.2017.06.087.

[59] V.L. Niranjani, K.C. Hari Kumar, V. Subramanya Sarma, Development of high strength Al-Mg-Si AA6061 alloy through cold rolling and ageing, *Mater. Sci. Eng. A.* 515 (2009) 169–174. DOI:10.1016/j.msea.2009.03.077.

[60] A. Sarkar, K. Saravanan, N. Nayan, S.V.S. Narayana Murty, P. Ramesh Narayanan, P.V. Venkitakrishnan, J. Mukhopadhyay, Microstructure and mechanical properties of cryorolled aluminum alloy AA2219 in different thermomechanical processing conditions, *Metall. Mater. Trans. A.* 48 (2016) 321–341. DOI:10.1007/s11661-016-3807-x.

[61] Q. Liu, X. Huang, D.J. Lloyd, N. Hansen, Microstructure and strength of commercial purity aluminium (AA 1200) cold-rolled to large strains, *Acta Mater.* 50 (2002) 3789–3802. DOI:10.1016/S1359-6454(02)00174-X

[62] N. Rangaraju, T. Raghuram, B.V. Krishna, K.P. Rao, P. Venugopal, Effect of cryo-rolling and annealing on microstructure and properties of commercially pure aluminium, *Mater. Sci. Eng. A.* 398 (2005) 246–251. DOI:10.1016/j.msea.2005.03.026.

[63] M. Weiss, A.S. Taylor, P.D. Hodgson, N. Stanford, Strength and biaxial formability of cryo-rolled 2024 aluminium subject to concurrent recovery and precipitation, *Acta Mater.* 61 (2013) 5278–5289. DOI:10.1016/j.actamat.2013.05.019.

[64] A. Dhal, S.K. Panigrahi, M.S. Shunmugam, Influence of annealing on stain hardening behaviour and fracture properties of a cryorolled Al 2014 alloy, *Mater. Sci. Eng. A.* 645 (2015) 383–392. DOI:10.1016/j.msea.2015.08.020.

[65] K. Chandra Sekhar, R. Narayanasamy, K. Velmanirajan, Experimental investigations on microstructure and formability of cryorolled AA 5052 sheets, *Mater. Des.* 53 (2014) 1064–1070. DOI:10.1016/j.matdes.2013.08.008.

[66] D.R. Satish, F. Feyissa, D.R. Kumar, Cryorolling and warm forming of AA6061 aluminum alloy sheets, *Mater. Manuf. Processes.* 32 (2017) 1345–1352. DOI:10.1080/10426914.2017.1317352.

[67] S.K. Panigrahi, R. Jayaganthan, Development of ultrafine grained high strength age hardenable Al 7075 alloy by cryorolling, *Mater. Des.* 32 (2011) 3150–3160. DOI:10.1016/j.matdes.2011.02.051.

[68] S.K. Panigrahi, R. Jayaganthan, Effect of annealing on thermal stability, precipitate evolution, and mechanical properties of cryorolled Al 7075 alloy, *Metall. Mater. Trans. A.* 42A (2011) 3208–3217. DOI:10.1007/s11661-011-0723-y.

[69] F. Taye, P. Das, D. Ravi Kumar, B. Ravi Sankar, Characterization of mechanical properties and formability of cryorolled aluminium alloy sheets, *5th International & 26th All India Manufacturing Technology, Design and Research Conference (AIMTDR)*, December 2014, IIT Guwahati, India.

[70] S. Cheng, Y.H. Zhao, Y.T. Zhu, E. Ma, Optimizing the strength and ductility of fine structured 2024 Al alloy by nano-precipitation, *Acta Mater.* 55 (2007) 5822–5832. DOI:10.1016/j.actamat.2007.06.043.

[71] P. Kumar, M. Kawasaki, T.G. Langdon, Review: Overcoming the paradox of strength and ductility in ultrafine-grained materials at low temperatures, *J. Mater. Sci.* 51 (2016) 7–18. DOI:10.1007/s10853-015-9143-5.

[72] S.K. Panigrahi, R. Jayaganthan, Effect of ageing on microstructure and mechanical properties of bulk, cryorolled, and room temperature rolled Al 7075 alloy, *J. Alloys Compd.* 509 (2011) 9609–9616. DOI:10.1016/j.jallcom.2011.07.028.

[73] T. Shanmugasundaram, B.S. Murty, V.S. Sarma, Development of ultrafine grained high strength Al-Cu alloy by cryorolling, *Scr. Mater.* 54 (2006) 2013–2017. DOI:10.1016/j.scriptamat.2006.03.012.

[74] A. Dhal, S.K. Panigrahi, M.S. Shunmugam, Deformation behaviour and fracture mechanism of ultrafine-grained aluminium developed by cryorolling in U.S. Dixit, R.G. Narayanan, *Strengthening and joining by plastic deformation*, Springer Nature Singapore Pte Ltd, 2019, 31–52.

[75] C.X. Huang, W. Hu, G. Yang, Z.F. Zhang, S.D. Wu, Q.Y. Wang, G. Gottstein, The effect of stacking fault energy on equilibrium grain size and tensile properties of nanostructured copper and copper-aluminum alloys processed by equal channel angular pressing, *Mater. Sci. Eng. A.* 556 (2012) 638–647. DOI:10.1016/j.msea.2012.07.041.

[76] H. Yu, A.K. Tieu, C. Lu, X. Liu, M. Liu, A. Godbole, C. Kong, Q. Qin, A new insight into ductile fracture of ultrafine-grained Al-Mg alloys, *Sci. Rep.* 5 (2015) 1–9. DOI:10.1038/srep09568.

[77] U.G. Kang, J.C. Lee, S.W. Jeong, W.J. Nam, The improvement of strength and ductility in ultra-fine grained 5052 Al alloy by cryogenic and warm-rolling, *J. Mater. Sci.* 45 (2010) 4739–4744. DOI:10.1007/s10853-010-4573-6.

[78] P.N. Rao, A. Kaurwar, D. Singh, R. Jayaganthan, Enhancement in strength and ductility of Al-Mg-Si alloy by cryorolling followed by warm rolling, *Procedia Eng. MRS Singapore-ICMAT Symposia Proceedings*, 75 (2014) 123–128. DOI:10.1016/j.proeng.2013.11.027.

[79] L. Mei, X.P. Chen, P. Ren, Y.Y. Nie, G.J. Huang, Q. Liu, Effect of warm deformation on precipitation and mechanical properties of a cryorolled Al-Zn-Mg-Cu sheet, *Mater. Sci. Eng. A.* 771 (2020) 138608. DOI:10.1016/j.msea.2019.138608.

[80] O.B. Bembalge, S.K. Panigrahi, Hot deformation behavior and processing map development of cryorolled AA6063 alloy under compression and tension, *Int. J. Mech. Sci.* 191 (2021) 106100. DOI:10.1016/j.ijmecsci.2020.106100.

[81] A.S. Taylor, M. Weiss, T. Hilditch, N. Stanford, P.D. Hodgson, Formability of cryo-rolled aluminium in uniaxial and biaxial tension, *Mater. Sci. Eng. A.* 555 (2012) 148–153. DOI:10.1016/j.msea.2012.06.044.

[82] F.T. Feyissa, D.R. Kumar, Enhancement of drawability of cryorolled AA5083 alloy sheets by hydroforming, *J. Mater. Res. Technol.* 8 (2018) 411–423. DOI:10.1016/j.jmrt.2018.02.012.

[83] D. R. Satish, F.T. Feyissa, D.R. Kumar, Formability of cryorolled aluminum alloy sheets in warm forming, *Int. J. Mater. Mech. Manuf.* 6 (2018) 123–126. DOI: 10.18178/ijmmm.2018.6.2.360.

[84] F. Vollertsen, Effects on the deep drawing diagram in micro forming, *Prod. Eng. Res. Devel.* 6 (2012) 11–18. DOI:10.1007/s11740-011-0355-5.

[85] A. Dhal, S.K. Panigrahi, M.S. Shunmugam, Achieving excellent microformability in aluminum by engineering a unique ultrafine-grained microstructure, *Sci. Rep.* 9 (2019) 10683. DOI:10.1038/s41598-019-46957-4.

6 Deformation Mechanism in Single Point Incremental Forming (SPIF) and Significance of Crystallographic Texture in Sheet Metal Forming Operations

Parnika Shrivastava
National Institute of Technology, Hamirpur, India

CONTENTS

6.1 INTRODUCTION

Metal framing is one of the foundations of manufacturing sectors other than being an astounding industry in itself (Groover 2007). For mass manufacturing of sheet metal components, conventional forming techniques are already playing a vital role. Conventional forming operations require a die to replicate the required shape. However, Incremental sheet forming (ISF) is one such type of non-conventional forming family that does not require die for shape formation. A ball nose-shaped tool moves in a predetermined trajectory and forms a sheet clamped around its periphery (Shrivastava and Tandon 2015). Conventional forming operations are well established and the process mechanism has been identified, however, ISF process is still under exploration and deformation mechanism is still not fully established. The chapter discusses the theories developed by various researches that uncover the dominant deformation mechanism of the process. In addition, significance of crystallographic texture in sheet metal industry is discussed. Crystal Plasticity Finite Element (CPFE) model is discussed as a future solution to predict microstructural behavior of the sheet metal components.

DOI: 10.1201/9781003226703-6

6.2 FORMING BEHAVIOR AND DEFORMATION MECHANISM IN SPIF

SPIF process is characterized by higher formability in comparison to conventional deep drawing or stamping processes (Emmens and van den Boogaard 2009; McAnulty et al. 2016; Shrivastava et al. 2018). The process permits to achieve much higher permissible strain values prior to failure (Shrivastava and Tandon 2018a). The same has been shown in Figure 6.1. However, in spite of various efforts, the mechanism of the deformation and the reason for this enhanced forming behavior is not yet fully established and is still under exploration. Material and constitutive models that are capable to predict the materials behavior in conventional forming operations fail to predict the behavior of the same material when subjected to SPIF (Gatea et al. 2018). Research had been carried out to identify the deformation mechanism of SPIF process. The enhanced formability was attributed to localized nature of the deformation but still there is a lack of explanation in support to that (Behera et al. 2017; Shrivastava and Tandon 2019).

Emmens and van den Boogaard (2009) reported that bending under tension is the governing mechanism in incremental sheet forming. They have performed Continuous Bending under Tension (CBT) test to validate that localized bending is the crucial deformation characteristics of SPIF process. The claimed major deformation modes are explained in Figure 6.2 Further, they have also studied the cause of enhanced formability in SPIF, i.e., difference between fracture limit and necking limit (absence of material localization prior to fracture). They have concluded that contact stress, bending-under-tension, shear, cyclic straining, geometrical inability to grow and hydrostatic stress are the major deformation mechanisms in SPIF. The first three of these are responsible for localized deformation behavior and the last three are responsible for delaying in necking phenomena. However, they were not able to comprehend the role of Hydrostatic pressure as far as stability above the FLC was concerned.

Later, Jackson and Allwood (2009) examined the deformation mechanism of SPIF process through experimentation. The copper plates with gridded cross-section were brazed and made as a single plate. Further, the brazed plate was formed and then separated to examine the distorted grid post forming. The schematic diagram of the procedure is shown in Figure 6.3. Deformation engineering strain and thickness distribution were examined to identify the dominant deformation mechanism in SPIF, which were plotted in global coordinate system. For strain measurement, they elicited information before and after deformation for the points of interaction of the gridlines in

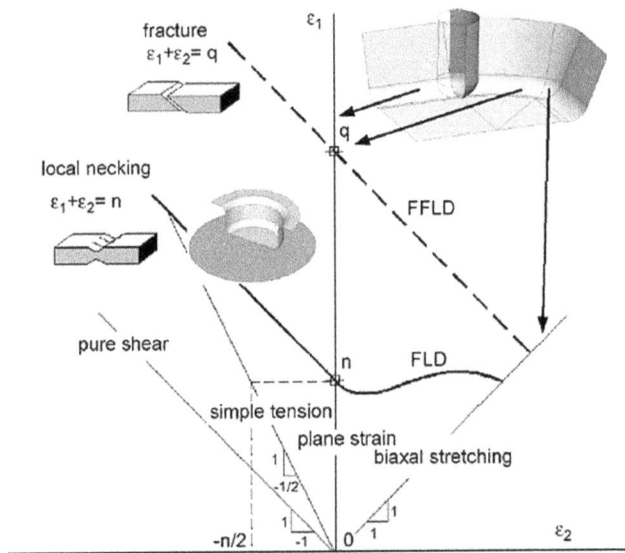

FIGURE 6.1 Forming limit diagram for different conventional and ISF process (Martins et al. 2008).

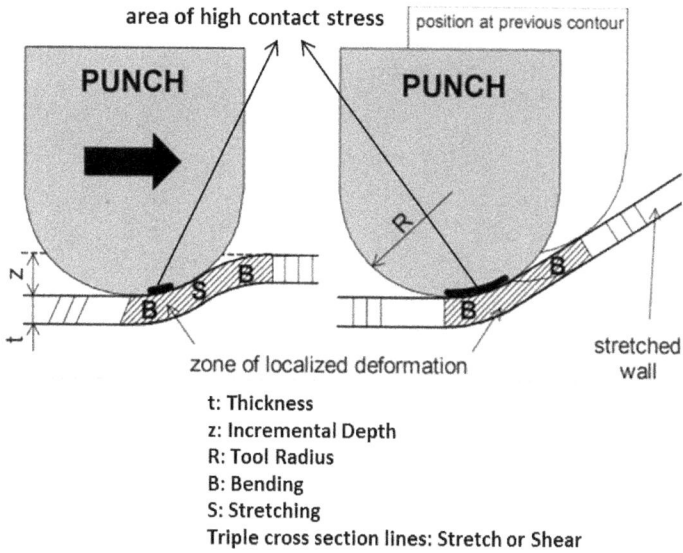

t: Thickness
z: Incremental Depth
R: Tool Radius
B: Bending
S: Stretching
Triple cross section lines: Stretch or Shear

FIGURE 6.2 Major deformation stresses in SPIF (Emmens and van den Boogaard 2009).

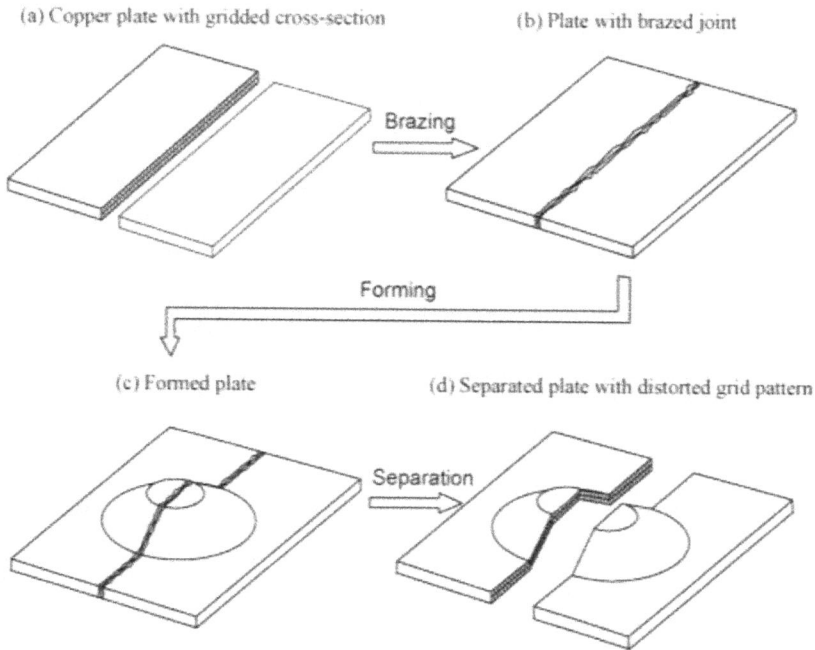

FIGURE 6.3 Procedure for measuring through the thickness deformation (Jackson and Allwood 2009).

the local coordinate system, as shown in Figure 6.4. Post forming, by analyzing the distribution of through the thickness strain values they have concluded that stretching and shear in the plane normal to the punch movement and shear in the plane parallel to the punch direction are the major deformation mechanism in SPIF process (Shrivastava and Tandon 2018b).

In order to understand formability and fracture behavior in SPIF, a different experimental approach adopted by Silva et al. (2011) led to the conclusion that it is the ratio of sheet thickness and

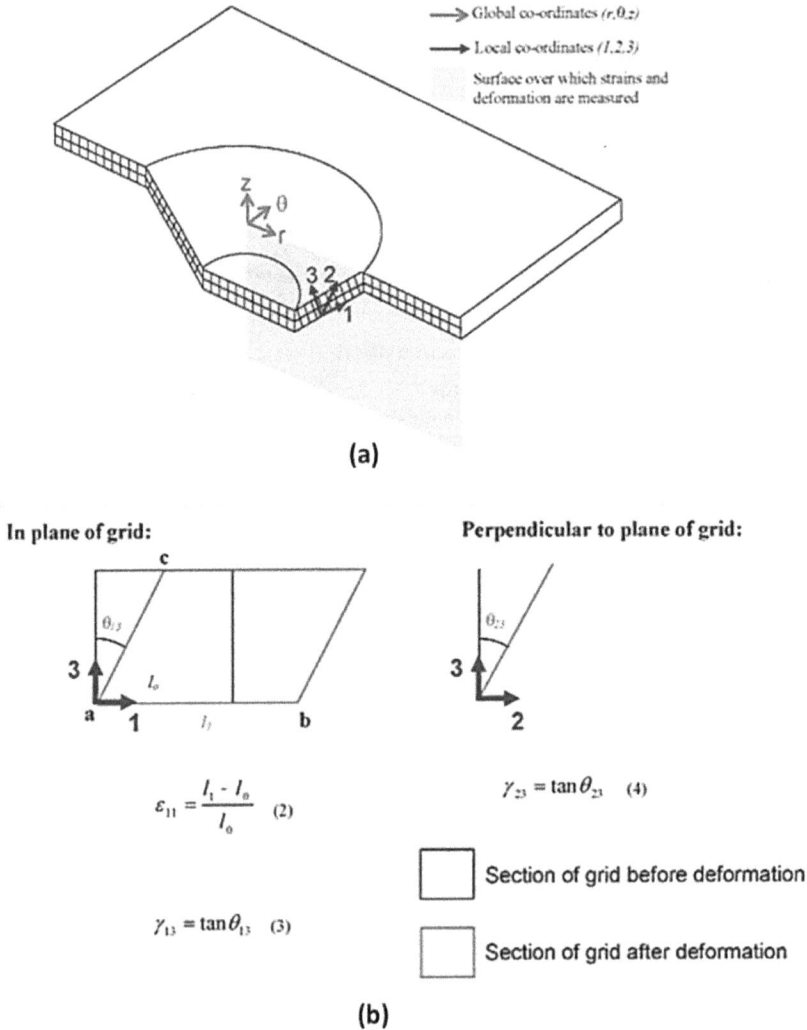

(a)

In plane of grid:

Perpendicular to plane of grid:

$$\varepsilon_{11} = \frac{l_1 - l_o}{l_o} \quad (2)$$

$$\gamma_{23} = \tan\theta_{23} \quad (4)$$

☐ Section of grid before deformation

☐ Section of grid after deformation

$$\gamma_{13} = \tan\theta_{13} \quad (3)$$

(b)

FIGURE 6.4 (a) Global and local coordinate system in three dimensional space and (b) strain calculation as per the deformed grids (Jackson and Allwood 2009).

tool radius that governs the occurrence of fracture. Tools with larger radii promote dynamic bending under tension which is responsible for enhanced formability, whereas tools having smaller radii lead to crack initiation with suppression of necking. They have also formulated the stress and strain in SPIF based on the tool radius and sheet thickness, as shown in Figure 6.5. Malhotra et al. (2012) studied deformation mechanism in SPIF and they have analyzed that both through-the-thickness shear and localized bending at the tool-sheet contact region determines the nature of deformation and failure in SPIF. A theory named as "noodle" theory was proposed in support of the enhanced formability by validating the localized nature of deformation in SPIF. The schematic representation of the noodle theory is presented in Figure 6.6. The theory utilizes the string fixed at one of the ends and stretched from the free end in the case of conventional forming and stretched at different positions in the case of SPIF.

	Plane strain conditions (flat and rotational symmetric surfaces)	$d\varepsilon_\varphi = -d\varepsilon_t > 0$ $d\varepsilon_\theta = 0$ $d\varepsilon_t < 0$	$\sigma_\varphi = \sigma_1 = \dfrac{\sigma_Y}{\left(1 + t/r_{tool}\right)} > 0$ $\sigma_\theta = \sigma_2 = \tfrac{1}{2}(\sigma_1 + \sigma_3)$ $\sigma_t = \sigma_3 = -\sigma_Y \dfrac{t}{\left(r_{tool} + t\right)} < 0$
	Equal bi-axial stretching (corners)	$d\varepsilon_\varphi = d\varepsilon_\theta > 0$ $d\varepsilon_t < 0$	$\sigma_\varphi = \sigma_\theta = \sigma_1 = \dfrac{\sigma_Y}{\left(1 + 2t/r_{tool}\right)} > 0$ $\sigma_t = \sigma_3 = -2\sigma_Y \dfrac{t}{\left(r_{tool} + 2t\right)} < 0$

ε	True strain	t	Thickness of the sheet
σ	True stress	r_{tool}	Radius of the SPIF tool
σ_θ	Circumferential stress	r_{part}	Radius of the SPIF part
σ_o	Meridional stress	r_{part}/r_{tool}	Incremental tool ratio
σ_t	Thickness stress		
σ_Y	Yield stress		

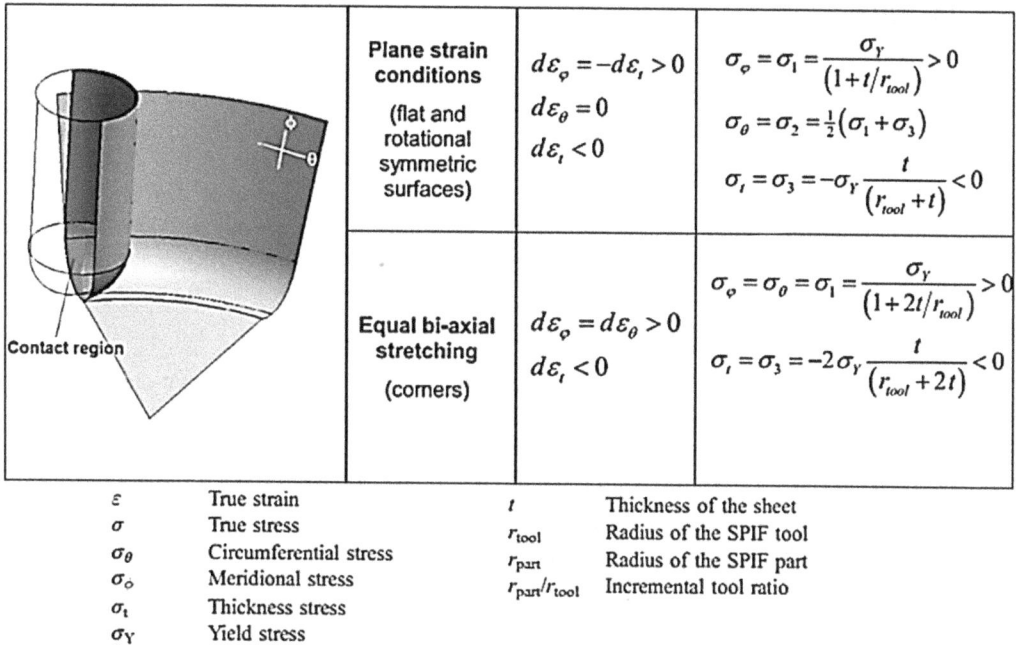

FIGURE 6.5 State of stress and strain in SPIF based on tool radius and sheet thickness (Silva et al. 2011).

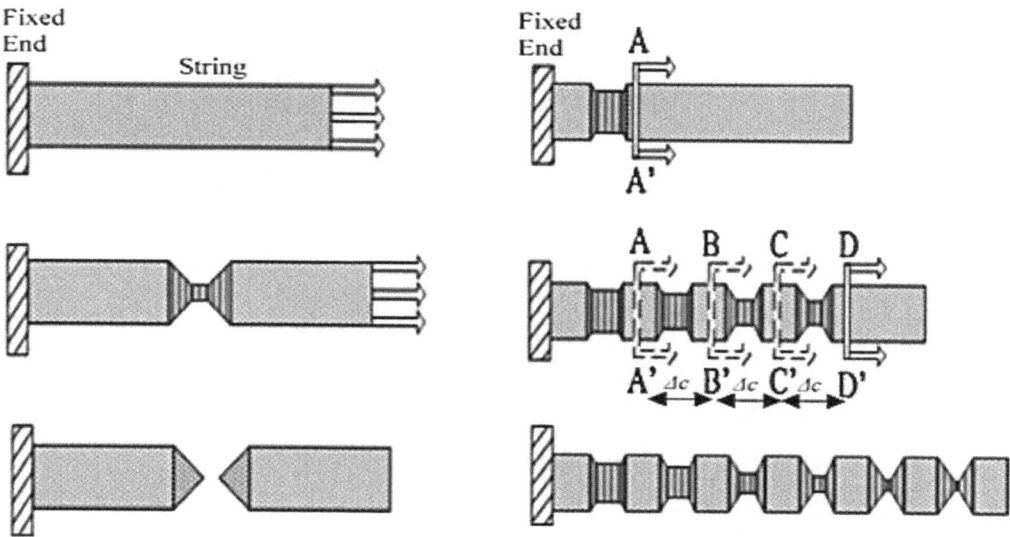

FIGURE 6.6 Deformation in conventional forming is limited by necking whereas incremental and localized deformation occur in SPIF (Noodle Theory) (Malhotra et al. 2012).

6.3 SIGNIFICANCE OF MICROSTRUCTURE AND CRYSTALLOGRAPHIC TEXTURE IN SHEET METAL FORMING

Complex microstructures and crystallographic textures developed during forming determines the distribution of grain boundaries and slip system, which in turn determines the dislocation movements responsible for the nature and extent of plastic deformation in metals (Satheesh Kumar and Raghu 2015; Hamid et al. 2017; Liao et al. 2017; Vargas et al. 2017; Singh et al. 2018).

The microstructural constituents which cumulatively govern the overall physical model are explained with the help of Figure 6.7.

6.3.1 Quantitative Description of Texture

Analysis of texture (crystallographic orientation) incorporates the concept of Orientation Distribution (OD), which parameterizes the crystal orientation, usually in the form of three Euler angles ($\varphi_1\Phi\varphi_2$). Euler angles signify a crystal orientation with respect to the sample axes, as shown in Figure 6.8. The three Euler angles forms a three-dimensional, orthogonal orientation space which is termed as Euler space as shown in Figure 6.9. Therefore, a point in Euler space corresponds to a particular orientation which is signified by triple Euler angles (Cho and Finocchiaro 2010).

OD is a normalized probability distribution which is associated with the frequency of occurrence of any given texture component. Continuous orientation density is described in the form of "orientation distribution function" (ODF) whereas, pole figure is the two-dimensional projection of the information presented by orientation distribution.

6.3.1.1 Texture Components in Face-Centered Cubic (FCC) Metals

Deformation of metal results in several common texture components that are classified on the basis of Euler angles. Texture components are named either after the scientist or by the material they

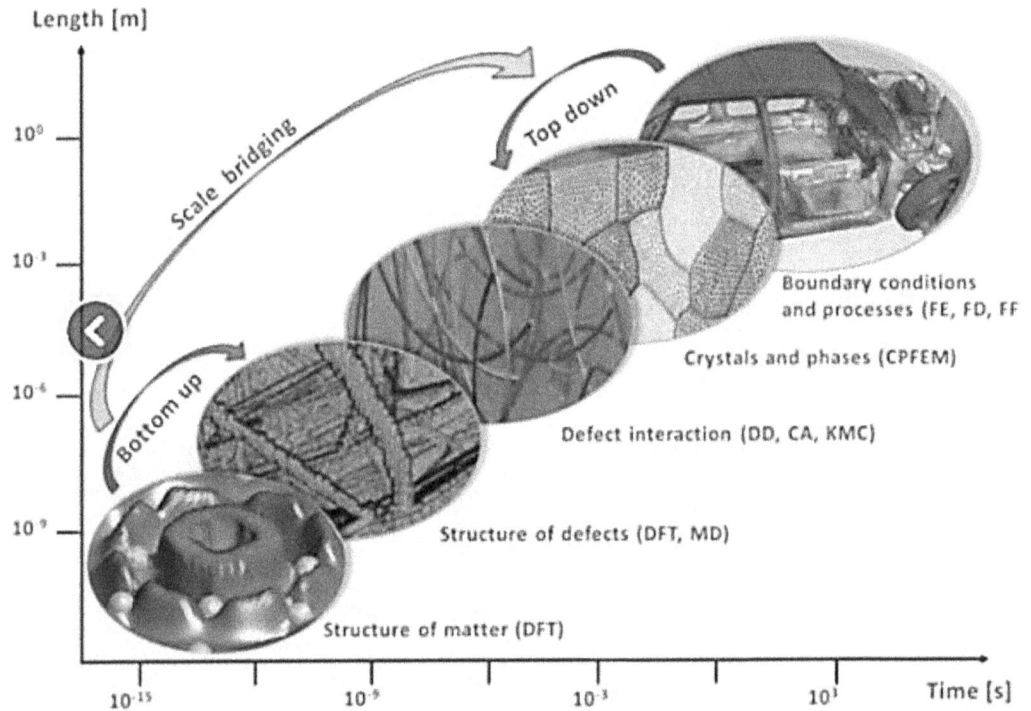

FIGURE 6.7   Microstructural constituents of any overall physical model (Zhao et al. 2008).

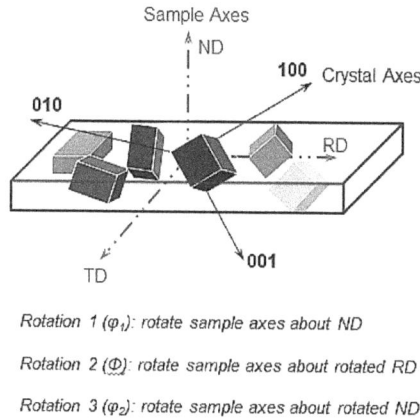

Rotation 1 (φ₁): rotate sample axes about ND

Rotation 2 (Φ): rotate sample axes about rotated RD

Rotation 3 (φ₂): rotate sample axes about rotated ND

FIGURE 6.8 Schematic representation of differently oriented crystals in a rolled sample having sample axes as rolling direction (RD), transverse direction (TD) and normal direction (ND) (Humphreys and Hatherly 2004).

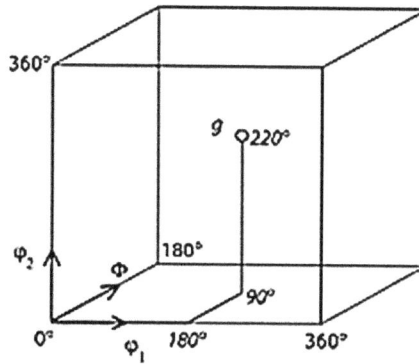

FIGURE 6.9 Cartesian Euler space (Humphreys and Hatherly 2004).

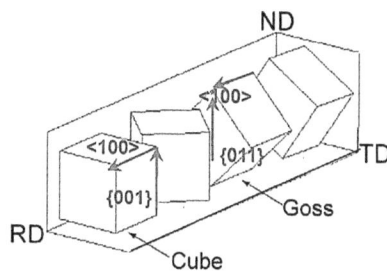

FIGURE 6.10 Orientation of typical texture components corresponding to sample axes.

are found in abundance. In face-centered cubic (FCC) material, texture components like, "Cube," "Copper," "Brass," "Goss," and "S" are most commonly found. The crystal orientation of typical texture components corresponding to sample axes are depicted in Figure 6.10, whereas miller indices and Euler angles are provided in Table 6.1.

For FCC material, the positions of important texture components and fibers in three-dimensional Euler space are shown in Figure 6.11. For the sake of understanding, critical texture components and their distribution in 3D Euler space are depicted individually in Figure 6.12.

TABLE 6.1

Miller Indices and Euler Angles of the Important Texture Components Present in FCC Metal (Cho and Finocchiaro 2010)

Texture Component	RD	ND	Miller Indices {hkl}<uvw>	Euler Angles (°)
Cube	<100>	{001}	{001}<100>	(0, 0, 0)
Goss	<100>	{011}	{011}<100>	(0, 45, 0)
Brass	<211>	{011}	{011}<211>	(35, 45, 0)
Copper	<111>	{112}	{112}<111>	(90, 30, 45)
S	<634>	{123}	{123}<634>	(59, 34, 65)

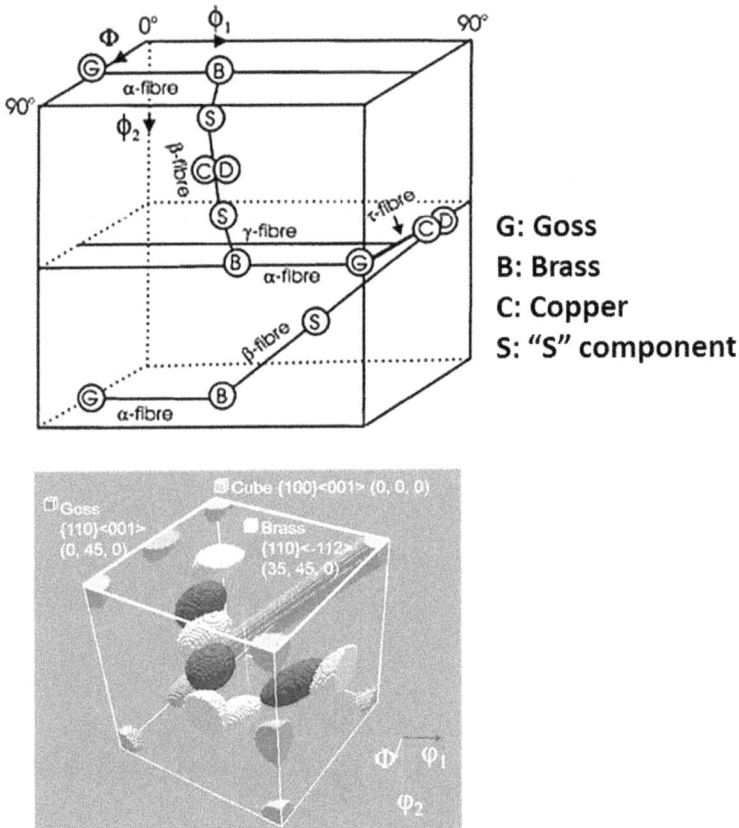

FIGURE 6.11   Texture components in three-dimensional Euler space (Humphreys and Hatherly 2004).

Due to the sake of understanding, ODF are presented in sections rather than three dimensional representations. The practice of extracting the three-dimensional information and presenting in two-dimensional section in the form of OD is explained in Figure 6.13. Each section (OD plot) is obtained by varying one Euler angle (ϕ_2 for FCC metals) keeping the other two as constant, which is shown in Figure 6.13 (a). OD corresponding to $\varphi_2 = 45^0$ is depicted in Figure 6.13 (b).

It is also noteworthy that orthorhombic sample and cubic crystal symmetry of FCC metals only a sub volume of Euler space (with $0 \leq \varphi_1$, Φ, $\varphi_2 \leq 90°$) is prerequisite to epitomize the whole texture. Thus, $\varphi_2 = 45°$, $\varphi_2 = 65°$ and $\varphi_2 = 90°$ sections are considered for texture representation for FCC metals (Kestens and Pirgazi 2016).

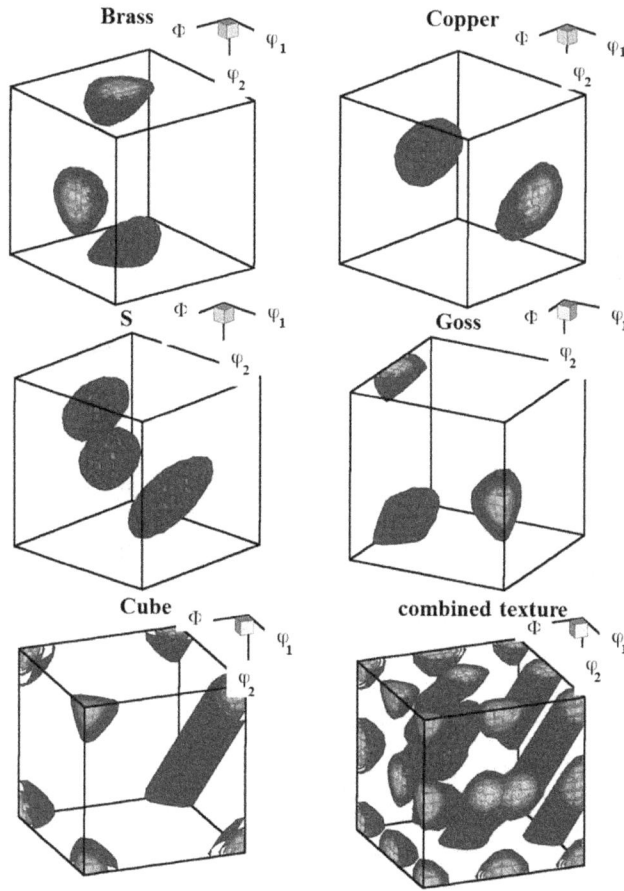

FIGURE 6.12 Distribution of important texture components in Euler space (Humphreys and Hatherly 2004).

6.3.1.2 Orientation Fiber

FCC metals like aluminium possess two dominant orientation fibers named as α and β fibers. The orientation spreads in a characteristic manner from one texture component to another lead to the formation of orientation tube in three-dimensional Euler space, as shown in Figure 6.13. However, for the sake of understanding the same can be shown with the help of connecting lines in Euler space as shown in Figure 6.14.

Where in Euler space, α-fiber stems from $\{011\}$ <100> Goss (Gs) to $\{011\}$ <211> Brass (Bs) texture component and β-fiber oriented from the $\{112\}$ <111> Copper (Cu) texture component to the $\{011\}$ <211> Brass texture component via $\{123\}$ <634> S texture component.

The position details of important texture components and fibers that may be present in deformed or recrystallized aluminium alloy sheet in $\varphi_2 = 45°$, $\varphi_2 = 65°$ and $\varphi_2 = 90°$ ODF sections, as shown in Figure 6.14.

6.3.2 DEFORMATION MECHANISM AND FORMING BEHAVIOR IN SPIF AS PER DEVELOPED TEXTURES

Barnwal et al. 2018, in an attempt to understand the macro and micro deformation behavior of AA 6061 aluminium alloy used techniques like XRD and EBSD. The role of anisotropy on deformation and effect of deformation on grain shape, size and crystallographic texture development was also

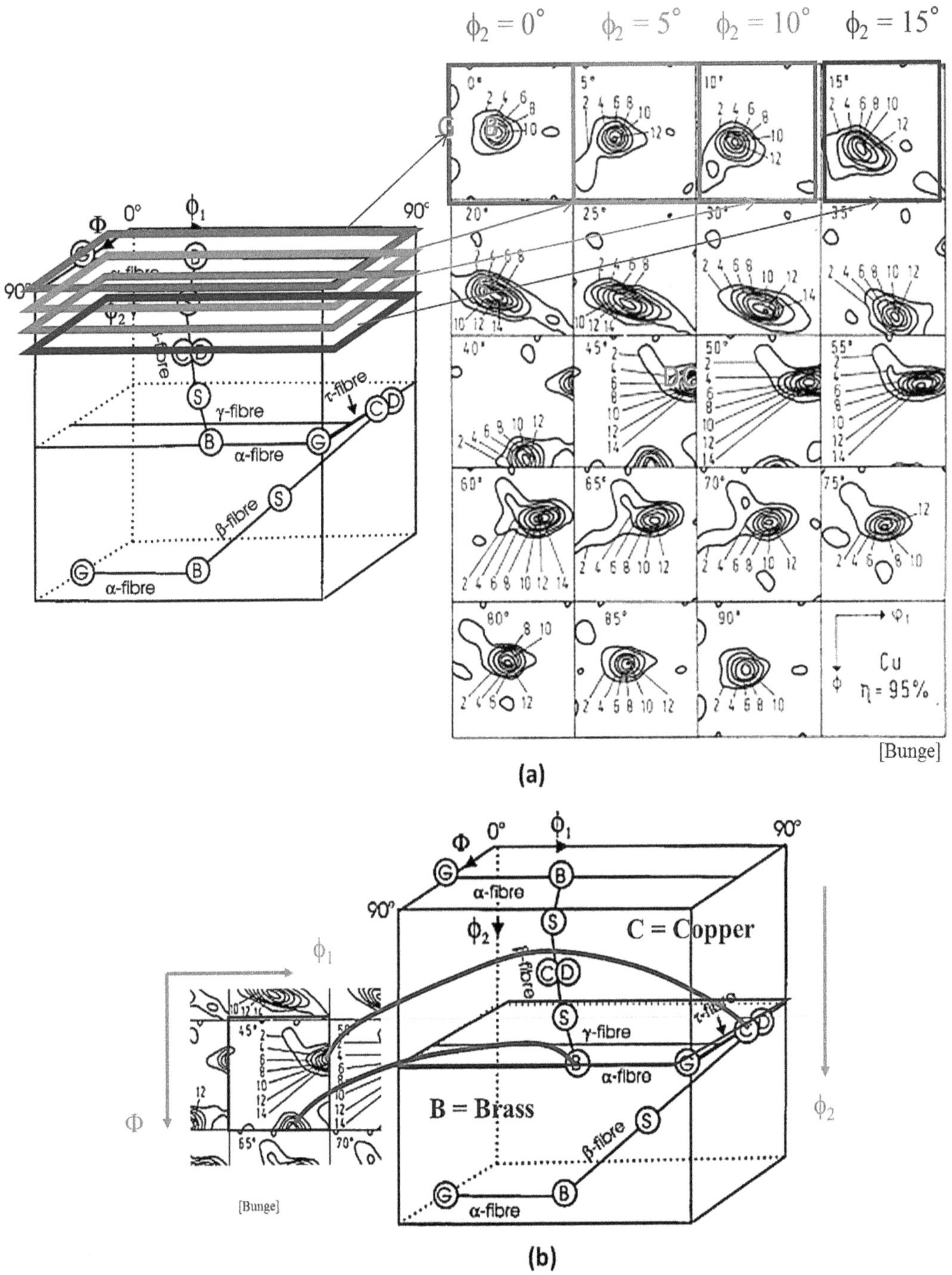

(a)

(b)

FIGURE 6.13 (a) Extraction of three dimensional representations of texture components and conversion into two dimensional OD sections and (b) detailed representation of $\varphi_2 = 45^0$, OD section (Humphreys and Hatherly 2004).

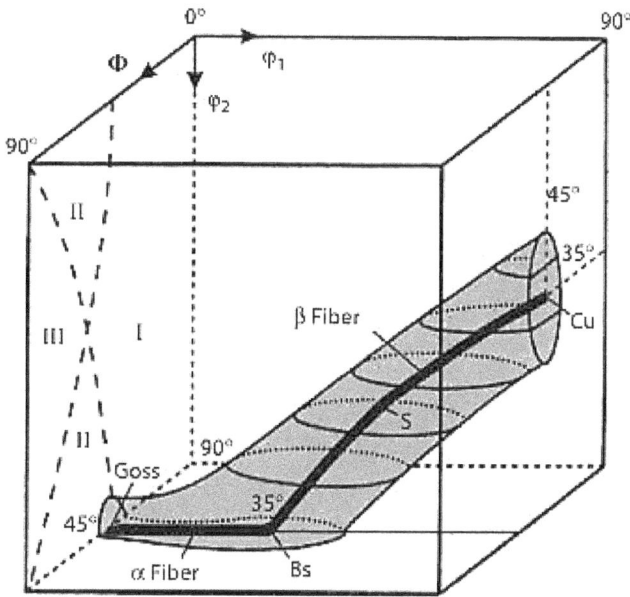

FIGURE 6.14 Distribution of α and β fibers in Euler space (Kestens and Pirgazi 2016).

studied. Microstructure analysis showed different resistance to deformation along different directions. Nath et al. 2018, compared surface finish and texture of samples formed by SPIF and deep drawing & bulge testing. The crystallographic texture comparison showed difference in texture of SPIF formed and deep drawn sample. Hussain et al. 2020, conducted a microstructure study pre and post ISF for two aluminium alloys and reported changes in microstructure being dependent upon the process parameters. Basak et al. 2020, tried to correlate microstructure evolution, crystallographic texture, and surface roughness. Grain refinement led to major increase in hardness and the amount of roughening was also related to crystallographic texture. A deformation mechanism study with insights from crystallographic texture and microstructure was conducted by Mishra et al. 2021, reported through thickness shear playing vital role in deformation through SPIF. Ghosh et al. 2021, used SEM and EBSD techniques to study the influence of temperature on the deformation of EN AW 6016 alloy.

Deformation behavior in SPIF process was unfolded by detailed analysis of the texture developed at different stages of the forming operation (Shrivastava and Tandon 2018b). Detailed analysis of the bulk texture evolution was carried out by investigating texture present in the undeformed sheet and the sheet deformed under imposed strain in SPIF. During texture measurements, attention was paid to ensure that area under bulk texture analysis and EBSD remained the same. Obtained crystallographic textures were characterized by standard pole figures and orientation density function (ODF) plots for $\varphi_2 = 45°$, $65°$ and $90°$. The pole figures and ODF plots obtained for undeformed sheet and SPIF samples pertaining to early, intermediate and final forming stages are provided in Figure 6.15 (a–d). Further, ODF data was used to quantify the volume fraction of dominant texture components evolved with the forming operation, and the same has been illustrated in Figure 6.15. For the sake of convenience, the peculiar findings from bulk texture analyses are categorized and discussed as per the strain levels the sheet material has undergone under the forming load.

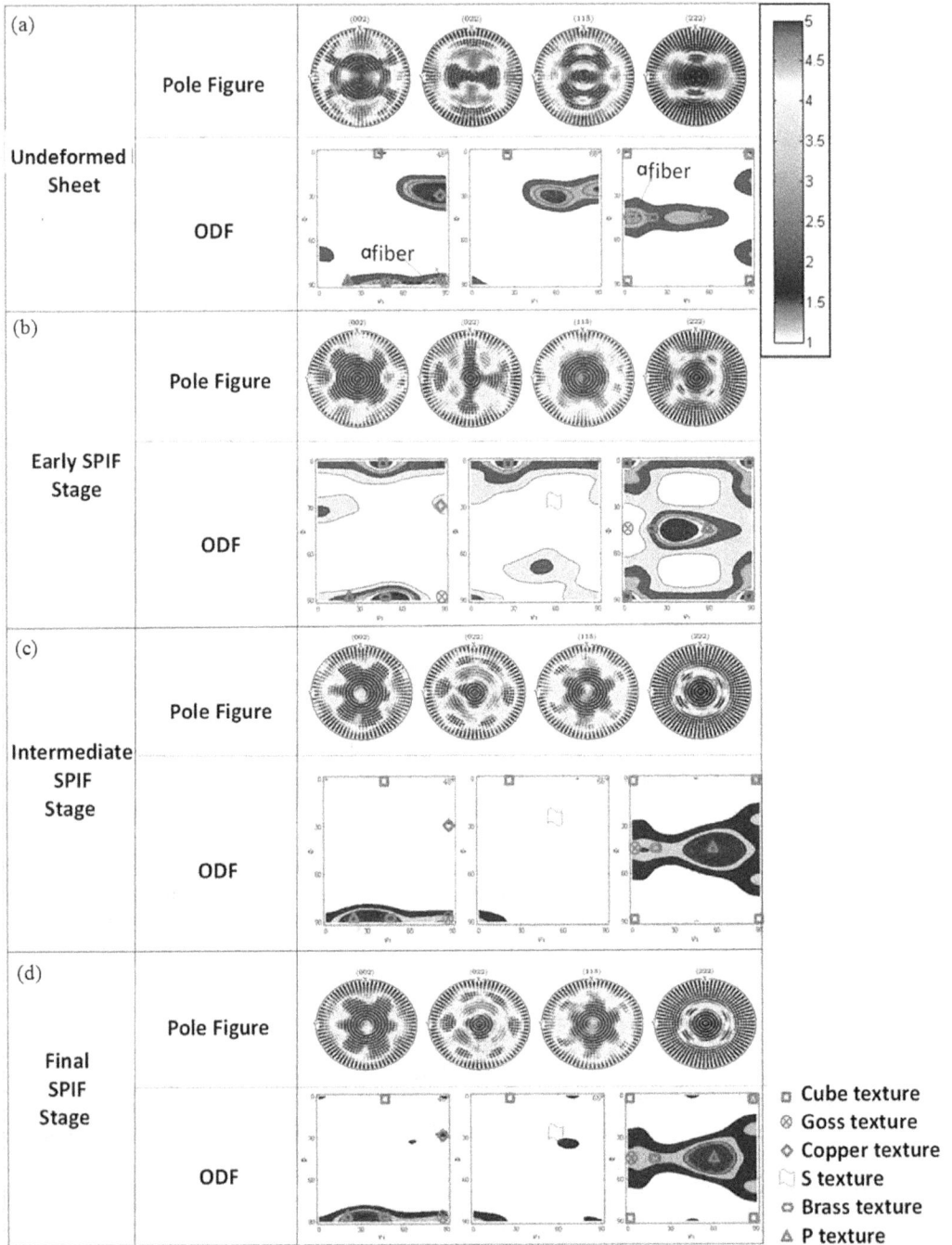

FIGURE 6.15 Experimental X-ray pole figures and φ_2 sections of orientation density functions in Euler space $(\varphi_1, \Phi, \varphi_2)$ for (a) undeformed sheet and different SPIF stages, (b) early, (c) intermediate, and (d) final (texture components are marked in ODFs as per their positions and intensity information is provided in color bar at the top right corner of the figure) (Shrivastava and Tandon 2018b).

6.3.2.1 Undeformed Sheet Material

From pole figures and ODF plots shown in Figure 6.15 (a), it is evident that the undeformed sheet possessed typical rolling texture having α-fiber along with the β-fiber. Where in Euler space, α-fiber stems from {011} <100> Goss (Gs) to {011} <211> Brass (Bs) texture component and β-fiber oriented from the {112} <111> Copper (Cu) texture component to the {011} <211> Brass texture component via {123} <634> S texture component. Presence of {011} <100> Goss, {011} <211> Brass, {112} <111> Copper, and {123} <634> S texture components signify that the AA1050 sheets have under gone rolling before SPIF (Souza et al. 2012). Along with Goss, Brass, Copper, and S texture components, {001} <100> Cube (C) texture components were also observed, albeit in minute quantity.

6.3.2.2 Early SPIF Stage

The texture developed in the early SPIF stage is shown with the help of pole figures and ODF plots shown in Figure 6.15 (b). It was observed that early stage of single point incremental forming developed {001} <100> Cube component as a dominant deformation texture. Being easy to reorient, it may be implied that Copper, Brass, and S components of β-fiber present in undeformed sheet reoriented to have to Cube texture (Cai et al. 2009). The formation and abundance of Cube texture components indicates that the shear force is dominant during early SPIF stage and leads to major deformation (Souza et al. 2012).

The shearing of the sheet metal in the early forming stage can be justified by the fact that transverse forming load, which is responsible for predominant bending in the early forming stage, as observed during experimental analysis, apart from causing the compression in the internal surface of the sheet in contact with the tool, and stretching on the external surface of the sheet, also results in shear stresses in the parallel as well as in perpendicular direction of the load (Whitney 1969). Thus, shearing can be concluded as the dominant deformation mechanism in the initial stage of SPIF. Volume fractions of all other texture components get reduced in this stage of forming in comparison to that present in undeformed material (Figure 6.16(a)).

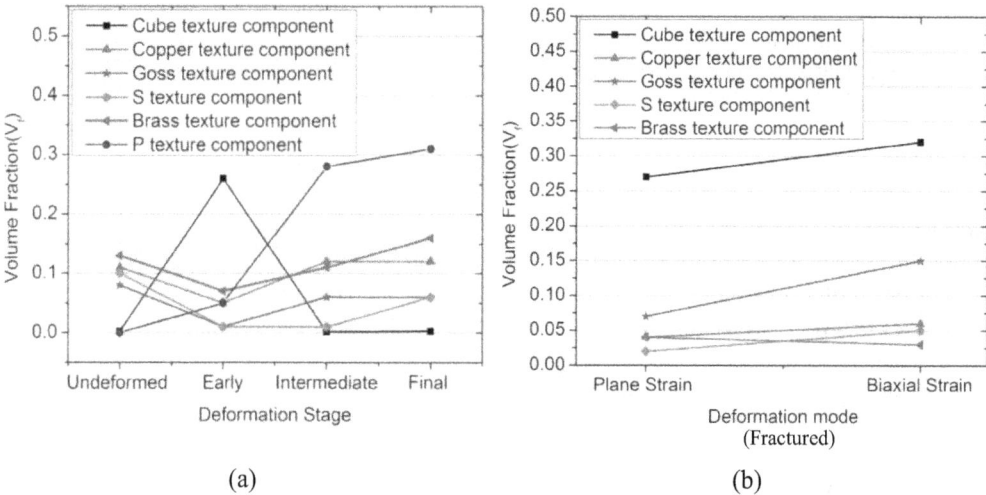

(a)   (b)

FIGURE 6.16   Volume fractions (V_f) of different texture components resulting with different (a) deformation stages and (b) deformation modes.

6.3.2.3 Intermediate and Final SPIF Stage

The texture developed in the intermediate and the final SPIF stages are shown in Figure 6.15(c) and Figure 6.15(d) respectively. It can be easily interpreted from the pole figures and ODFs that both the intermediate and final stage of forming leads to more or less identical textures. With the progression of forming, Cube (shear) texture which was present in abundance in the early forming stage diminished with increase in strain. This may be due to the fact that grains orientated along {001} <100> are comparatively unstable, when pure aluminium alloy undergoes heavy plane strain deformation at room temperature (Panchanadeeswaran and Field 1995). With increase in strain, cube texture reorients favorably and initially transforms to Goss orientation followed by Brass orientation. Therefore, the volume fraction of Goss and Brass texture components increase in comparison to that in the early SPIF stage (Figure 6.16). However, the volume fraction of Goss texture component increases in comparison to early forming stage but still the amount is less in comparison to that present in undeformed sheet. At this stage, Copper texture component regains its amount which was decreased during the early stage of forming (Figure 6.16(a)). Texture analysis also showcase that during SPIF of AA1050 sheet, the intermediate stage develops dominant {011} <111> P texture, and this remains intact even up to the final forming stage. This indicates the attainment of textural stability in SPIF, where initial orientations present in the sheet material converges to a single dominant texture, i.e., P texture here. In fully formed component, Brass and Copper texture components also exist (Figure 6.16(a)) but in lesser amount in comparison to P texture. P texture component generally arises, when material undergoes recovery and/or recrystallization at elevated temperature (LI et al. 2016). However, recovery and/or recrystallization is not possible in this case because SPIF (at normal feed and rotational speed) was performed at room temperature and friction between tool and sheet is not capable to significantly increase the sheet temperature (Ambrogio et al. 2013). Thus, there must be some other phenomenon which is responsible for P texture occurrence. It was found that when a material having high stacking fault energy (like aluminium) is subjected to excessive high strains, the grain refinement and shear band occur, which simulates kind of recovery or recrystallization process, even at room temperature (Sidor et al. 2012).

Excessive deformation in incremental forming at room temperature led to high energy storage around aluminium alloy particles (which are usually of large size), which eventually leads to increase in the amount of P texture. At such a high strain, apart from large sized particles, deformation inhomogeneity, like accumulated dislocations and/or shear bands, act as initiating sites for P texture. The forming and sustaining capabilities of P texture at high strain values evolved during SPIF is due to their special boundary orientation relationship with the surrounding deformed matrix (Savoie et al. 1996; Hirsch 2005).

The origination and presence of P texture in SPIF indicates that Particle Stimulated Nucleation (PSN) and/or strain induced boundary migration (SIMB) had taken place in the sheet material due to excessive strains (Savoie et al. 1996; Hirsch 2005; Sidor et al. 2012). The localized deformation behavior of the SPIF leads to dislocation accumulation near the grain boundaries. Subsequent forming increases the density of accumulated dislocations as well as shear bands inside the metal structure. SPIF of aluminium causes the grain refinement and shear band formation, which simulates kind of recovery or recrystallization process. Thus, PSN at high strains in SPIF may be reason behind enhanced formability. Further, increase in volume fraction of Brass texture component also facilitates formability of sheet material (Barnwal et al. 2017).

6.4 FINITE ELEMENT ANALYSIS

Analysis of aluminium sheet microstructures evolved during conventional forming reports the presence of dominant beta fibers consisting of dominant Cube, Goss, S, Brass, and Copper texture components (Barnwal et al. 2018). Fang et al. (2017) analyzed the microstructural differences that resulted post conventional and electromagnetic pulse assisted incremental drawing of 5052 aluminium alloy. They have correlated the formability of the parts with resulted grains and formed slip bands. Shore et al. (2018) claimed that asymmetric rolling of sheet introduces shear components and thus, crystallographic texture of the formed sheet alters. Evolved textures during single pass of asymmetric cold rolling were investigated to analyze the effect on macroscopic anisotropy of the sheet material. They further successfully established the relationship between volume fractions of shear texture, texture heterogeneity and plastic anisotropy behavior. Satheesh Kumar and Raghu (2015) had performed conventional and cross constrained groove pressing (CGP) technique. They confirmed the influence of strain path change on the dislocation interactions and evolution of boundary characteristics. Further, development of misorientation distribution was correlated with grain refinement during forming. Thus, to analyze the deformation and forming behavior of SPIF from the core, there is a need to investigate the sheets deformation at the microstructural level.

Usually, experimental execution of SPIF for both small components with complex features and large components with simpler geometry are time-consuming and cost-ineffective. Thus, numerical simulation of the process is practiced beforehand to extensively analyze the process characteristics and evaluate the complex state of stress and strain evolved in the process (Shrivastava et al. 2018). The overall SPIF process performance is the result of complicated stresses such as shear, compression, and tension during the forming operation, which is difficult to evaluate during actual forming (Do et al. 2017; Narayan 2015; Yue et al. 2017).

Researches have been done on modeling and numerical simulation of the process to predict thickness reduction, formability limits, and geometrical accuracy in SPIF (Benedetti et al. 2017; Formisano et al. 2017). Li et al. (2017) had implemented the FE simulation of SPIF process. Their objective was to investigate the deformation mechanism in SPIF through finite element (FE) simulation approach. For the purpose, they have utilized fine solid elements and reported that the deformation mechanism in SPIF process includes shearing, bending, and stretching, with varying influence in each direction.

During plastic deformation, polycrystalline materials possess anisotropic behavior due to heterogeneous tendency of crystallographic microstructure (Dwivedi 2016; Kumar 2016). Recently, a few numerical studies were carried out considering the microstructure and texture evolution of polycrystalline material during the plastic deformation by Crystal Plasticity Finite Element (CPFE) model. The method is capable to predict formability in polycrystalline material by involving the fundamental mechanism of dislocation movements and crystallographic slip systems. Farukh et al. (2016) had developed physically-based crystal plasticity models to predict the behavior of individual grains in polycrystalline materials. In addition, crystal plasticity when combined with finite element method is capable of predicting the global and local stresses, and strain response of crystalline materials subjected to different kinds of loading Conditions. To predict the material response at microstructural level, Farukh et al. (2016) utilized representative volume element (RVE) including adequate quantity of grains. The technique was capable to predict the global behavior by employing real microstructural arrangement in comparison to virtual microstructures created by Voronoi tessellation technique. The overall technique is shown in Figure 6.17 which includes a basic step to procure real microstructures from scanning electron microscope and to determine the coordinates of the grain boundaries by using image-processing code developed within MATLAB. The generated geometry through the calculated coordinates was utilized as an input to ABAQUS CAE software which in conjugation to FE model generated the realistic grain microstructure. A flow chart depicting the procedure to produce RVE is shown in Figure 6.18. The model is grounded on two dimensional (surface) images of the material and thus, neglects the material behavior in the third dimension.

CPFE models are extremely complex and demands enormous computational time and sophisticated hardware because they work on the principle of sub model technique to predict the behavior on global and macro level. In addition, finite element homogenization causes the calculation cumbersome. The schematic representation of the components and the procedure of CPFE modeling are shown with the help of Figure 6.19.

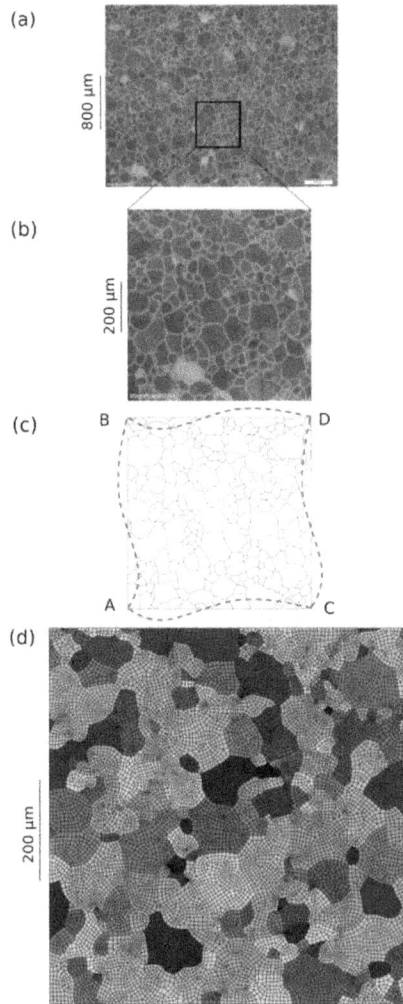

FIGURE 6.17   (a) SEM image, (b) grains in RVE region, (c) illustration of the RVE and periodic boundary Conditions, and (d) finite element mesh (Farukh et al. 2016).

Efforts have been made to apply CPFE model on the macro level (Zhao et al. 2008). Zhao et al. (2008) investigated the origin of plastic strain localization and surface roughening in polycrystalline aluminium sample subjected to uniaxial loading conditions. The mesh of the finite element model was created by mapping the geometry of the grain boundaries according to the Electron Back Scattered Diffraction (EBSD) measurement as shown in Figure 6.20 (a). The surface strain distribution resulted after FEA and experimental observations were compared with each other and shown in Figure 6.20 (b).

The study suggested that the grain topology and micro texture have a significant influence on the origin of strain heterogeneity. Moreover, it also suggested that the final surface roughening profiles are related both to the macro strain localization and intra-grain interaction. However, discrepancies between experiment and simulation results indicated the inefficiency of CPFE model to predict the actual grain scale behavior of the macro components. The method is appropriate for small components where micro-level studies are involved but seems impractical for analyzing the macro parts formed by SPIF (Khalatbari et al. 2015; Kim et al. 2017).

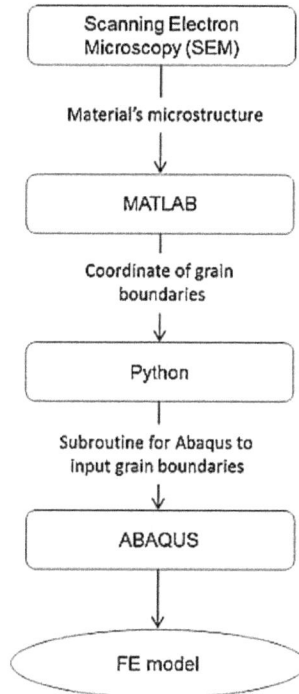

FIGURE 6.18 Flow chart of developing RVE-based on realistic microstructure (Farukh et al. 2016).

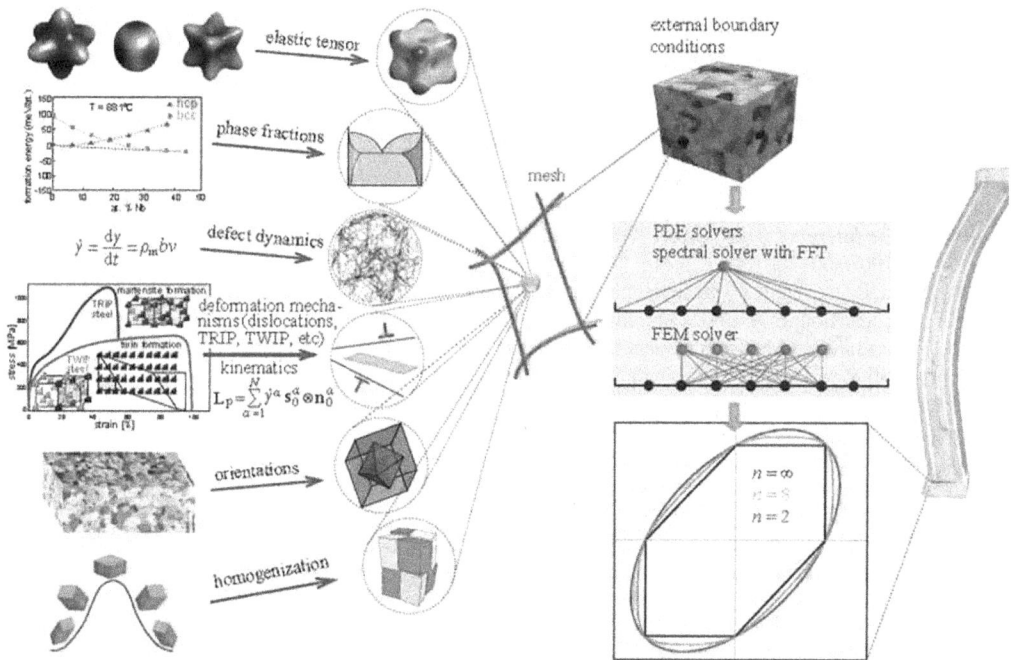

FIGURE 6.19 Key elements and working methodology of crystal plasticity finite element model (Ma et al. 2007).

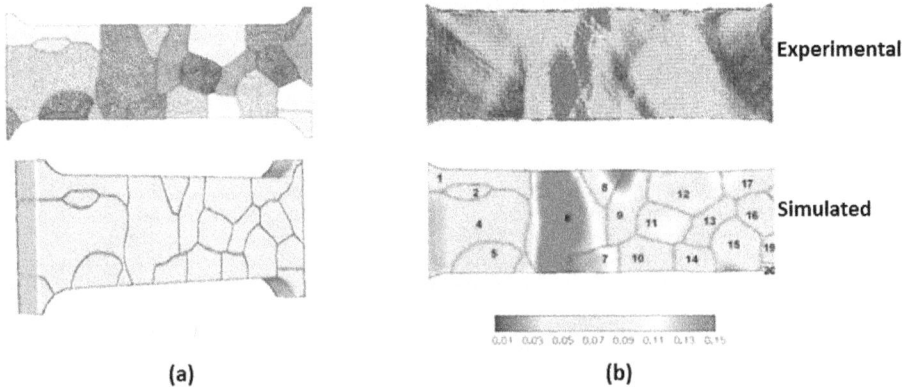

FIGURE 6.20 (a) Creation of FE mesh by mapping the grain boundaries according to the EBSD measurements and (b) resulted experimental and simulated strain localization (Zhao et al. 2008).

6.5 CONCLUSION

From the literature, it can be concluded that analysis done from the root level i.e., microstructural and crystallographic texture perspective can unfold the dominant deformation mechanism of sheet metal forming processes. Complex microstructures and crystallographic textures which developed during forming determine the distribution of grain boundaries and slip system, which in turn determines the dislocation movements responsible for the nature and extent of plastic deformation in metals. It was inferred from the analyses that in Incremental sheet forming, shearing, bending, and stretching are the major deformation mechanisms. The influence of strain path change on the dislocation interactions and evolution of boundary characteristics was also confirmed. In spite of all the efforts prediction/simulation of SPIF components on grain scale is still under exploration. Crystal Plasticity FE model are proven inefficient to predict the actual grain scale behavior of the macro components.

REFERENCES

Ambrogio, G., Gagliardi, F., Bruschi, S., & Filice, L. (2013). On the high-speed Single Point Incremental Forming of titanium alloys. *CIRP Annals*, *62*(1), 243–246. https://doi.org/10.1016/J.CIRP.2013.03.053

Barnwal, V. K., Chakrabarty, S., Tewari, A., Narasimhan, K., & Mishra, S. K. (2018). Forming behavior and microstructural evolution during single point incremental forming process of AA-6061 aluminum alloy sheet. *The International Journal of Advanced Manufacturing Technology*, *95*(1–4), 921–935. https://doi.org/10.1007/s00170-017-1238-5

Barnwal, V. K., Raghavan, R., Narasimhan, K., & Mishra, S. K. (2017). Effect of microstructure and texture on forming behaviour of AA-6061 aluminium alloy sheet. *Materials Science and Engineering: A*, *679*, 56–65.

Basak, S., Prasad, K. S., Mehto, A., Bagchi, J., Ganesh, Y. S., Mohanty, S., … Panda, S. K. (2020). Parameter optimization and texture evolution in single point incremental sheet forming process. *Proceedings of the Institution of Mechanical Engineers, Part B: Journal of Engineering Manufacture*, *234*(1–2), 126–139. https://doi.org/10.1177/0954405419846001

Behera, A. K., de Sousa, R. A., Ingarao, G., & Oleksik, V. (2017). Single point incremental forming: An assessment of the progress and technology trends from 2005 to 2015. *Journal of Manufacturing Processes*, *27*, 37–62. https://doi.org/10.1016/J.JMAPRO.2017.03.014

Cai, C., Ji, Z., Zhang, H., & Wang, G. (2009). Study on the microstructure and texture of 3003 aluminum sheets rolled by laser-textured roll. *Journal of Metallurgy*, *2009*, 1–6. https://doi.org/10.1155/2009/587938

Cho, S., & Finocchiaro, E. T. (2010). *Handbook of prebiotics and probiotics ingredients: Health benefits and food applications*. CRC Press. Retrieved from https://books.google.co.in/books?id=mpLq_0Bkn6cC&pg=PA142&lpg=PA142&dq=cartesian+euler+space&source=bl&ots=zPNF_uOBtX&sig=n5fPU9pvLZ61JAZl-xHEcNbAkps&hl=en&sa=X&ved=2ahUKEwj6iKORhc7cAhUTSI8KHcQrAfsQ6AEwCHoECAEQAQ#v=onepage&q=cartesianeuler space&f=true

Dwivedi, D. (2016). *Numerical investigations of coupled and uncoupled material models in incremental sheet forming process*. PDPM, IIITDM Jabalpur.

Emmens, W. C., & van den Boogaard, A. H. (2009). An overview of stabilizing deformation mechanisms in incremental sheet forming. *Journal of Materials Processing Technology*, *209*(8), 3688–3695. https://doi.org/10.1016/j.jmatprotec.2008.10.003

Fang, J., Mo, J., & Li, J. (2017). Microstructure difference of 5052 aluminum alloys under conventional drawing and electromagnetic pulse assisted incremental drawing. *Materials Characterization*, *129*, 88–97. https://doi.org/10.1016/J.MATCHAR.2017.04.035

Farukh, F., Zhao, L. G., Jiang, R., Reed, P., Proprentner, D., & Shollock, B. A. (2016). Realistic microstructure-based modelling of cyclic deformation and crack growth using crystal plasticity. *Computational Materials Science*, *111*, 395–405. https://doi.org/10.1016/J.COMMATSCI.2015.09.054

Gatea, S., Xu, D., Ou, H., & McCartney, G. (2018). Evaluation of formability and fracture of pure titanium in incremental sheet forming. *The International Journal of Advanced Manufacturing Technology*, *95*(1–4), 625–641. https://doi.org/10.1007/s00170-017-1195-z

Ghosh, A., Roy, A., Ghosh, A., & Ghosh, M. (2021). Influence of temperature on microstructure, crystallographic texture and mechanical properties of EN AW 6016 alloy during plane strain compression. *Materials Today Communications*, *26*, 101808. https://doi.org/10.1016/j.mtcomm.2020.101808

Groover, M. (2007). *Fundamentals of modern manufacturing: Materials processes, and systems, 2Nd Ed - Mikell P. Groover - Google Books*. John Wiley & Sons. Retrieved from https://books.google.co.in/books?hl=en&lr=&id=yFMlBb68KsIC&oi=fnd&pg=PA2&dq=mass+manufacturing+and+shet+metal+forming&ots=3ki5sJCP3s&sig=bOLkBWnR4DABy_grZvc621ap3SE#v=onepage&q=mass manufacturing and shet metal forming&f=false

Hamid, M., Lyu, H., Schuessler, B., Wo, P., & Zbib, H. (2017). Modeling and characterization of grain boundaries and slip transmission in dislocation density-based crystal plasticity. *Crystals*, *7*(6), 152. https://doi.org/10.3390/cryst7060152

Hirsch, J. (2005). Texture and anisotropy in industrial applications of aluminium alloys. *Archives of Metallurgy and Materials*, *50*(1), 21–34. Retrieved from http://www.imim.pl/files/archiwum/Vol1_2005/art.nr.2.pdf

Humphreys, F. J., & Hatherly, M. (2004). *Recrystallization and related annealing phenomena*. Elsevier.

Hussain, G., Ilyas, M., Lemopi Isidore, B. B., & Khan, W. A. (2020). Mechanical properties and microstructure evolution in incremental forming of AA5754 and AA6061 aluminum alloys. *Transactions of Nonferrous Metals Society of China (English Edition)*, *30*(1), 51–64. https://doi.org/10.1016/S1003-6326(19)65179-4

Jackson, K., & Allwood, J. (2009). The mechanics of incremental sheet forming. *Journal of Materials Processing Technology*, *209*(3), 1158–1174. https://doi.org/10.1016/J.JMATPROTEC.2008.03.025

Kestens, L. A. I., & Pirgazi, H. (2016). *Materials science and technology texture formation in metal alloys with cubic crystal structures*. https://doi.org/10.1080/02670836.2016.1231746org/10.1080/02670836.2016.1231746

Khalatbari, H., Iqbal, A., Shi, X., Gao, L., Hussain, G., & Hashemipour, M. (2015). High-speed incremental forming process: A trade-off between formability and time efficiency. *Materials and Manufacturing Processes*, *30*(11), 1354–1363. https://doi.org/10.1080/10426914.2015.1037892

Kim, J. H., Lee, M.-G., Kang, J.-H., & Oh, C.-S. (2017). Crystal plasticity finite element analysis of ferritic stainless steel for sheet formability prediction. *International Journal of Plasticity*, *93*, 26–45. https://doi.org/10.1016/J.IJPLAS.2017.04.007

Kumar, A. (2016). *Estimation of material model parameters and finite element analysis of ISF process*. PDPM, IIITDM Jabalpur.

Li, D., Zhang, D., Liu, S., Shan, Z., Zhang, X., Wang, Q., & Han, S. (2016). Dynamic recrystallization behavior of 7085 aluminum alloy during hot deformation. *Transactions of Nonferrous Metals Society of China*, *26*(6), 1491–1497. https://doi.org/10.1016/S1003-6326(16)64254-1

Liao, J., Sousa, J. A., Lopes, A. B., Xue, X., & Pereira, A. B. (2017). Mechanical, microstructural behaviour and modelling of dual phase steels under complex deformation paths. *International Journal of Plasticity*, *93*, 269–290. https://doi.org/10.1016/J.IJPLAS.2016.03.010

Ma, A., Roters, F., & Raabe, D. (2007). A dislocation density based constitutive law for BCC materials in crystal plasticity FEM. *Computational Materials Science*, *39*(1 SPEC. ISS.), 91–95. https://doi.org/10.1016/j.commatsci.2006.04.014

Malhotra, R., Xue, L., Belytschko, T., & Cao, J. (2012). Mechanics of fracture in single point incremental forming. *Journal of Materials Processing Technology*, *212*(7), 1573–1590. https://doi.org/10.1016/J.JMATPROTEC.2012.02.021

Martins, P. A. F., Bay, N., Skjoedt, M., & Silva, M. B. (2008). Theory of single point incremental forming. *CIRP Annals*, *57*(1), 247–252. https://doi.org/10.1016/J.CIRP.2008.03.047

McAnulty, T., Jeswiet, J., & Doolan, M. (2016). Formability in single point incremental forming: A comparative analysis of the state of the art. *CIRP Journal of Manufacturing Science and Technology*, *16*, 43–54. https://doi.org/10.1016/j.cirpj.2016.07.003

Mishra, S., Yazar, K. U., Kar, A., Lingam, R., Reddy, N. V., Prakash, O., & Suwas, S. (2021). Texture and microstructure evolution during single-point incremental forming of commercially pure titanium. *Metallurgical and Materials Transactions A: Physical Metallurgy and Materials Science*, *52*(1), 151–166. https://doi.org/10.1007/s11661-020-06000-y

Nath, M., Shin, J., Bansal, A., Banu, M., & Taub, A. (2018). Comparison of texture and surface finish evolution during single point incremental forming and formability testing of AA 7075. *Minerals, Metals and Materials Series, Part F4*, 225–232. https://doi.org/10.1007/978-3-319-72284-9_31

Panchanadeeswaran, S., & Field, D. P. (1995). Texture evolution during plane strain deformation of aluminum. *Acta Metallurgica et Materialia*, *43*(4), 1683–1692. https://doi.org/10.1016/0956-7151(94)00316-A

Satheesh Kumar, S. S., & Raghu, T. (2015). Strain path effects on microstructural evolution and mechanical behaviour of constrained groove pressed aluminium sheets. *Materials & Design*, *88*, 799–809. https://doi.org/10.1016/J.MATDES.2015.09.057

Savoie, J., Zhou, Y., Jonas, J. J., MacEwen, S. R. (1996). Textures induced by tension and deep drawing in aluminum sheets. *Acta Materialia*, *44*(2), 587–605. Retrieved from https://ac.els-cdn.com/1359645495002146/1-s2.0-1359645495002146-main.pdf?_tid=4baa5700-6f88-4d5e-bc9f-4a8f13e882aa&acdnat=1525596359_5d7583d1c3ad4d94d302f6f39407a32b

Shore, D., Kestens, L. A. I., Sidor, J., Van Houtte, P., & Van Bael, A. (2018). Process parameter influence on texture heterogeneity in asymmetric rolling of aluminium sheet alloys. *International Journal of Material Forming*, *11*(2), 297–309. https://doi.org/10.1007/s12289-016-1330-7

Shrivastava, P., Kumar, P., Tandon, P., & Pesin, A. (2018a). Improvement in formability and geometrical accuracy of incrementally formed AA1050 sheets by microstructure and texture reformation through preheating, and their FEA and experimental validation. *Journal of the Brazilian Society of Mechanical Sciences and Engineering*, *40*(7), 335. https://doi.org/10.1007/s40430-018-1255-9

Shrivastava, P., Kumar, P., Tandon, P., & Pesin, A. (2018b). Improvement in formability and geometrical accuracy of incrementally formed AA1050 sheets by microstructure and texture reformation through preheating, and their FEA and experimental validation. *Journal of the Brazilian Society of Mechanical Sciences and Engineering*, *40*(7). https://doi.org/10.1007/s40430-018-1255-9

Shrivastava, P., & Tandon, P. (2015). Investigation of the Effect of Grain Size on Forming Forces in Single Point Incremental Sheet Forming. *Procedia Manufacturing*, *2*, 41–45. https://doi.org/10.1016/J.PROMFG.2015.07.008

Shrivastava, P., & Tandon, P. (2018a). Enhancement of process capabilities and numerical prediction of geometric profiles and global springback in incrementally formed AA 1050 sheets. *Transactions of the Indian Institute of Metals*, 1–11. https://doi.org/10.1007/s12666-018-1346-4

Shrivastava, P., & Tandon, P. (2018b). Microstructure and texture based analysis of forming behavior and deformation mechanism of AA1050 sheet during Single Point Incremental Forming. *Journal of Materials Processing Technology*. https://doi.org/10.1016/J.JMATPROTEC.2018.11.012

Shrivastava, P., & Tandon, P. (2019). Effect of preheated microstructure vis-à-vis process parameters and characterization of orange peel in incremental forming of AA1050 sheets. *Journal of Materials Engineering and Performance*, *28*(5), 2530–2542. https://doi.org/10.1007/s11665-019-04032-z

Sidor, J. J., Decroos, K., Petrov, R. H., & Kestens, L. A. I. (2012). Particle stimulated nucleation in severely deformed aluminum alloys. *Materials Science Forum*, *706*(389–394). https://doi.org/10.4028/www.scientific.net/MSF.706-709.389

Silva, M. B., Nielsen, P. S., Bay, N., & Martins, P. A. F. (2011). Failure mechanisms in single-point incremental forming of metals. *The International Journal of Advanced Manufacturing Technology*, *56*(9–12), 893–903. https://doi.org/10.1007/s00170-011-3254-1

Singh, J., Kim, M.-S., & Choi, S.-H. (2018). The effect of initial texture on micromechanical deformation behaviors in Mg alloys under a mini-V-bending test. *International Journal of Plasticity*. https://doi.org/10.1016/J.IJPLAS.2018.01.008

Souza, F. M., Plaut, R. L., Lima, N. B. de, Fernandes, R. do C., & Padilha, A. F. (2012). Recrystallization and crystallographic texture in AA4006 aluminum alloy sheets produced by twin roll caster and direct chill processes. *Rem: Revista Escola de Minas*, *65*(3), 363–370. https://doi.org/10.1590/S0370-44672012000300013

Vargas, M., Lathabai, S., Uggowitzer, P. J., Qi, Y., Orlov, D., & Estrin, Y. (2017). Microstructure, crystallographic texture and mechanical behaviour of friction stir processed Mg-Zn-Ca-Zr alloy ZKX50. *Materials Science and Engineering: A*, *685*, 253–264. https://doi.org/10.1016/J.MSEA.2016.12.125

Whitney, J. M. (1969). The effect of transverse shear deformation on the bending of laminated plates. *Journal of Composite Materials*, *3*(3), 534–547. https://doi.org/10.1177/002199836900300316

Zhao, Z., Ramesh, M., Raabe, D., Cuitiño, A. M., & Radovitzky, R. (2008). Investigation of three-dimensional aspects of grain-scale plastic surface deformation of an aluminum oligocrystal. *International Journal of Plasticity*, *24*(12), 2278–2297. https://doi.org/10.1016/j.ijplas.2008.01.002

7 Tribological Behavior in Bulk and Sheet Forming Processes

Vishal Bhojak and Jinesh Kumar Jain
Malaviya National Institute of Technology Jaipur, Jaipur, India

CONTENTS

7.1 INTRODUCTION

In 2015, the United Nations set 17 sustainable development goals aimed at eradicating poverty, reducing inequality, and encouraging responsible consumption and production. Friction, wear, and lubrication are all aspects of tribology that have a significant impact on industrial and manufacturing processes. Tribology is also important for achieving sustainable development goals. A lot of work has gone into developing new environmentally friendly lubricants and ensuring the long-term viability of tribology [1].

Tribology word is derived from the Greek word tribos, which means "rubbing" [2]. The tribology is the study, where we study about surfaces which are in relative motion and various phenomenon like friction, wear and tears applied on them during the relative motion. Interaction of tribology with different science and technology are shown in Figure 7.1.

Friction is a tribological component that acts as resistance when one body moves over another in motion during sliding or rolling. Friction can be static or kinetic, sliding, and rolling between the surfaces. Friction theories are defined in dry friction on the basis of adhesion and deformation between two surface. Lubrication is defined as the process, where contraction between the surfaces

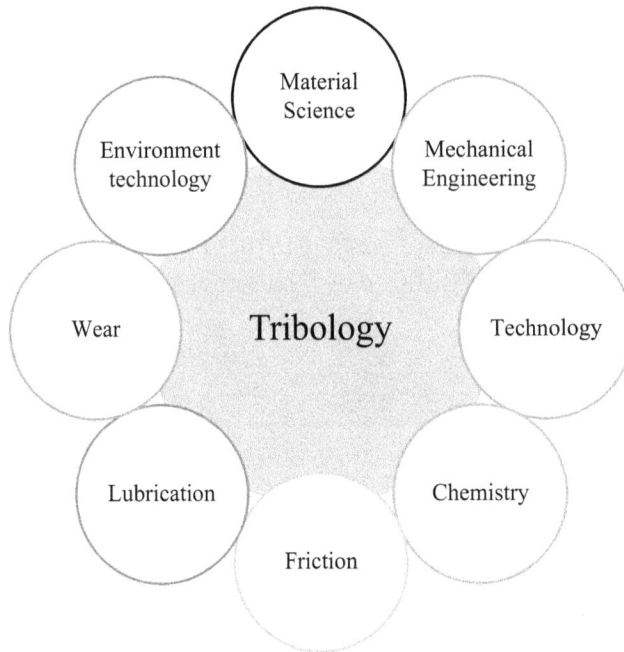

FIGURE 7.1   Tribology interacting with other sciences [3].

can be reduced by minimizing the coefficient of friction. Lubrication can come in a variety of forms, including solids (graphite, graphene, molybdenum disulfide), liquids (oil, water, etc.), gaseous (air), and semisolid (grease). It can be categorized based on their primary function.

- Anti-wear additives (AW): These additive reduce wear by forming a protecting layer over them.
- Extreme pressure additives (EP): A coating layer has been formed on the surface to prevent them against seizing (graphite, molybdenum disulfide).
- Friction modifiers: These are solid particles that are used to control friction (graphite, molybdenum disulfide, tungsten disulfide, etc.).
- Corrosion inhibitors: They are used to form a corrosion-resistant coating on surfaces to protect them against chemical attack.
- Viscosity Index Improvers: These additives are used to prevent or reduce the viscosity index of lubricants when the temperatures have been raised.

Wear can be defined as a phenomenon which is caused by friction and a lack of appropriate lubrication. The wear rate depends upon various parameters such as type of loading, the nature of the motion between the contact surfaces, and the temperature. The phenomenon of tribology in aspects of bulk and sheet metal forming have been discussed in further section.

7.2   METAL FORMING

Metal forming is a critical manufacturing process, due to its cost effectiveness, improved mechanical qualities, flexible operations, increased productivity, and significant material savings, it has gained industrial importance among diverse manufacturing activities. Man-made, engineered parts are obtained from some raw material through some manufacturing process to create the items and articles we utilize in our daily lives.

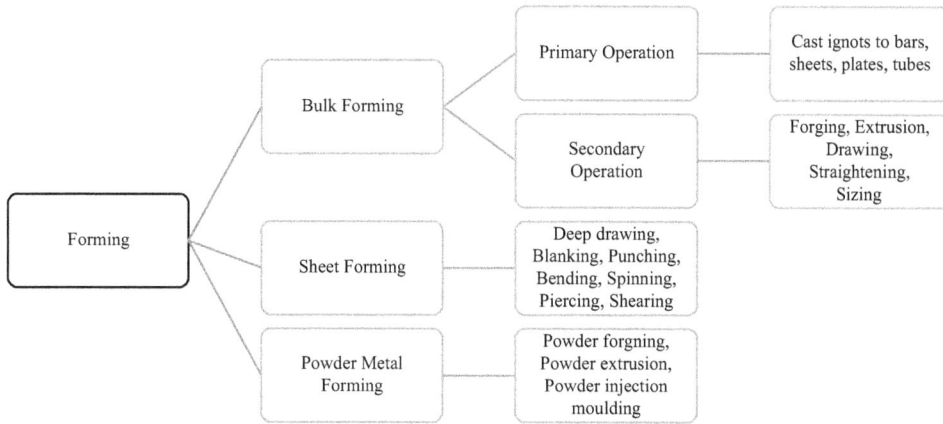

FIGURE 7.2 Metal forming processes classification [4].

Metal forming techniques can be in two categories: (i) bulk forming and (ii) sheet metal forming. Bulk deformation process has been used for shaping raw materials with a low surface area to volume ratio. The operation such as rolling, forging, extrusion, and drawing are mainly coming in the category of such processes. The forces such as compressive, compressive and tensile, shear, or a combination of these forces are applied on the raw material to deform their shapes. A set of tool and die are used for bulk forming of such materials. The components such as Gears, bushed, valves, connecting rods, hydraulic valves can be produced easily by using metal forming process.

Sheet metal forming process mainly used shear force to convert a raw material in the form of sheets, plates, and strips. The hydraulic and pneumatic process have been used for performing the sheet metal operation. Sheet working procedures are performed with a set of tools known as die and punch. Bending, drawing, shearing, blanking, punching is example of sheet metal operations.

Apart from that, Powder forming's method has gained attention of several researchers due to its ability to generate parts with very close tolerance and minimal material waste. The Figure 7.2 represent the classification chart of forming processes.

7.2.1 ROLLING

Rolling is the most frequent metal forming technique, which transforms ingots into blooms and slabs, and then into bars, billets, rods, plates, sheets, strips, and coils. The work material has been deformed elastically due to the compressive force which act between rollers.

Friction is very important aspect of rolling. It is not possible to draw the material between the roller if there is no friction, and the roll will try to slide over the work piece. As a result, pure hydrodynamic films are uncommon, however the mechanics of hydrodynamic lubrication are still important to understand in tribology The capacity to forecast lubricant film thickness at the contact zone's entry, such as via the Wilson-Walowit equation [5], is the key benefit which is given by:

$$h = \frac{6\eta\gamma U}{\tan\theta\left(1 - e^{-\gamma\sigma}\right)} \tag{7.1}$$

where h is the film thickness, η is the lubricant viscosity, γ is the viscosity pressure coefficient, U is the mean inlet rolling speed, and θ is the contact angle between roll and work piece.

Oil droplets floating in water are a most frequent lubricant in metal rolling. Schmid et al. [6] highlighted a number of modelling methodologies, mixing theory of water and oil, including plate out of oil, dynamic concentration of oil, and surface tension based theories, in a recent summary. Each of them appears to be dominant at a different speed range, greatly complicating emulsion mathematical and experimental studies.

7.2.1.1   Wear During Rolling Process

The work roll surface is successively loaded, heated, and cooled by contacting hot steel during the rolling process. Large compressive as well as tensile stresses are generated, during high frequency of the roll rotation, resulted in mechanical and thermal surface fatigue. As a result, the rolls must have great wear resistance properties and high impact toughness resistance [7, 8]. Work roll wear is either uniform over the surface which are in contact during rolling or localized in deeper wear bands.

Abrasion is a kind of micro chipping which caused by thermal induced surface fatigue, that is the main cause for uniform wear [7]. The thermal fatigue induced over the roll surface is the reason for critical wear during bending operation [9]. Cracks are generated perpendicular to roll surface due to induced thermal fatigue [10]. Contact fatigue cracks run parallel to the surface and are caused by high-pressure contact with the backup rolls.

To improve the wear resistance during hot rolling process, HSS rolls were used instead of HCr [11]. HSS rolls have better wear resistance properties from HCr rolls, according to Park et. al [12] and Ziehenberger and Windhager [13].

7.2.2   Forging

Forging is also metal deformation process, where compressive stresses are generated during plastic deformation process. Forging can be classified as open and closed die process, during the closed die forging process work pieces are operated in cold condition while hot metal work pieces are used for open die forging. Extrusion of small discrete components is frequently done in dies, and the lubricating difficulties are similar to those addressed above for rolling.

7.2.2.1   Hot Forging

As mentioned earlier, hot forging process are used with open die forging to make desired shape and size on cast billets. Die sets are typically made of hardened tool steel and have simple forms. Due to high temperature and the surroundings condition, formation of oxide scale over the billet and surfaces are common. These oxide layer of scale will be removed thought forging process, which results in non-uniform surface profile. Friction is a major issue that has an impact on the stress that the work piece is subjected to. Friction has identified as a major factor which influences the upsetting force and maximum die pressure. From a slab method analysis, it can be proven that the die pressure distribution in plane strain upsetting of a work piece is provided by:

$$P = \frac{2}{\sqrt{3}} Y exp\left(\frac{2\mu\left(x-a\right)}{h} \right) \tag{7.2}$$

where P is the die pressure at location x, x is the distance from the work piece's edge, Y is the material uniaxial flow stress, a is the semi-width of the work piece, and h is the thickness of the work piece. A friction factor can be used to get a comparable expression, but the importance of friction is evident. The hot forging operation are associated with open die process so the course action could be hammer drop on work piece, mechanical operated presses, and hydraulic operated presses. Due to high impact forces are induced during the process, the material for inserts and die are made of high alloy steels. Glass or graphite can be used as solid lubricant in hot open die forging process.

7.2.2.2   Cold Forging

Cold forging processes are employed where advances surface characteristics or surface treatments are required. As name suggest, temperature of the work piece and during the process are comparable low so this can be utilized as heading, orbital forging, impression die forging, and extrusion. In this process high forces and stresses are induced, so reduced the effect of forces lubricants used. A number of lubricants with different additives are used for this process because there is no limitation of temperature rather hot forging process.

7.2.2.3 Lubricants for Forging Processes

The lubricant applied a thermal resistance between the die and the work piece. This accomplishes two goals. The lubricant can slow down the heat loss because cooler work parts have a higher flow strength. Moreover, a robust heat barrier reduces die wear.

Among the lubricants used in forging are:

- Metal coatings.
- Solid lubricants.
- Polymer coatings.
- Liquid lubricants.
- Phosphate conversion coating.

Orange peel formation during the process is due to formation of thick lubrication film which is major disadvantage of lubrication process. The mechanics of surface roughening have been described by Wilson et al. [14]. There are also reverse effect of lubricants like powder soaps, which try to stick and foul the tool arrangement. This could be an reason for dimensional inaccuracy.

7.2.3 EXTRUSION AND DRAWING

Extrusion and drawing are the similar drawings phenomenon but the measure difference between them is, one pushed through dies (extrusion) and another dragged/pulled through dies (drawing). In reality, there are significant distinctions between the processes, the most significant of which is that extrusion is a type of batch process, whereas drawing is a type of continuous process. Extrusion, on the other hand, uses significantly greater area reductions per pass than drawing and, as a result, is often executed at a higher temperature to get better work piece ductility. Saha [15], Schey [7], and Laue et al. [16] are good generic extrusion sources. Schey [7] and Tassi [17] are two general drawing sources.

7.2.3.1 Extrusion

Extrusion can be used in conjunction with forging as a press working activity. This section is about the more sophisticated process of producing semi-fabricated goods. A cast or already rolled billet is used as the beginning material, which is placed in a container and driven through a die by a hydraulic press. Extrusion can be further defined in three processes, direct, indirect, and hydrostatic extrusion. High friction forces are generated during direct extrusion process because the direction of material flows and pressing stem are in same direction. During the indirect or reverse extrusion process, work piece is at rest and the friction forces are generated at die/work piece interface.

During the extrusion process smoothness of the surface is a concern, which can be improved by proper utilization of friction and lubrication. Oil peel is a type of work piece roughening caused by extremely thick film thicknesses Work piece scoring and abrasion can occur when land lengths are too long.

7.2.3.2 Lubrication in Extrusion Processes

Glass powder is measure component of lubricants which used with binders like sodium silicate or bentonite during the hot extrusion process which is discussed earlier in the hot forging operation. Glass as powder create cushion in front of billet, which make easier the process. Spraying glass particles on the billet before inserting it into the cylinder is an another option for lubrication process.

Due to this glass particles are stick and melt over the surface of the billet and create insulating layer between billet and container. But control over the lubrication process need to made to avoid thick lubrication which cause orange peel formation. Other molten solids, such as polymers and waxes, can be used instead of glass, though they are less common and better utilization of waste material as lubrication material.

Instead of solid lubricants, soft metals, polymers, or lattice layer materials can be employed during hot extrusion process. It is also identified through previous stated, metal coatings over surfaces can reduce friction effect which improved the heat resistance for the surface. Cladding is the process of create coating on the surface, in extrusion process cladding is provided to coat the surface of work piece. In both hot and cold extrusion, graphite and polymers are employed. In cold extrusion process surface finish is main concern after the process, so liquid organic lubricants are the best option in this process.

7.2.3.3 Drawing

Sometime drawing can be misunderstanding with deep drawing operation that is related to sheet metal forming and drawing is associated with bulk metal forming process. The purpose of drawing is to pull or drag the billet/bar into continuous lengths of tube and wire form. Basic criteria of drawing are better surface finish and smooth surface, due to which most of the drawing process are done with cold process. In the drawing of tube, an internal mandrel or a plug mandrel can be used for smooth movement of flowing material. The only time friction is useful is when using a plug mandrel, and lubrication is normally pursued to reduce drawing strains. With the goal of achieving a mixed film regime, liquid lubricants are frequently utilized. According to theory it is mention viscous fluid/liquid used to create hydrodynamic lubrication film, but when it comes to real life implementation different die features are used to make thicker hydrodynamic lubrication that reduce the friction. According to Schey [7], isothermal sketching has been the subject of the majority of research.

7.2.3.4 Wear of Drawing Dies

Wear of the drawing dies due to many reason like heat induce during friction, pulling/dragging of the wire. The measure effect of abrasion and adhesion wear mechanism are observed during wire drawing processes. The repeating motion of dies and work piece will have resulted in surface fatigue that cause the wear in the drawing die. Due to the friction sticking and chipping phenomenon occurs which resulted in adhesive shear and wear of the surface. Coated oxide bits or impurities on wire's profile might damage the drawing die's surface, causing abrasive wear. [18].

The surface topology of the die material should be kept at low level to reduce friction forces and interaction between the drawing die and the wire rod. Polishing is the another way to improve the surface quality of the work piece. By the polishing, the performance and wear resistance of drawing die will be improved [19].

7.2.4 Tribology in Sheet Metal Forming

In this section the sheet metal deformation processes and sheet metal forming tribology have been discussed. The characteristics of sheet metal, tool material, and tribological conditions at the tool/steel sheet contact, the extent of plastic deformation all influence the performance of sheet metal forming operations [20, 21]. The optimization of the process parameters is necessary for avoiding part failures and reducing manufacturing costs [22, 23]. Steel sheet metal's exceptional formability allows items to be formed into a broad variety of shapes.

7.2.4.1 Tribological Systems

Tribo-systems play vital role in those applications where extreme tribological conditions affect the performance of the system. Tribology is primarily concerned with the interaction of surfaces in motion. Tribology is defined as the science and technology of interacting surfaces in relative motion, according to Michael [24]. It entails the investigation of the various elements involved in friction and wear processes, as well as the existing relationships between them and their properties.

Interacting elements constitute a single system, according to Bhushan [25], which is subject to external examination and perception.

Tribological parameters have the same impact on sheet metal forming operations as other forming system parameters, according to Srbislav et al. [26]. They have emphasized the importance of explicitly defining the limits of tribological systems in the analysis of processes involving them. Several studies, such as Bhushan [27] and Kenneth et al. [28], reveal that wear resistance features cannot be assigned to a single member of the tribological system since they are dependent on the attributes of all members of the tribo-system. Friction is required in many manufacturing processes to keep the process under control. Emmens' research [29] suggests that in deep drawing process, friction forces in the blank holder is required to control the tool movement. Additional restraining devices, such as draw beads, are added in the tool when the friction forces in the blank holder is insufficient.

Given that the majority of component failures occur at the surface, tribological aspects of materials, coatings, and lubricants must be taken into account when designing system components.

After these evident of friction, one can conclude that only changing the single parameter at a given instant, not to be an ideal approach. Peter Blau [30] explained the need of increasing the force; otherwise, the wear rate of material will increase, modifying the surface roughness and bringing debris particles into the interface. The temperature in the interface can change while evaluating the impact of sliding speed across a large range. In light of such synergistic influences, Peter Blau [31] contends that isolating the influence of merely one variable on tribological changes is difficult.

The load, velocity, type of motion, temperature, time, and materials flow are the operating parameters. Every friction interaction can be broken down into four tribological components shown also in Figure 7.3:

- Main body.
- Counter body.
- Intermediate lubrication film.
- Working condition.

The operating settings and the qualities of each constituent, including the lubricant and the environment, control tribological behaviors in this system.

Lubrication is provided as an intermediate medium, which separates the main and counter bodies to increase the load capacity and minimize the wear of the surface. Air, as well as any unintended splashes of water or the presence of chemicals, are frequently present in the environment. Solvents or varnishes, as well as a vacuum or any type of radiation load, should all be considered "environmental" and included in this unit. In such cases, the lubricant's resistance to ambient conditions is commonly required.

7.2.4.2 The Stribeck Curve in Sheet Metal Forming

Friction in sheet metal forming activities is challenging to account for in process models. Low friction, on the other hand, causes the sheet to move over the tooling and disperse strains over a broader area, reducing maximum strain and boosting formability. In order to create durable tooling, it is desirable to forecast sheet strains when designing tooling. A Coulomb friction model produces extremely erroneous predictions, necessitating a more sophisticated approach to friction.

During the sheet metal forming process, the lubrication regimes like boundary, thin film, thick film and mixed can occur at multiple places in the tool/work piece interface at the same time. It can be concluded through the curve that friction coefficient is low in the full film regime and the friction coefficient will increase when the system transformation from the mixed to the boundary film regimes. It can be concluded the same results for the wear rates also. Emmens (1988) studied friction for strips/sheet of various surface quality finishes drawn between flat platens with various lubricants and sliding speeds, as shown in Figure 7.4. The abscissa variable is the dimensionless speed specified by:

$$S_S^* = \frac{\mu U x}{p R^2} \tag{7.3}$$

where μ, is the friction coefficient, U is the sliding speed, x is the contact length, p is the interface pressure, and R is the roughness (maximum peak height).

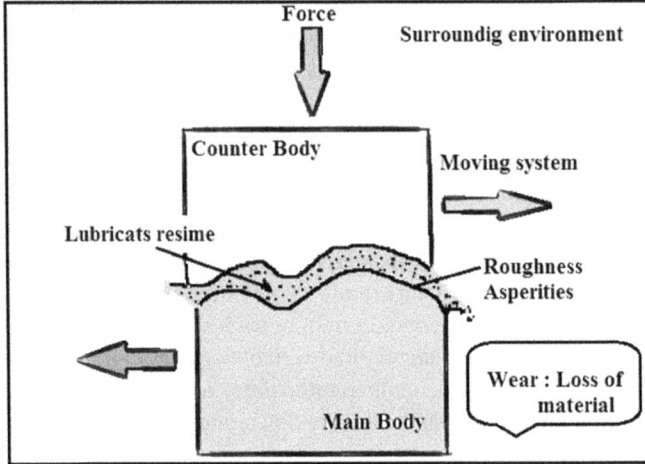

FIGURE 7.3   Tribological system [31].

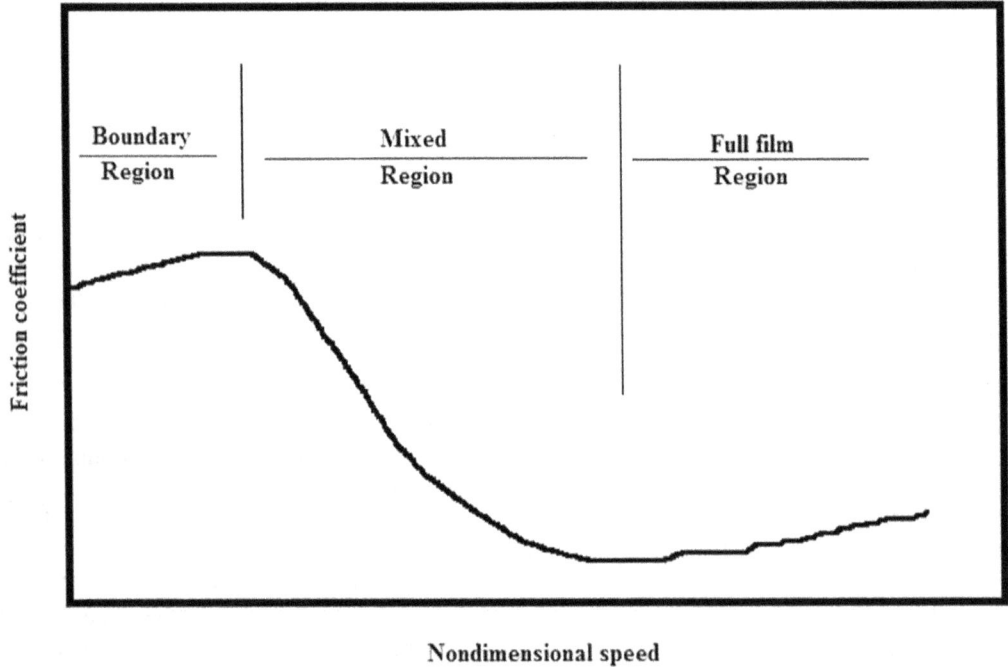

FIGURE 7.4   Deep drawing friction coefficient as a function of dimensionless speed [5, 29].

7.2.4.3 Lubrication Influence in Sheet Metal Forming

In many studies it has been identified, the friction coefficient of an unlubricated forming operation is rarely less than 0.5, [32–34]. The value of friction coefficient is high that would cause in high frictional forces, resulting in fracture surfaces and frictional energy losses in the sheet metal formation. So use of different lubricants in sheet metal forming industries to reduce and control over the frictional force between the surfaces [3, 35]. Friction is reduced when the adhesion component of friction is lower than the abrasion component of friction.

Tapering on the input clearance in lubricated sliding contacts and usually on curved surfaces involves the liquid lubricant [36], which is forced to enter the contact gap under certain conditions like load, speed, geometry of surfaces, and elastic properties thereby supporting the imposed load. If there is an excess of oil (more than what is required for the production of the boundary film), the wear boundary film is restored. Under the influence of boundary lubrication, the friction process continues to flow. Rudi Haar [37] claimed in his study that increasing the amount of oil in the distance between the mating parts causes a mixed regime of friction.

Demand for lubricants in forming processes, as well as quality criteria and accumulated knowledge about the processes that occur in the working environment. The selection of lubricant and its application is influenced by a variety of factors should be addressed from three angles [35]:

- The status of type of lubrication and the preform for forming process, as well as their practical applications.
- Behavior throughout the forming process.
- Anti-corrosion protection.

The following criteria should be addressed while choosing lubricants for sheet metal forming [38]:

- Application methods.
- Additive kinds.
- Viscosity of lubricant/desired film thickness.
- Corrosion management.
- Cleaning procedures.
- Compatibility with the pre-lubricants and pre-applied oils.
- Post-metal forming activities.
- Environmental safety issues and recycling of the material.

Economic, environmental, hygienic, and toxicological considerations are all given a lot of weight nowadays. An important topic is the use of pre lubrication in rolling stand to reduce the number of lubrication sites necessary for stamping operations. Pre lubrication saves money not only because it saves lubricants but also because it saves time. The true benefit of employing pre lubrication instead of special oils in each stage of the entire production chain is a large reduction in the cost of the process, making it crucial in the optimization of all manufacturing stages.

Appropriate oils for pre lubrication allow for better cost savings. This is especially true when multifunctional lubricants are utilized, as they are entirely compatible and may be utilized for a variety of applications. Pre-lubrication systems are now modular, allowing you to employ oils of various viscosities.

Table 7.1 demonstrates the different types of lubricants available for the forming process, as well as the most appropriate applicable regions. Lubricant can be applied to the metal sheet manually, by spray, by dip, by rotating roller, or by grease lubrication in the event of deep drawing.

Cleaning fluids and emulsions used to clean large parts of car bodies before forming should be compatible with milling and pre-lubrication oils applied to the sheet metal previously, and ideally, they should be the same, chemically similar [39]. Spraying is the most convenient technique of application since, like manual application, it allows the lubricant to be applied to the appropriate spot. Spraying can be used to apply oils, emulsions, and lubricants for forming operations.

TABLE 7.1
Various Lubricants for the Metal Forging Process [39]

Lubricant	Main Application	Note
a milling lubricant	Stamping, extrusion, shaping, and deep drawing are some of the techniques used, Calibration and Punching	The flashpoint is between 40 and 65 degrees Celsius.
polar additives in the lubricant	Punching thin plates and circular blanks with a punching machine. Embossing, stamping, shaping, and pressing are some of the techniques used. Deep drawing is used in the production of automotive bodywork.	Apply lubricants with a low viscosity, example mineral oils without additives.
Anti-scuffing additives in the lubricant (without chlorine)	Deep pressing, pressing Blank operations that are exact. When stretched, the sheet material undergoes a deep drawing and thinning process.	Copper may develop corrosion spots as a result of lubricants.
Forming lubricants that do not contain solid components	The formation of a complex	Steels with low and high carbon content
Forming lubricants with solid components or anti-scuffing additives	Deep drawing application	They're similar to pasty lubricants.
Lubricants	For forming, emulsions or solutions are used. Hydraulic power hydraulic forming fluids	Synthetic, semi-synthetic, or equivalent lubrication
Forming lubricants that do not contain solid lubricant components	Deep drawing application	Pasty lubricants are sometimes used.
Forming lubricants that contain solid lubricant components	Forming and deep drawing are the most hardest tasks.	Pasty lubricants are frequently used.

Spraying requirements for lubricants can be specified specially to ensure compliance with industrial shop standards.

The complexity of the shaping process has a big impact on the lubricant choice. The lubricant should make it easier to process the workpiece by reducing friction and ensuring the formation of an effective separation layer between the workpiece and the tool. The lubricant must contain additives that produce an efficient separation layer throughout the process due to the rarity of the hydrodynamic friction regime in the forming process. Erdemir's previous research [40] showed that additives can be either polar or chemicals that can modify the metal surface as a result of chemical interactions. For example, fine solids can be introduced into the lubricant as a solid lubricant component or as an inert filler to generate a separation effect.

7.2.5   POWDER FORGING PROCESS

Powder forging is a natural extension of the press and sinter (P/M) process, which has been proven to be a reliable method for creating a wide range of parts in net or near-net shape. The powder forging process is depicted in Figure 7.5. In essence, a porous preform is densified in a single stroke via forging. Heat aids the process of developing a net or near-net shape by facilitating densification and deformation. Typically, forging is done in heated, completely enclosed dies with little to no flash [41].

7.2.5.1   Lubrication of Powder Producing Dies

Before each blow, grease the dies. In most cases, a spray of colloidal graphite in water or oil is sufficient. Lubricants made of synthetic materials are also commonly used. Dies are normally sprayed by hand, although in press forging, automatic sprays synchronized with the press stroke are occasionally employed. In press forging, glass is occasionally utilized as a lubricant or billet coating. Glass is applied to the heated forging by dipping it in molten glass or dusting it with glass frit. Glass is a good lubricant, but its viscosity needs to be suitable with the forging temperature [42].

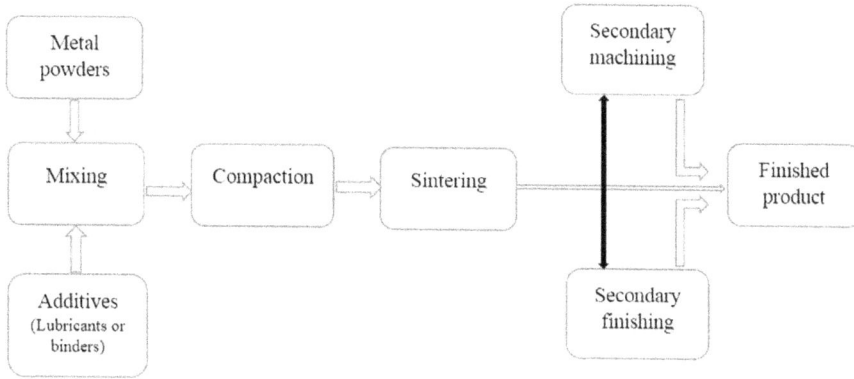

FIGURE 7.5   Powder forging flow process diagram [41].

7.3   CONCLUSION

Bio-lubricants, which are often manufactured from vegetable oil feed stocks and are almost fully biodegradable, have received a lot of attention in recent years. The use of tribology to mechanical systems can assist reduce global carbon emissions by reducing friction, lowering wear, and extending the life of resources. By reducing energy loss, machinery can run more effectively, and reducing wear minimizes the need for equipment replacement, saving money, resources, and time.

Forming is a plastic deformation technique for metals that employs various engineering mechanics. Tribology has an impact on forces during the process, product quality, vibration, and sound effect, and other factors due to the relative effect of roll, die, metal surface, and other equipment. The forming methods and available lubrication for each process have been discussed throughout the chapter.

Tribology in the bulk and metal forming processes has been discussed throughout the chapter. On behalf of that process flow chart for metal forming process defines lubrication, including some of the most modern lubrication processes in Figure 7.6. To make the process smoother and reduce the wear of the instruments, proper study has been made before implementing the process.

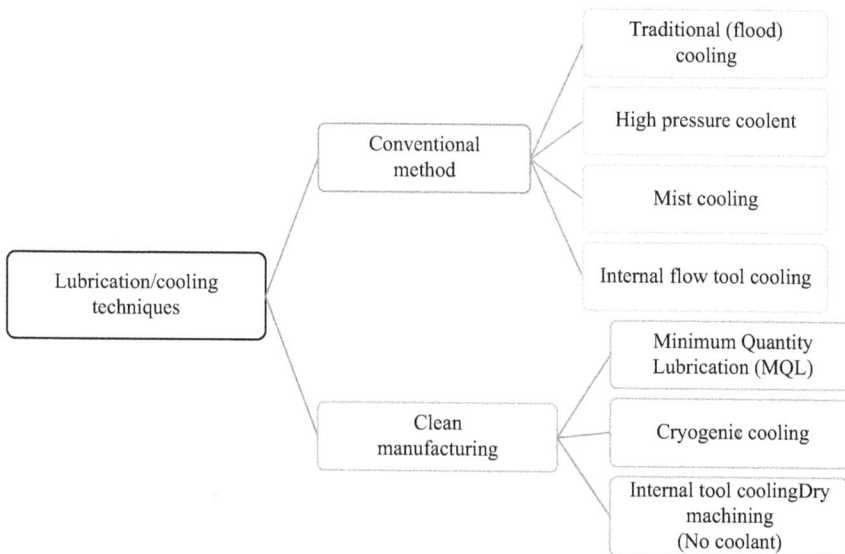

FIGURE 7.6   Lubrication for manufacturing possesses.

REFERENCES

[1] Enrico Ciulli, "Tribology and industry: From the origins to 4.0," *Hypothesis and Theory*, vol. 5, 2019. https://doi.org/10.3389/fmech.2019.00055

[2] M. Woydt, "The importance of tribology for reducing CO_2 emissions and for sustainability," *Wear*, Vols. 474–475, 2021. https://doi.org/10.1016/j.wear.2021.203768

[3] K. C. Ludema, *Friction, wear, lubrication, A textbook in tribology*, CRC Press LLC, 1996.

[4] D. J. Schipper and E. Heide, "Friction and wear in lubricated sheet metal forming processes," in *Modern tribology handbook*, CRC Press LLC, 2001.

[5] W. R. D. Wilson and J. A. Walowit, "An isothermal lubrication theory for strip rolling with front and back tension," in *Trib. Conv. 1971, institution of mechanical engineers*, London, 1971. https://doi.org/10.1243%2FPIME_PROC_1990_204_064_02

[6] S. R. Schmid and W. R. D. Wilson, "Lubrication mechanisms for oil-in-water emulsions," in *Lubrication Engineering*, 1996, pp. 168–175.

[7] J. Schey, *Tribology in metalworking*, American Society for Metals, Metals Park, OH, 1984. https://doi.org/10.1016/S0007-8506(07)60191-7

[8] K. Hwang, S. Lee, H. Lee, "Effects of alloying elements on microstruc-ture and fracture properties of cast high speed steel rolls Part I: Micro-structural analysis," *Materials Science and Engineering, A*, vol. 254, pp. 282–295, 1998.

[9] L. C. Erickson, S. Hogmark, "Analysis of banded hot rolling rolls," *Wear*, vol. 165, pp. 231–235, 1993. https://doi.org/10.1016/0043-1648(93)90340-R

[10] G. Walmag, "Mechanisms of work roll degradation in HSM," in *Proceedings of the 44DG Seminario de Laminacao Processos e Prudutos Laminados e Revestidos*, Jordao, Brasil, 2007.

[11] L. Caithness, S. Cox, and S. Emery, "Surface behaviour of HSS in hot strip mills," in *Proceedings of the Rolls 2000+ Conference*, Birmingham, UK, 1999.

[12] J. W. Park, H. C. Lee, S. Lee, "Composition, microstructure, hardness, and wear properties of high-speed steel rolls," *Metallurgical and Materials Transaction A*, vol. 30, 399–409, 1999.

[13] K. H. Ziehenberger, M. Windhager, "State of the art work rolls for hot rolling flat products," in *Proceedings of the CONAC 2007 – 3rd Steel Industry Conference and Exposition*, Monterrey, Mexico, 2007.

[14] W. R. D. Wilson, and W. Lee, "Mechanics of surface roughening in metal forming processes," in *1st Int. Conf. Tribol. Manufacturing Processes*, Gifu, Japan, 1997.

[15] P. K. Saha, *Aluminum extrusion technology*, American Society for Metals, Park, OH, 2000. ISBN: 978-0-87170-644-7

[16] K. Laue and H. Stenger, "Extrusion," in *American society for metals*, Metals Park, OH, 1981. ISBN: 978-0-87170-644-7.

[17] O. J. Tassi, "Nonferrous wire handbook," in *The wire association*, Guilford, CT, 1981.

[18] I. M. Hutchings, *Tribology – friction and wear of engineering materials*, Butterworth-Heinemann, Oxford, 1992. ISBN: 9780081009512.

[19] B. Podgornik and S. Hogmark, "Surface modification to improve friction and galling properties of forming tools," *Journal of Materials Pro-cessing Technology*, vol. 174, pp. 334–341, 2006. https://doi.org/10.1016/j.jmatprotec.2006.01.016

[20] D. J. Schipper and E. Heide, *Friction and wear in lubricated sheet metal forming processes*, CRC Press LLC, 2001.

[21] R. Gedney, *Sheet metal testing guide*, ADMET, Inc. 51 Morgan Drive, Norwood, 2013.

[22] R. Padmanabhan, M. C. Oliveira, J. L. Alves, and L. F. Menezes, "Influence of process parameters on the deep drawing of stainless steel," *Finite Element Analysis Design*, vol. 43, no. 14, p. 1062–1067, 2007.

[23] A. Danel and M. Buard, "Estimation of coining force on sheet metal," in *IDDRG, CETIM – Technical Centre for Mechanical Industries*, Paris, France, 2014.

[24] M. J. Neale, *The tribology handbook* (second edition), Butterworth Heinemann, Farnham, Surrey, UK, 2001. ISBN: 9780080519661.

[25] B. Bhushan, *Introduction to tribology* (second edition), Wiley: JohnWiley & Sons, Ltd, 2013. ISBN: 978-1-118-40322-8.

[26] S. Aleksandrovich, M. Stefanovich and T. Vujinovich, "Variable tribologcal conditions on the blank holder as significant factor in deep drawing process," in *8th International Tribology Conference, Belgrade*, Serbia, 2003.

[27] B. Bhushan, "Friction, scratching/wear, indentation, and lubrication using scanning probe microscopy," in *Modern tribology handbook*, CRC Press LLC, 2001. ISBN: 978-1-118-40322-8.

[28] K. Holmberg and A. Matthews, "Tribological properties of metallic and ceramic coatings," in *Modern tribology handbook*, CRC Press LLC, 2001. https://doi.org/10.1201/9780849377877

[29] W. C. Emmens, "Tribology of flat contacts and its application in deep drawing," PhD thesis, University of Twente, the Netherlands, 1997.

[30] P. J. Blau, "Effects of tribosystem variables on friction (second edition)," in *Friction science and technology*, Taylor & Francis Group LLC, 2009, p. 269–313.

[31] P. J. Blau, "Introductory mechanics approaches to solid friction," in *Friction science and technology*, Taylor & Francis Group LLC, 2009, pp. 17–41.

[32] H. Sofuoglu and J. Rasty, "On the measurement of friction coefficient utilizing the ring compression test," *Tribology International*, vol. 32, no. 6, p. 327–335, 1999. https://doi.org/10.1016/S0301-679X(**99**)00055-9

[33] W. Wang, Y. Zhao, Z. Wang, M. Hua, and X. Wei, "A study on variable friction model in sheet metal forming with advanced high strength steels," in *Tribology international*, University of Hong Kong, China, 2015. https://doi.org/10.1016/j.triboint.2015.09.011

[34] L. Figueiredo, A. Ramalho, M. C. Oliveira, and L. F. Menezes, "Experimental study of friction in sheet metal forming," *Wear, CEMUC*, vol. 271, no. 12, pp. 1651–1654, 2011.

[35] T. Mang and W. Dresel, *Lubricants and lubrication*, Wiley-VCH Verlag GmbH & Co.KGaA, Weinheim, 2007. DOI:10.1002/9783527610341

[36] L. J. Gschwender, D. C. Kramer, B. K. Lok, S. K. Sharma, J. C. E. Snyder, and M. L. Sztenderowicz, "Liquid lubricants and lubrication," in *Modern tribology handbook*, CRC Press LLC, 2001.

[37] R. ter Haar, "Friction in sheet metal forming, the influence of (local) contact conditions and deformation," PhD thesis, University of Twente, the Netherlands, 1996.

[38] H. Kim and N. Kardes, "Friction and lubrication," in *Sheet met. forming – fundamentals*, ASM International, 2012.

[39] G. E. Totten, "Handbook of lubrication and tribology (second edition)," in *Aplication and maintenance*, Taylor & Francis Group LLC, CRC, 2006.

[40] A. Erdemir, "Solid lubricants and self-lubricating films," in *Modern tribology handbook*, CRC Press LLC, 2001.

[41] B. Lynn Ferguson, "Powder forging," in *Metalworking: bulk forming*, ASM International, Materials Park, OH, 2005, pp. 205–220.

[42] George Mochnal, "Forging of stainless steels," in *Metalworking: bulk forming*, ASM International, Materials Park, OH, 2005, pp. 261–275. https://doi.org/10.31399/asm.hb.v14a.a0003992

8 Hole Expansion Ratio (HER) for Automotive Steels

Surajit Kumar Paul
Indian Institute of Technology Patna, Patna, India

CONTENTS

8.1 INTRODUCTION

In recent times, advanced high strength steels (AHSS) are progressively adopted by the automobile industry to reduce emission and to improve fuel efficiency by reducing the weight of the car without sacrificing passenger's safety [1, 2]. However, edge cracking of AHSS is the major constraint in their use [1, 2]. Such edge cracking does not arise for soft steel grades like low carbon and mild steels [3, 4]. The edge cracking may visible for two conditions (a) during the regular sheet metal forming operation; the blank free edge might get stretched, and (b) manufacturing of stretch flange; the material stretched tangential to the free edge and simultaneously shrinks along the perpendicular direction to the free edge. Casellas et al. [5] shown the edge cracking in cold-formed AHSS automotive parts for both conditions i.e., at normal free edge during forming and during manufacturing of stretch flange. Figure 8.1 schematically explain the stretch flanging operation. Normally stretch flange is introduced in the design of sheet metal parts to provide rigidity of panels or fasten the parts together. Common instances of stretch flanging in the automobile body parts include cut-outs in automotive inner panels, corners of the window panel and hub-hole of wheel discs, etc. Traditional forming limit diagram (FLD) is not able to predict edge cracking in stretch flanging [1]. Hole expansion ratio (HER) is successfully used to address such edge cracking. As a consequence, the HER becomes one of the crucial formability parameters for sheet metal forming operation.

HER is normally determined from the hole expansion test (HET). A schematic illustration of the HET is explained in Figure 8.2. Normally a square-shaped (100 mm × 100 mm) specimen with a central hole of 10 mm diameter is used for the HET. A conical punch with a cone angle of 60° is used to expand the central hole of 10 mm initial diameter (according to the ISO 16630 standard). The diameter of central hole is measured before and after the execution of HET. The test stop criterion is selected as the through-thickness formation of edge crack at the central hole edge. However, flat-bottom and hemispherical punches (non-standard tests) are also adopted by researchers [1, 2, 6, 7].

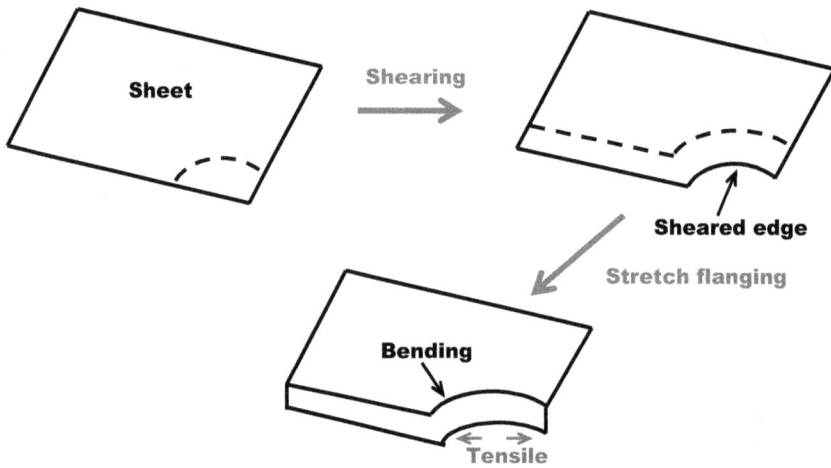

FIGURE 8.1   Schematic example of stretch flanging.

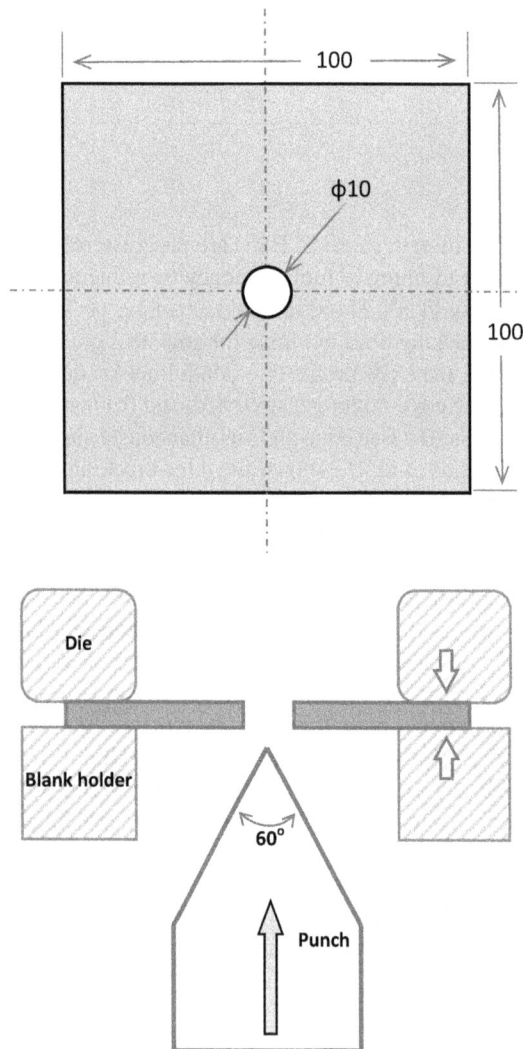

FIGURE 8.2   Schematic diagram of specimen and tool for hole expansion test (all dimensions are in mm).

HER can be calculated from Equation (8.1).

$$HER = 100\frac{\left(d_f - d_0\right)}{d_0}$$ (8.1)

Where d_f and d_o are the final and initial diameter of the central hole.

Extensive investigation on stretch flangeability (HER) was started in the last decade. Taylor [8], Yoahida [9], and Wang and Wenner [10] were numerically investigate the hole expansion. Yamada et al. [11] reported that uniaxial tensile deformation happens at the hole edge during hole expansion. Latter Karima et al. [12] also observed uniaxial tensile deformation at the hole edge during the hole expansion with a flat-bottomed punch. Matsuzu et al. [13] noticed that punched hole edge property (burr shape and damage/crack) and finally HER depends upon the microstructure of metals. After that number of research papers were published on HER. In this chapter, published literature on HER is reviewed and a comprehensive discussion is done section wise.

8.2 DIFFERENCE BETWEEN HOLE EXPANSION AND UNIAXIAL TENSILE DEFORMATION

The stress state at the central hole edge is approximately uniaxial tensile during the HET [1, 2, 6]. The finite element (FE) modelling of HET was done by Paul et al. [14] for EDD steel and Paul [15] for DP steel, and confirms roughly uniaxial tensile stress state present at the central hole edge. However, a slight difference exists initially among three different planes at the central hole edge i.e., inner, mid, and outer edges. Uniaxial tensile deformation path can be easily understood from the in-plane minimum principal strain (i.e., radial strain, ε_2) versus in-plane maximum principal strain (i.e., hoop strain, ε_1) plot. For isotropic material following von Mises yield criteria, Equation (8.2) can be followed.

$$\frac{\varepsilon_2}{\varepsilon_1} = -\frac{1}{2}$$ (8.2)

For material following normal anisotropy can be described by Equation (8.3) with Hil-48 yield criteria [16].

$$\frac{\varepsilon_2}{\varepsilon_1} = \frac{-r}{\left(1+r\right)}$$ (8.3)

Where r is the normal anisotropy. If $r = 1$ in Equation (8.3), then the material turns to isotropic and following Equation (8.2). The stress state at the central hole edge during the HET can be easily determined from ε_1 vs. ε_2 plot. After collecting strain (ε_1 and ε_2) data with time at the central hole edge, a similar diagram like Figure 8.3 can be obtained. Figure 8.3 is a schematic diagram of major and minor principal strains. Forming limit curve (FLC) and fracture forming limit curve (FFLC) are also positioned in that diagram. When the loading point reaches FLC localized necking starts and after reaching FFLC fracture happens. Uniaxial tensile loading paths with Equations (8.2) and (8.3) are also shown in Figure 8.3. In Figure 8.3, $r > 1$ is assumed. Therefore, with increasing r the uniaxial tensile loading path shifts right side and necking is delayed as it reaches FLC at higher ε_1. In the case of incremental forming, localized necking can be suppressed and superior HER can be achieved [18]. As localized necking is suppressed in case of incremental forming, the material can deform up to fracture (FFLC) and hence HER is improved. Therefore, the highest HER can be achieved from the incremental forming operation.

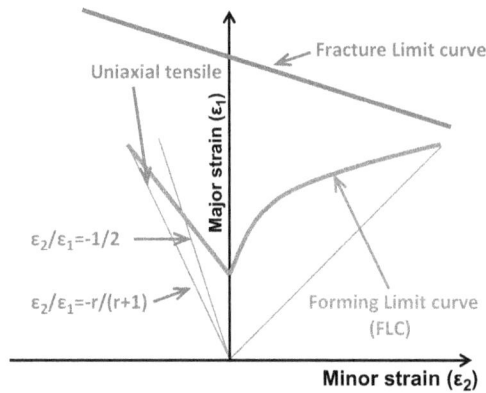

FIGURE 8.3 Schematic representation of uniaxial tensile loading path, and position of FLC and FFLC.

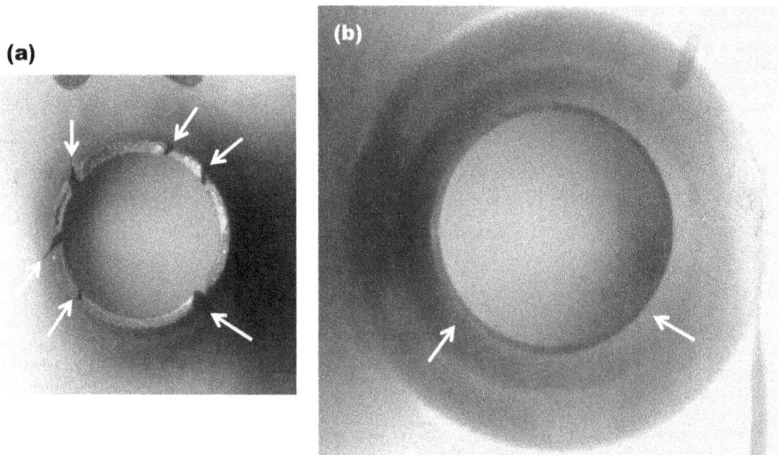

FIGURE 8.4 Hole expansion test samples of DP steel after completion of hole expansion test for (a) punched hole and (b) EDM machined hole.

Though uniaxial tensile deformation takes place at the central hole edge, tensile properties are not sufficient to predict HER [19]. Reasons behind this can be stated as follows: (1) There is a difference in sample geometry. Along the width uniaxial tensile test sample, two free edges are present. While for the HET sample, only one free edge (i.e., central hole edge) is present, and upper-lower dies to hold the other side of the central hole edge. (2) Existence of a prominent deformation gradient in the HET sample is observed. The highest deformation is noticed at the central hole edge and deformation reduces as moving away from the central hole edge. No such deformation gradient is present along the width of the uniaxial tensile test sample (up to uniform elongation). (3) In the course of the HET, first sheet sample bends about the punch radius and then bends about the die radius. But purely in-plane deformation (no bending) is observed for a uniaxial tensile test sample. Because of the above mentioned three reasons, multiple cracks are visible at the central hole edge after completion of the HET (Figure 8.4). However, only a single crack is evident in the necked zone after the uniaxial tensile test. For the same three reasons, the necking behaviour in HET is different from the uniaxial tensile test. In the case of HET (only for smooth central hole edge), the diffuse necking is delayed/absent [20]. Figure 8.5 schematically differentiates the necking in the uniaxial tensile test and HET (only for smooth central hole edge).

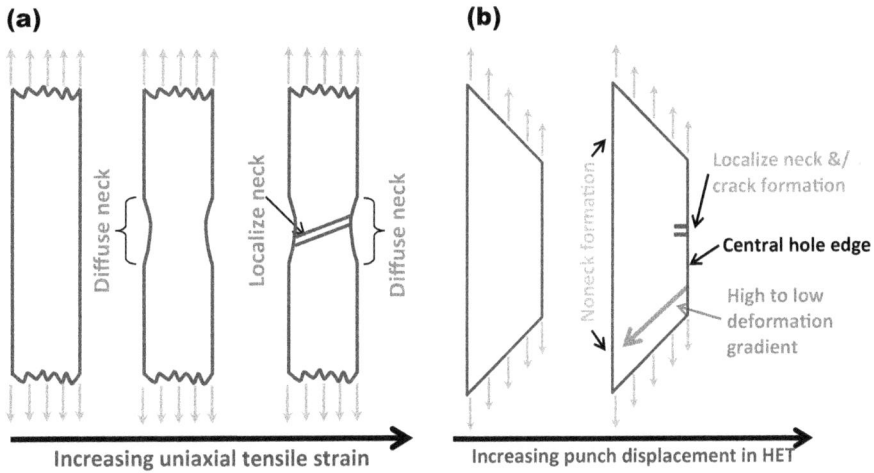

FIGURE 8.5 Schematic diagram of necking and failure in (a) uniaxial tensile test: diffuse neck followed by the localized neck and (b) hole expansion test: only localize neck and (or) crack propagation.

8.3 AFFECTING FACTORS OF HER

Paul [19] summarized that the controlling factors of HER of steels are microstructure, mechanical properties, hole edge preparation method, and punch geometry (Figure 8.6). Each of the factors is discussed in the below section wise.

8.3.1 MICROSTRUCTURE

The vital microstructural characteristics of high strength steel (HSS) that affect the HER are illustrated in Figure 8.7. High HER can be achieved in single-phase equiaxed ferritic steel microstructure [21]. But single-phase equiaxed ferritic steel without any precipitation hardening is normally soft. Nowadays, the weight reduction of a car to improve fuel efficiency and carbon emission is feasible by the adoption of HSS grades. Without much deterioration the ductility of steel, strength can be

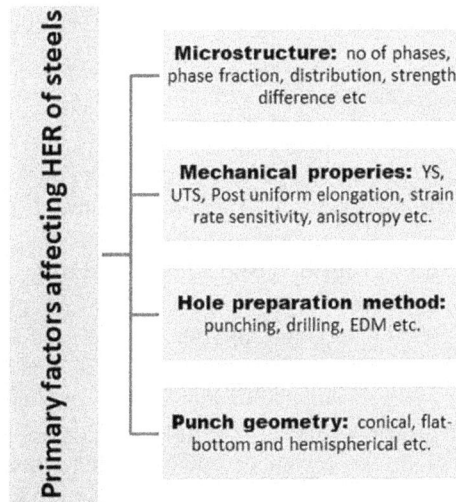

FIGURE 8.6 Primary factors affecting HER of steels.

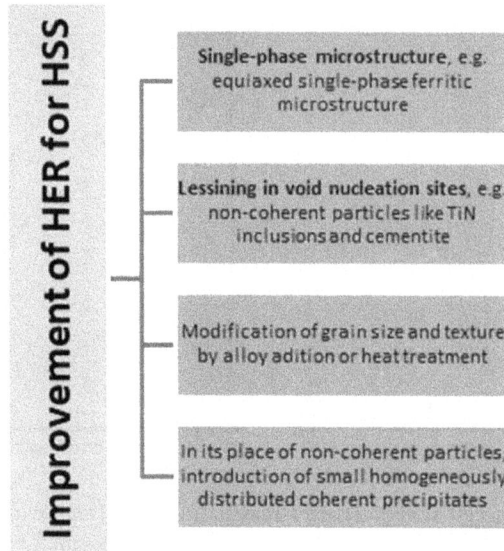

FIGURE 8.7   Vital microstructural features that can control the HER of high strength steel (HSS).

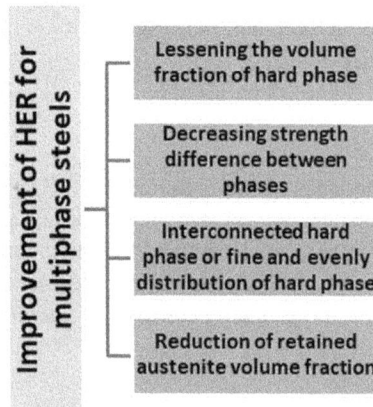

FIGURE 8.8   Main microstructural factors that can affect HER of multiphase steel.

improved by precipitation strengthening. At the same time, secondary particles like cementite and TiN inclusions act as void nucleation sites [21]. Therefore, fine evenly distributed coherent precipitates (precipitation strengthening) inside the grain can outcome in strength enhancement and lessening in void nucleation locations i.e., without or slight ductility scarifies.

In the case of multiphase steel, usually two or more phases are present. Normally soft phase ensures ductility and the dispersed hard phase improve strength, and strain hardening is improved by restricting the movement of dislocation. The strength-ductility and also HER of the multiphase steel can be manipulated by varying the volume fraction and morphology of the hard phase, strength difference between soft and hard phases, the volume fraction of metastable phase e.g., retained austenite etc. [19]. Figure 8.8 explains the key microstructural features that can control HER of multiphase steel. HER of DP steel depends upon the martensite volume fraction [22], ferrite-martensite hardness (strength) difference [23–25], martensite carbon content [23–26], and martensite morphology [27]. Terrazas et al. [24] reported that fine and evenly distributed martensite colonies in DP steel

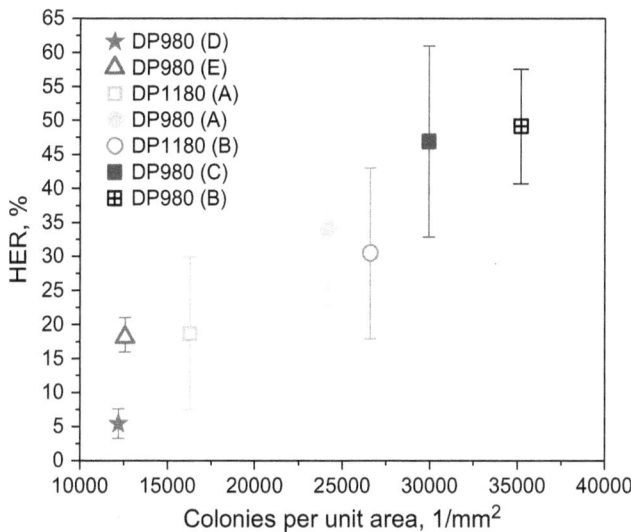

FIGURE 8.9 Alteration of HER with martensite colonies per unit area. Adapted from experimental work by [24].

result in an improvement in HER because such martensite colonies restrict the crack propagation. The martensite colonies per unit area in DP steel has an excellent correlation with HER (Figure 8.9). All experimental data related to Figure 8.9 are collected from Terrazas et al. [24]. Interconnected martensite colonies also improve HER as plastic deformation happens in martensite colonies and less strain localization in the ferrite phase [26, 28]. The HER of ferrite-martensite steel is inferior to ferrite-bainite steel as the strength difference between ferrite-martensite is higher than the strength difference between ferrite-bainite [29, 30]. Because of the same reason, the HER of CP steel is better than DP steel [31]. Normally bainite and tempered martensite are present as a hard phase in CP steel, and which is softer than martensite in DP steel. Generally, the TRIP effect is detrimental for HER [32, 33] as a transformation of retained austenite takes place at the time of hole preparation (punching process). But, HER deterioration lessens with increasing prior austenite grain size. HER of TWIP steel is also poor as strain rate sensitivity is low and the presence of elongated MnS particles [27, 34, 35]

8.3.2 Mechanical Properties

As uniaxial tensile stress state exists at the central hole edge during HET, many researchers try to correlate HER with uniaxial tensile properties. Similarly, as pre-existing defects (crack and void) present at the punched hole edge, so researchers also attempt to correlate HER with fracture parameters. The correlation of HER with other mechanical properties is tabulated in Table 8.1.

Paul [15] collected HER data from literature and plotted different correlations with HER and uniaxial tensile properties. Figure 8.10 is plotted with the same set of data. HER shows a definite correlation with UTS, YS, r, TEL, and PUEL. Paul [15] also reported that the HER has no definite correlation with YS/UTS, n, and UEL. A non-linear correlation between uniaxial tensile properties of steel and HER was reported by Paul [15]. Equation (8.4) describes the non-linear correlation between HER and uniaxial tensile properties.

$$HER = -48 + 302\,EXP(-0.0035\sigma_{UTS} + 144\left(1 - EXP\left(-0.6r\right)\right) + 0.1\varepsilon_t \tag{8.4}$$

where, HER and total elongation (ε_t) are in %, and ultimate tensile stress (σ_{UTS}) is in MPa.

TABLE 8.1

Correlation of HER with Mechanical Properties

Mechanical Properties	Authors
yield stress (YS)	[2, 15]
ultimate tensile stress (UTS)	[2, 15, 36]
YS/UTS	[37]
total elongation (TEL)	[15, 38]
post uniform elongation (PUEL)	[15, 38–41]
reduction of area (ROA)	[1]
strain hardening exponent (n)	[40]
strain rate sensitivity (m)	[39, 40, 42]
normal anisotropy (r)	[15, 40, 42]
fracture toughness (K1C, J1C)	[5, 43]
notch mouth opening displacement (NMOD)	[44]

Paul [19] summarized that among all uniaxial tensile properties, m and r have a massive effect on necking resistance. So, PUEL is likely to improve with an increase in r &/m. Therefore, high PUEL delays necking and crack formation during HET and HER improves accordingly. The important uniaxial tensile properties that control HER is schematically illustrated in Figure 8.11.

Punched hole edge normally contains defects i.e., micro-cracks and dimples, and macro-cracks are generated from those defects with the progression of deformation. Therefore, HER can be correlated to the resistance of defect growth i.e., crack growth in the metal. As a result, researchers are also attempted to correlate HER with fracture toughness [5, 43] or notch mouth opening displacement (NMOD) [20]. Figure 8.12 shows that the HER is linearly related to NMOD of different steel grades (TWIP, DP and HIF steels).

Paul et al. [45] examined the effect of planar anisotropy on HER. They have found that crack initiation site during HET is dependent upon the planar anisotropy of the metal for smooth hole edge i.e., central hole prepared by W-EDM machining. However, no such correlation is found for punched hole edge i.e., hole edge contains defects. Paul et al. [45] did FE simulation of HET of DP 590 steel. They have used Hill-48 yield criteria [16] to define the planar anisotropy of sheet metal. Material constants of Hill-48 yield criteria are determined from the Lankford coefficients (r-values) determined at 0°, 90° and 45° to the rolling directions of the steel sheet. Figure 8.13 illustrates the variation of sheet thickness (central hole edge) with different angles to rolling direction during the HET. Due to the planar anisotropy of the DP 590 steel, sheet thickness reduction at the central hole edge is not uniform during the HET. Higher thickness reduction at a certain angle to rolling direction results in higher straining and finally initiation of crack from that location only.

8.3.3 HOLE PREPARATION METHOD

Experimental FLC and HER for reamed and punched hole are shown in Figure 8.14. Experimental FLC and HER data are collated from Pathak et al. [1]. HER with a reamed hole edge is located above the FLC, while HER with a punched hole edge is positioned under the FLC. Therefore, FLC unable to explain the HER and hole edge condition plays a crucial role in HER.

The central hole in HET sample can be manufactured by various methods like punching, milling, wire-EDM cutting, laser cutting, etc. Different levels of damage at the central hole edge are introduced during different hole preparation methods [46]. The maximum damage takes place during the punching process as shear failure happens. Researchers [43, 47–50] are observed voids and cracks

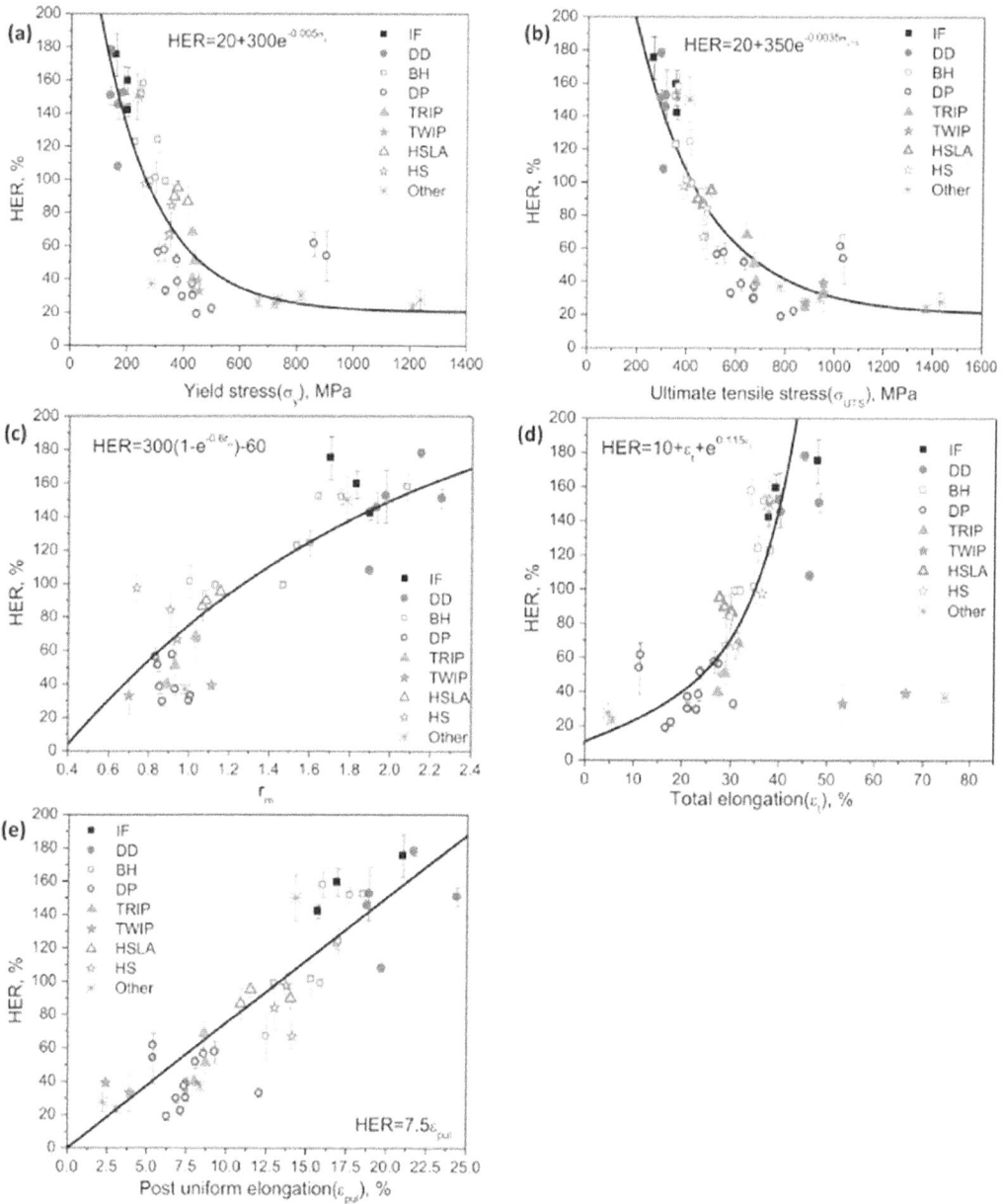

FIGURE 8.10 Correlation between HER (%) and tensile properties: (a) yield stress, MPa, (b) ultimate tensile stress, MPa, (c) normal anisotropy, (d) total elongation, %, and (e) post uniform elongation, % [15].

at the sheared fracture surface. Dalloz et al. [51] detected voids within 200 μm from the sheared face for DP steel.

From the top to bottom of the sheared edge is not symmetric. A burr is usually evident at the fractured surface bottom. Defects (voids) are normally positioned near the bottom surface (burr and fracture surface), while no such defects are visible at the top surface (shear drop region). As a result, lesser HER displays in burr up position (placement of sheet sample during HET), while high HER shows in burr down position [1].

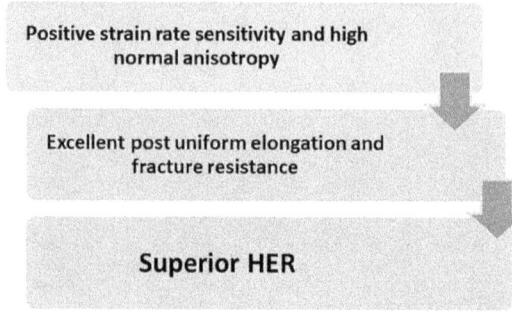

FIGURE 8.11   Vital tensile properties that can affect HER.

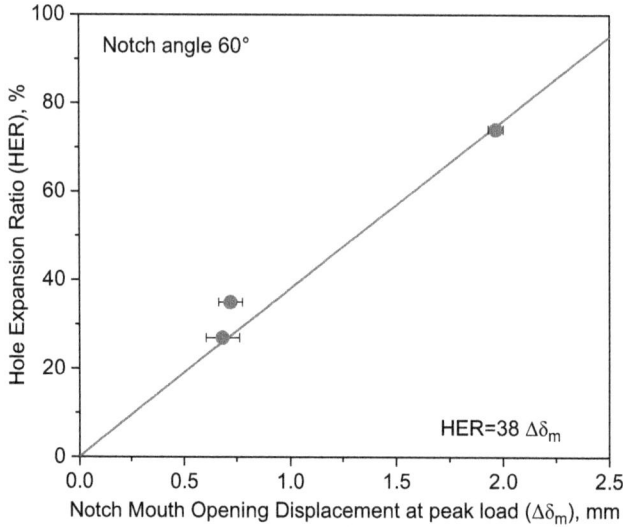

FIGURE 8.12   Correlation between HER and notch mouth opening displacement at peak load ($\Delta\delta m$) [20].

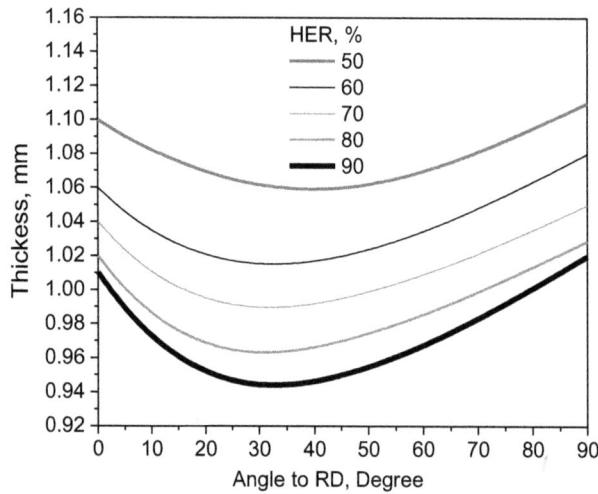

FIGURE 8.13   Variation of sheet thickness of DP 590 steel at different angles to rolling direction (RD). Adapted from experimental work by [45].

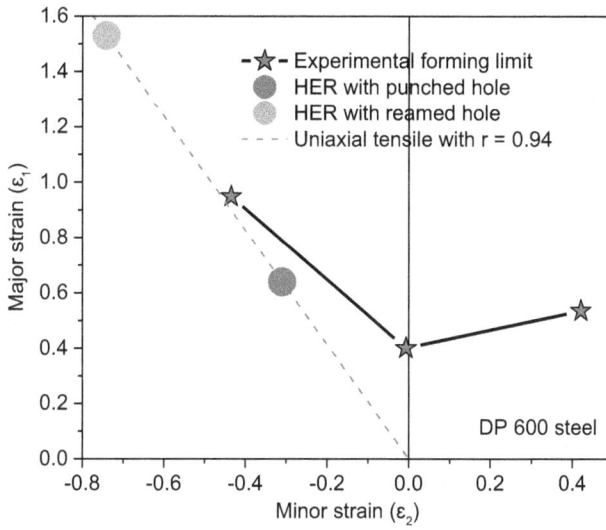

FIGURE 8.14 The position of HER with punched and reamed hole on the FLD. Experimental data are collated from Pathak et al. [1].

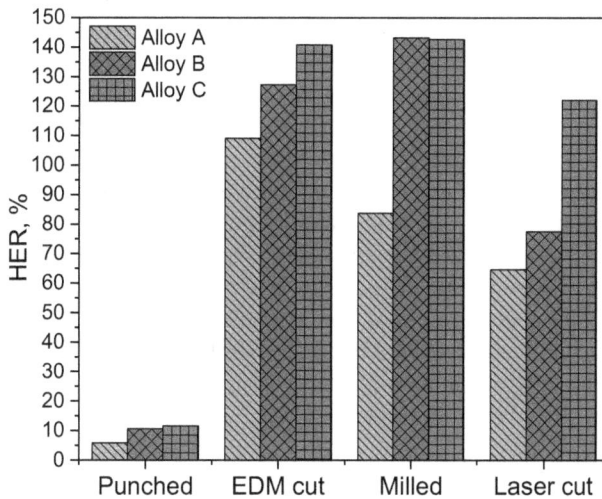

FIGURE 8.15 Variation in HER in different hole edge preparation methods. Adapted from experimental work by Branagan et al. [52].

On three third-generation AHSS, Branagan et al. [52] conducted HET with various hole edge conditions (punching, wire-EDM cutting, milling, laser cutting). Experimental data reported by Branagan et al. [52] is replotted in Figure 8.15. Branagan et al. [52] noticed that the lowest HER is observed for punched hole edge, and maximum HER for EDM cut and milled hole edge. They finally concluded that the size and severity of the damage zone created during hole preparation finally control HER. Paul [53] also performed HET on DP steel with two (punching, and w-EDM cutting) hole edge conditions. He reported 30% HER for the punched hole, whereas 110% HER for w-EDM cut hole.

8.3.4 PUNCH GEOMETRY

As per the ISO 16630 [54] standard, a conical punch with a cone angle of $60°$ is normally used to perform the HET. However, hemispherical and flat-bottom punches are also used by researchers [1, 2, 6, 7]. Konieczny and Henderson [55] reported maximum HER for conical punch, intermediate HER for hemispherical punch, and lowermost HER for flat-bottom punch. For aluminium alloys, Stanton et al. [7] also observed higher HER by using a conical punch than a flat-bottom punch. A similar observation was also reported by Madrid et al. [22] for DP steels, Pathak et al. [1] for DP and CP steels. Pathak et al. [1] also noticed that the hole edge condition (edge preparation method) does not influence the HER determined from a flat-bottom punch.

Figure 8.16 shows three different punch geometries: conical, flat-bottom and hemispherical punches. Paul [56] investigated the reasons for varying HER with punch geometries. FLC of DP 600 steel is collected from Pathak et al. [1] work. FE simulation results are collected from Paul [56] work. Paul [56] considered dies as a discrete rigid body and DP 600 steel sheet as a deformable body. Von Mises yield function (isotropic material) is considered to model the DP 600 steel sheet. Tensile true stress-strain curve (isotropic hardening) is provided as a material input. The DP 600 steel sheet has meshed with shell elements, and five integration points are selected along with its thickness. Major strain (ε_1) and minor strain (ε_2) data at the hole edge for the entire loading path are collected for three punch geometries (conical, flat-bottom and hemispherical), and illustrated in Figure 8.17. As von Mises yield criteria is adopted in this current investigation, the strain ratio (β) = 0.5 for uniaxial tensile loading path, $\beta = 0$ for plane strain tensile loading path, and $\beta = 1.0$ for equi-biaxial tensile loading path. The strain ratio (β) can be defined as the ratio of in-plane minimum and maximum principal strains i.e., $\beta = \varepsilon_2/\varepsilon_1$. Paul [56] noticed that uniaxial tensile deformation occurs at the hole edge irrespective of punch geometry. But failure happens in different locations for different punch geometries, like at the central hole edge for conical punch, slightly away from the central hole edge for flat-bottom and hemispherical punches. Identical experimental observation is also reported for flat-bottom punch [1, 57], and conical punch [2, 6, 7, 14, 15]. The causes of such observation are explained in Figure 8.17. Strain path data (ε_1 versus ε_2) are collected at the failure locations for various punch geometries. A purely uniaxial tensile strain path (at the failure location: at the central hole edge) is noticed for conical punch (Figure 8.17), and crack initiates from the central hole edge. Plane strain tensile strain path (slightly away from the central hole edge) is noticed for flat-bottom and hemispherical punches, and necking followed by cracking takes place. As crack initiates from the sheet inside (not from the hole edge), so HER is independent of central hole edge condition i.e., hole preparation method [56]. In the case of HET with hemispherical and flat-bottom punches, the failure condition is closely comparable to the Nakajima test (sheet metal forming experiment) with a plane strain condition [56].

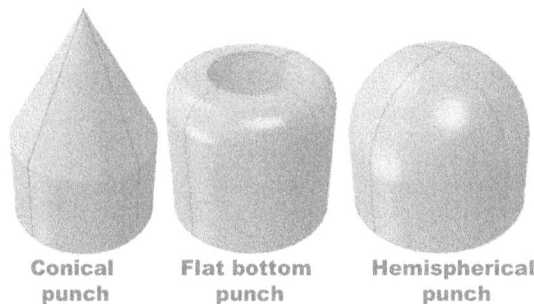

Conical punch   Flat bottom punch   Hemispherical punch

FIGURE 8.16 Different punch geometries to expand the central hole in hole expansion test.

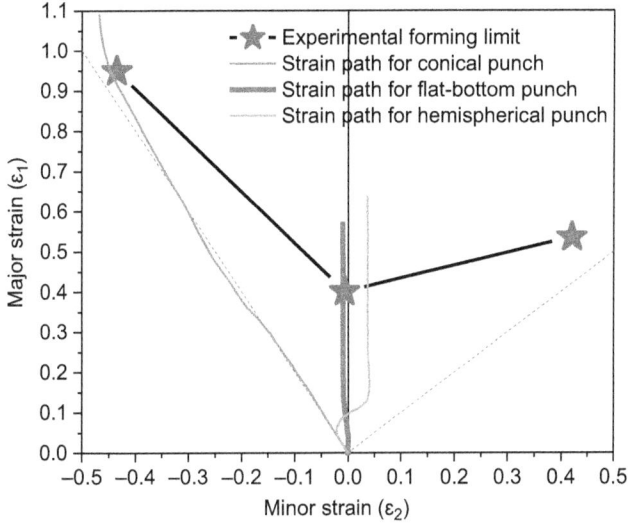

FIGURE 8.17   Loading paths at the central hole edge during HET for different punch geometries. Experimental FLC data collected from Pathak et al. [1].

8.4   CONCLUSIONS

The effect of various parameters affecting HER of automotive steels is summarized in this chapter. Materials microstructure and uniaxial tensile properties control HER. Hole edge preparation method and punch geometry play a crucial role in determining HER.

Equiaxed single-phase ferritic microstructure displays higher HER than multi-phase steels. In single-phase steels, HER can be improved by reducing void nucleating sites, i.e., non-coherent secondary precipitates i.e., TiN precipitate and cementite. Small size evenly dispersed coherent precipitates (precipitation strengthening) can provide higher HER with optimized strength. Inferior HER is noticed for steels with TRIP or TWIP effect. HER of multi-phase steels can be improved by the decrease in volume fraction of hard phase, reduce the strength difference between hard and soft phases, reducing the size of the hard phase, evenly distributing the hard phase. Better HER is noticed for ferrite-bainite steel over ferrite-martensite steel.

Uniaxial tensile deformation takes place at the central hole edge during the hole expansion test. Instead of uniaxial tensile stress state at the central hole edge, deformation and damage evolution and finally failure in hole expansion test is quite different than a uniaxial tensile test. Researchers establish prominent correlations among HER and uniaxial tensile properties like yield stress, ultimate tensile stress, total elongation, post uniform elongation, reduction of area, normal anisotropy, strain rate sensitivity etc. A definite correlation is also evident between HER and fracture parameters like fracture toughness and NMOD.

The hole edge preparation method introduces defects (void and cracks) at the hole edge. Hole edge condition significantly affects the HER determined by using a conical punch as cracks initiate from the hole edge only. However, the hole edge condition does not affect HER determined by using hemispherical or flat-bottom punch as the cracks initiate not from the hole edge but slightly inside the sheet. Highest HER (conical punch) is obtained for w-EDM cut hole and the lowest HER for the punched hole.

REFERENCES

[1] Pathak N, Butcher C, Worswick M. Assessment of the critical parameters influencing the edge stretchability of advanced high-strength steel sheet. *J Mater Eng Perform* 2016; 25: 4919–4932.

[2] Sadagopan S, Urban D. *Formability characterization of a new generation of high strength steels.* DOE Report No. 0012, 2003.

[3] Wu X, Bahmanpour H, Schmid K. Characterization of mechanically sheared edges of dual phase steels. *J Mat Proc Technol* 2012; 212: 1209–1224.

[4] Luo M, Wierzbicki T. numerical failure analysis of stretch-bending test on dual-phase steel sheets using a phenomenological fracture model. *Int J Solids Struct* 2010; 47: 3084–3102.

[5] Casellas D, Lara A, Frómeta D, Gutiérrez D, Molas S, Pérez L, Rehrl J, Suppan C. Fracture toughness to understand stretch- flangeability and edge cracking resistance in AHSS. *Metall and Mat Trans A* 2017; 48: 86–94.

[6] Larour P, Freudenthaler J, Grunsteidl A, Wang K. Evaluation of alternative stretch flangeability testing methods to ISO 16630 standard. *IDDRG Conf Proc.* 2014; pp. 188–193.

[7] Stanton M, Bhattacharya R, Dargue I, Aylmore R, Williams G. Hole expansion of aluminum alloys for the automotive industry. *AIP Conf Proc*, 2011; 1488–1493.

[8] Taylor GI. The formation and enlargement of a circular hole in a thin plastic sheet. *Q J Mech appl Math* 1948; 1: 103–124.

[9] Yoshida K. *Inst Phys Chem Res*, 1959; 53: 126.

[10] Wang NM, Wenner ML. An analytical and experimental study of stretch flanging. *Int J Mech Sci* 1974; 16: 135–143.

[11] Yamada, Koide. Analysis of the bore-expanding test by the incremental theory of plasticity. *Int J Mech Sci* 1968; 10: 1–14.

[12] Karima M, Chandrasekaran N, Tse W. Process signature in metal stamping: Basic concepts. *J Mater Shaping Technology* 1989; 7: 169–183.

[13] Matsuzu N, Itami A, Koyama K. Stretch-flange formability of high-strength steel. *SAE Technical Paper* 1991; 910513.

[14] Paul SK, Mukherjee M, Kundu S, Chandra S. Prediction of hole expansion ratio for automotive grade steels. *Comput Mater Sci* 2014; 89: 189–197.

[15] Paul SK. Non-linear correlation between uniaxial tensile properties and shear-edge hole expansion ratio. *J Mater Eng Perform.* 2014; 23: 3610–3619.

[16] Hill R. A theory of the yielding and plastic flow of anisotropic metals. *Proc Roy Soc London* 1948; 193: 281–297.

[17] Martínez-Donaire AJ, Borrego M, Morales-Palma D, Centeno G, Vallellano C. Analysis of the influence of stress triaxiality on formability of hole-flanging by single-stage SPIF. *Int J Mech Sci* 2019; 151: 76–84.

[18] Borrego M, Morales-Palma D, López-Fernández JA, Martínez-Donaire AJ, Centeno G, Vallellano C. Hole-flanging of AA7075-O sheets: Conventional process versus SPIF. *Proc Manuf* 2020; 50: 236–240.

[19] Paul SK. A critical review on hole expansion ratio. *Materialia* March 2020; 9: 100566.

[20] Paul SK. Fundamental aspect of stretch-flangeability of sheet metals. *Proc Inst Mech Eng, Part B: J Eng Manuf* 2019; 233(10): 2115–2119.

[21] Kaijalainen A, Kesti V, Vierelä R, Ylitolva M, Porter D, Kömi J. The effect of microstructure on the sheared edge quality and hole expansion ratio of hot-rolled 700 MPa steel. *J Phys: Conf Series* 2017; 896: 012103.

[22] Madrid M, Van Tyne CJ, Sadagopan S, Pavlina E, Hu J et al. Effects of testing method on stretch-flangeability of dual-phase 980/1180 steel grades. *JOM* 2018; 70: 918–923.

[23] Hasegawa K, Kawamura K, Urabe T, Hosoya Y. Effects of microstructure on stretch-flange-formability of 980 MPa grade cold-rolled ultra high strength steel sheets. *ISIJ Int* 2004; 44: 603–609.

[24] Terrazas O, Findley KO, Van Tyne CJ. Influence of martensite morphology on sheared-edge formability of dual-phase steels. *ISIJ Int* 2017; 57: 937.

[25] Taylor MD, Choi KS, Sun X, Matlock DK, Packard CE, Xu L, Barlat F. Correlations between nanoindentation hardness and macroscopic mechanical properties in DP980 Steels. *Mater Sci Eng A* 2014; 597: 431–439.

[26] Kim J, Lee MG, Kim D, Matlock DK, Wagoner RH. Hole-expansion formability of dual-phase steels using representative volume element approach with Boundary-smoothing technique. *Mater Sci Eng A* 2010; 527: 7353–7363.

[27] Chen L, Kim JK, Kim SK, Chin KG, Cooman BC De. On the stretch-flangeability of high Mn TWIP steels. *Mater Sci Forum* 2010; 278; 654–656.

[28] Miura M, Nakaya M, Mukai Y. *Cold-rolled, 980 MPa Grade Steel-sheets with Excellent elongation and stretch flangeability.* http://www.kobelco.co.jp/english//ktr/pdf/ktr_28/008-012.pdf. Accessed 15 2017.

[29] Takahashi Y, Kawano O, Tanaka Y. Fracture mechanical study on edge flange-ability of high tensile-strength steel sheets. *Proc of MS&T* 2009; 1317–1328. 2009/10/25.

[30] Sudo M, Hashimoto SI, Kambe S. Niobium bearing ferrite-bainite high strength hot-rolled sheet steel with improved formability. *Transactions of ISIJ* 1983; 23: 303–311.

[31] Karelova A, Krempaszky C, Werner E, Tsipouridis P, Hebesberger T, Pichler A. Hole expansion of dual-phase and complex-phase AHS steels – effect of edge conditions. *Steel Res Int* 2009; 80; 71–77.

[32] Jin X, Want L, Speer JG. Hole expansion in Q&P, DP, and TRIP steel sheets. *Int Symposium of Automotive Steels Conf Proc* 2013: 60–67.

[33] Chen X, Jiang H, Cuit Z, Lian C, Lu C. Hole expansion characteristics of ultra high strength steels. *Proce Eng* 2014; 1: 718–723.

[34] Cooman BC De, Kim HS, Chen L, Estrin Y, Voswinckel H, Kim SK. Hole expansion performance of TWIP steel, in: DY Yang (Ed.), *ICTP 2008*, Gyeongju, Korea, 2008; 554–555.

[35] Xu L, Barlat F, Lee MG. Hole expansion of twinning-induced plasticity steel. *Scripta Mater* 2012; 66: 1012–1017.

[36] Chen X, Jiang H, Cui Z, Lian C, Lu C. Hole expansion characteristics of ultra high strength steels. *Proc Eng* 2014; 81: 718–723.

[37] Fang X, Fan Z, Ralph B, Evans P, Underhill R. The relationships between tensile properties and hole expansion property of C-Mn steels, *J Mater Sci* 2003; 38: 3877–3882.

[38] Kim JH, Seol EJ, Kwon MH, Kang S, Cooman BC De. Effect of quenching temperature on stretch flangeability of a medium Mn steel processed by quenching and partitioning. *Mater Scie Eng A* 2018; 729: 276–284.

[39] Yoon JI, Jung J, Lee HH, Kim GS, Kim HS. Factors governing hole expansion ratio of steel sheets with smooth sheared edge. *Met Mater Int* 2016; 22: 1009–1014.

[40] Chen L, Kim J, Kim SK, Chin KG and Cooman BC De. On the stretch-flangeability of High Mn TWIP steels. *Mater Sci Forum* 2010; 654–656: 278–281.

[41] Lee J, Lee SJ, Cooman BC De. Effect of micro-alloying elements on the stretch-flangeability of dual phase steel. *Mater Sci Eng A* 2012; 536: 231–238.

[42] Chen L, Kim JK, Kim SK, Kim GS, Chin KG, Cooman BC De. Stretch-flangeability of high Mn TWIP steel. *Steel Res Int* 2010; 81(7): 552–568.

[43] Yoon JI, Jung J, Joo S-H, Song TJ, Chin K-G, Seo MH, Kim S-J, Lee S, Kim HS. Correlation between fracture toughness and stretch-flangeability of advanced high strength steels. *Mater Lett* 2016; 180: 322–326.

[44] Paul SK. Correlation between hole expansion ratio (HER) and notch tensile test. *Manuf Lett* 2019; 20: 1–4.

[45] Chinara M, Paul SK, Chatterjee S, Mukherjee S. Effect of planar anisotropy on the hole expansion ratio of cold-rolled DP 590 steel. Submitted for publication.

[46] Hance B, Comstock R, Scherrer D. The influence of edge preparation method on the hole expansion performance of automotive sheet steels. *SAE Tech Pap* 2013; 2013/04/08.

[47] Levy BS, Van Tyne CJ. Review of the shearing process for sheet steels and its effect on sheared-edge stretching. *J Mater Eng Perform* 2012; 21: 1205–1213.

[48] Karelova A, Krempaszky C, Werner E, Tsipouridis P, Hebesberger T. Pichler A. Hole expansion of dual-phase and complex-phase AHS steels - effect of edge conditions. *Steel Res Int* 2009; 80(1): 71–77.

[49] Mukherjee M, Tiwari S, Bhattacharya B. Evaluation of factors affecting the edge formability of two hot rolled multiphase steels. *Int J Miner Metall Mater* 2018; 25: 199–215.

[50] Wu X, Bahmanpour H, Schmid K. Characterization of mechanically sheared edges of dual phase steels. *J Mater Process Tech* 2012; 212: 1209–1224.

[51] Dalloz A, Besson J, Gourgues-Lorenzon AF, Sturel T, Pineau A. Effect of shear cutting on ductility of a dual phase steel. *Eng Fract Mech* 2009; 76: 1411–1424.

[52] Branagan DJ, Frerichs A, Meacham BE, Cheng S, Sergueeva AV. "New mechanisms governing local formability in 3rd generation AHSS," *SAE Tech Pap* 2017. 2017/03/28.

[53] Paul SK. Effect of deformation gradient in necking/failure during hole expansion test. *Manuf Lett* 2019; 21: 50–55.

[54] ISO. Metallic Materials - Method of Hole Expanding Test, http://www.iso.org (2009). (accessed May 17, 2016).

[55] Konieczny A, Henderson T. On formability limitations in stamping involving sheared edge stretching. *SAE Int Tech Pap* 2007; 20–29. 2007-01-01-0340.

[56] Paul SK. Effect of punch geometry on hole expansion ratio. *Proc. Inst. Mech. Eng., Part B: J. Eng. Manuf.* 2020; 234(3): 671–676.

[57] Suzuki T, Okamura K, Capilla G, Hamasaki H, Yoshida F. Effect of anisotropic evolution on circular and oval hole expansion behavior of high-strength steel sheets. *Int J Mech Sci* 2018; 146: 556–570.

9 Forming and Fracture Limit Diagrams of Inconel 718 Alloy at Elevated Temperatures

Gauri Mahalle, Nitin Kotkunde, and Amit Kumar Gupta
Birla Institute of Technology and Science,
Pilani Hyderabad Campus, India

Swadesh Kumar Singh
Gokaraju Rangaraju Institute of Engineering and Technology,
Hyderabad, India

CONTENTS

9.1 INTRODUCTION

Inconel 718 alloy (IN718) is an indispensable material for various critical applications due to its excellent blend of material properties. Despite its advantages, IN718 has a narrow forming process map, and complex microstructures. Thus, it required more deformation load at room temperature (RT) [1, 2]. From the literature, it is observed that IN718 can be formed at RT but requires an excessively high forming load. Also, high strength-to-weight ratio requires excessive deformation force to form a product, which increases the tendency of elastic recovery during forming processes [1, 2]. It is reported that noticeable localized necking or thinning tendency is not perceived during the stretch forming operation of Inconel alloy. Hence, it is challenging to define forming limits at the onset of necking, as fracture takes place without a noticeable necking in the blank specimen [3–5].

Warm or hot forming is the most prominent solution to overcome these concerns. Deformation resistance decreases during warm/hot forming because of thermal activation of dislocation motion, which results in easier plastic deformation at elevated temperature [6, 7]. This reduction in deformation load reduces spring back. It permits maximum plastic deformation with minimum annealing between forming operations by suitable overheating care [8, 9].

DOI: 10.1201/9781003226703-9

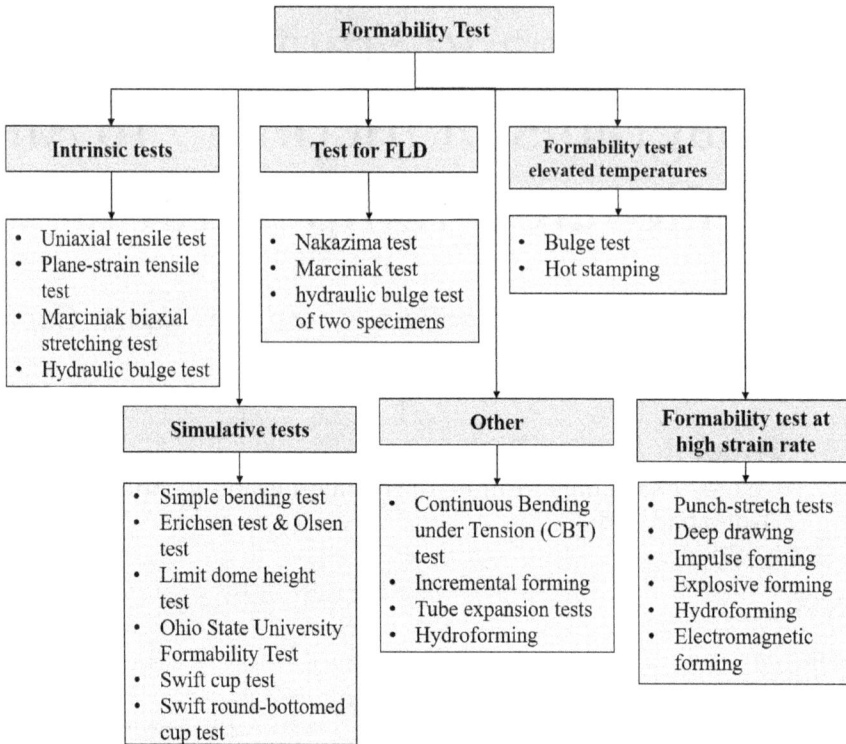

```
                              ┌─────────────────────┐
                              │   Formability Test  │
                              └─────────────────────┘
```

Intrinsic tests	Test for FLD	Formability test at elevated temperatures
• Uniaxial tensile test • Plane-strain tensile test • Marciniak biaxial stretching test • Hydraulic bulge test	• Nakazima test • Marciniak test • hydraulic bulge test of two specimens	• Bulge test • Hot stamping

Simulative tests	Other	Formability test at high strain rate
• Simple bending test • Erichsen test & Olsen test • Limit dome height test • Ohio State University Formability Test • Swift cup test • Swift round-bottomed cup test	• Continuous Bending under Tension (CBT) test • Incremental forming • Tube expansion tests • Hydroforming	• Punch-stretch tests • Deep drawing • Impulse forming • Explosive forming • Hydroforming • Electromagnetic forming

FIGURE 9.1   Experimental prediction of formability of sheet metal [10].

Generally, the ability of a metal to undergo plastic deformation without actual damage is defined as formability. Some of the performance index to measure formability are forming limit diagrams (FLDs), thickness distribution, limit dome height (LDH), and limit dome ratio (LDR). FLD is a popular performance index to represent forming process maps. In order to define FLDs, formability test can be broadly divided into three main groups, simulative tests, intrinsic tests, and tests developed to determine FLDs, as listed in Figure 9.1. Intrinsic tests give comprehensive information of a basic mechanical properties of sheet metals which relates to formability characteristics independent of surface conditions and sheet thickness. However, it reproduces strain states much easier than that industrial processes characteristics. It completely excludes the processing variables effect. Some of the popular intrinsic tests are uniaxial and plane-strain tensile tests, Hydraulic bulge test, and Marciniak biaxial stretching test. The uniaxial tensile test is most frequently used intrinsic test [10]. This test generates a stress state in tension–compression region (typically the drawing region), where minor strains are negative. This test can be easily conducted on universal testing machines where data can be rapidly extracted. Other advantages are the absence of friction effects, less scatter observed in experimental data and the optical device can be used for strain measurement. Marciniak biaxial stretching test and plane-strain tensile test are less popular methods. In plane-strain tensile test, observed minor strain values are equal to zero, material response ensures plane-strain state. Specimen geometry was modified with the decrease in gauge length and an increase in its width. In order to create a uniform in-plane biaxial strain at the specimen centre, Marciniak biaxial stretching test was conducted on a central hole specimen with a cylindrical punch [11]. In the hydraulic bulge test, a pressurized fluid allows biaxial stretching of specimen into a dome. Mainly out-of-plane strains and stresses generated in the specimen [12].

Simulative tests enforce stress and strain states that closely replicate stress states which particularly occur in a forming operation due to parameter effects. Some of the most popular simulative tests are the simple bending test, Erichsen test and Olsen test, LDH test, Ohio State University Formability Test (OSUFT), Swift cup test, and Swift round-bottomed cup test. These tests are categorized based on forming operations such as stretching, bending, drawing, and a combination of stretch & drawing. In the simple bending test, specimen with a specific thickness allows to form a 180° bend without failure [13]. The Erichsen test and Olsen test were developed to estimate metal stretch-ability especially during the stretch forming operation. A pre-hardened steel ball is used in these tests to stretch specimen and formed dome height defines stretchability of material [14]. LDH test, similar to above stretch forming test, is performed with hemispherical punch ($\phi = 100$ mm). Draw bead present in die to prevent slipping/drawing of specimens. This test is especially dedicated to produce plane strain stretching [15]. In order to overcome limitation of LDH test, OSUFT is introduced. In this test, an optimized geometry punch, obtained from FE simulation is used to generate plane strain conditions [10]. Where, Swift cup test consists of flat-bottom punch which is used to draw different diameters specimens.

In the sheet metal forming (SMF) process, FLD is widely used to describe materials deformation limit without fracture or necking. FLD gives a comprehensive picture of the material forming process map. Keeler [16] and Goodwin [17] introduced the FLD concept in 1960s. From time, it is extensively used to analyze sheet metal formability. Keeler defined strains only on right-hand side of FLD, and Goodwin expanded FLD by determining strain in left-hand side (negative minor strains). Test proposed by Keeler and Goodwin requires to use different radii punches to vary stress state. Main disadvantages of these tests are large amount of experimental work, high tooling cost, and position of FLC is highly influenced by punch radii. Furthermore, Hecker developed simplified techniques for evaluating FLD experimentally where sheet specimens are clamped at periphery and stretched over a hemispherical punch until failure [18]. This friction effect was overcome by varying lubrication or reducing the width of the original specimen

All over these years, experimental FLDs are established using various suggested experimental techniques, in which stretch forming using a hemispherical punch is mostly popular [19]. Fundamentally, there are two different tests types, namely stretch forming tests (for out-of-plane deformation) and in-plane stretching test (for in-plane deformation) which used to draw FLDs. Grid marking technique is used to measure strain in both tests. Major and minor surface strain values are measured in necking or fracture regions. Notable development was suggested by Nakazima et al. [20] for experimental FLD. Important features of Nakazima test are ease while using test setup and specimen dimensions. Different width Rectangular specimens are stretch formed by a hemispherical punch to define limiting strains. Both tension–tension (T–T) and tension–compression (T–C) regions of FLD were explored by this test [21]. Nakazima test set was improved with an optical system and an induction heating system by Turetta et al. [22]. An optical system is well equipped with an infrared thermo-camera to accurately measure surface strains for FLD during tests, even at high temperatures.

Marciniak test, among in-plane deformation tests, is most popular [23]. Different cross-sections namely elliptical, circular, and rectangular punches are used with different width specimens. This test provides much better measurement accuracy. However, the main disadvantage of Marciniak test is usage of complex shape tools, need of support blank and the limited thickness specimen possible. Comparison of Nakazima and Marciniak tests results shows good agreement for negative minor strain values, but slightly higher in-plane strain and positive minor strain values are given by Nakazima test. Banabic et al. [19] proposed a new test with two specimens based on the hydraulic bulging test. Main advantages include the capability to investigate all deformation regions specific to SMF operation, simple equipment setup, and reduction in parasitic effects, generated by friction and necking occurrence in the pole region.

Over the years, many non-conventional testing was proposed to evaluate formability. Some of formability tests are continuous bending under tension (CBT) test, Incremental forming, Tube expansion tests, and Hydroforming. CBT test was developed to review various failure mechanisms developed from enforced stress state and generate cyclic stretch-bending. Further, incremental sheet forming gives an interesting characteristic related to formability. Deformation mechanism in incremental forming vary from conventional deep drawing process. In different incremental techniques were developed to attain distinct strain paths and states. However, obtained FLDs are reasonably different from conventionally forming curves. Further, the tube expansion test is also studied for sheet metal formability. A multiaxial tube expansion test is established where different true stress or strain paths can be recognized to control internal pressure and axial force [24].

Based on a thorough literature survey, it has been perceived that very few research available on forming, and fracture prediction are reported for Ni-Cr-Fe alloy. Thus, main aim of the present study is to understand the complete forming behavior of alloy in different regions such as the safe region, and fracture region to produce a defect free-formed product.

9.2   MATERIALS AND METHODS

9.2.1   MATERIAL

Commercially available IN718 thin sheet (Grade: N07718) of thickness 1 mm is used for testing. The chemical composition of IN718 sheets is found using spark emission spectrometer. Standard ASTM E3047-16 test for emission vacuum spectrometric analysis has been used to determine the material composition of IN718 [25]. Chemical composition of IN718 is listed in Table 9.1.

Figure 9.2 shows optical micrographs of IN718 in different orientations, i.e., rolling direction (RD), diagonal direction (ND), and transverse direction to rolling direction (TD). Average diameter is 15.9 µmm with difference in morphology mostly elongated and compressed grain structures can be perceived in RD and ND. This can be accredited to large-scale rolling to achieve sheet material. Specimens confirm mostly fine grain size with considerable carbide stringers/intermetallic phases in an austenitic matrix consisting of fine equiaxed grains.

9.2.2   NAKAZIMA TEST

Stretch forming tests were performed on 40-ton hydraulic press displayed in Figure 9.3 (a). Stretch-forming tooling setup with hemispherical punch, die, and blank holder is shown in Figure 9.3 (b). Induction heating setup with a temperature controller having a K-type thermo-couples system used

TABLE 9.1
Chemical Composition of IN718

Element	Ni	Fe	Cr	Nb	Mo	Ti	Al	Cu
wt.%	51.463	20.441	18.279	5.0122	2.87	1.09	0.5611	0.0306
Element	**Mn**	**C**	**Si**	**Co**	**P**	**S**	**Zr**	**B**
wt.%	0.0616	0.0271	0.0505	0.0925	0.001	0.002	0.0091	0.0024

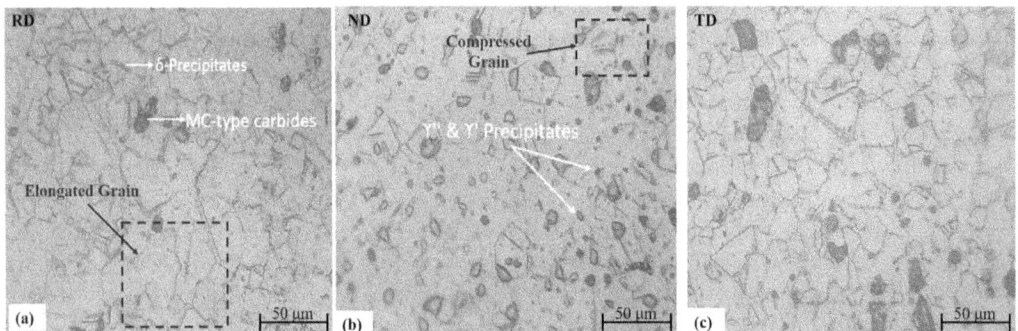

FIGURE 9.2   Microstructure of parent IN718 at (a) RD, (b) ND, and (c) TD orientations.

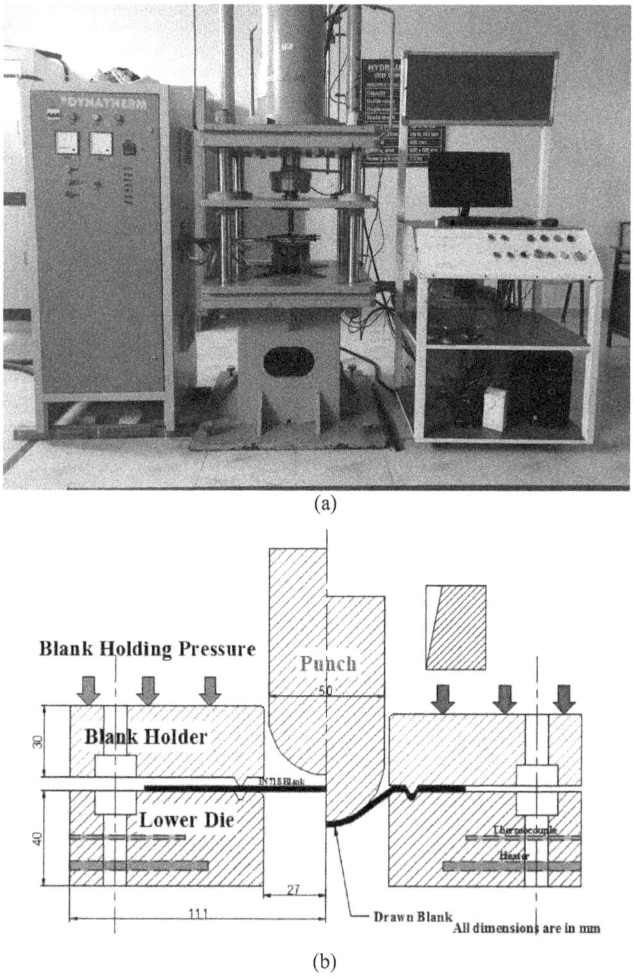

(a)

(b)

FIGURE 9.3 (a) Hydraulic press of 40-ton capacity used for stretch forming test and (b) Schematic diagram of stretch forming setup for FLD and FFLD evaluation.

to heat specimens and die at preferred temperature. K-type thermo-couples were located at appropriate locations of die and punch to measure temperature. All stretching tests were carried out using molybdenum-based lubricant (MOLYKOTE). In literature, many researchers used molybdenum-based lubricant for high temperature applications. ASTM E2218-15 standard is used to prepare different geometry and dimension blanks to plot FLDs and FFLDs [26]. Different geometry of specimens are shown in Figure 9.4. Hasek specimens (ISO12004-2 standard) were considered for lower width rectangular specimens to prevent draw bead failure. Draw bead was present in blank holder for holding blank tightly without any slipping action. It also stops the easy flow of flat flange part into the die cavity. So as to set optimal process parameters, few trials were initially conducted. Stretch forming tests were carried out at different temperatures (RT, 300°C, 500°C & 700°C) with fixed 2 mm/min punch speed and 2.5 MPa blank holding pressure.

For measuring major and minor strains, all specimens were laser-etched by 2.5 mm circular grid. Stretched specimens were distinguished by necking, safe, and fracture states. Representative stretched specimens at 300°C temperature are shown in Figure 9.5. Precise and accurate measurement of the deformed grid is one of the critical issues to get accurate limiting strains in FLD. Major

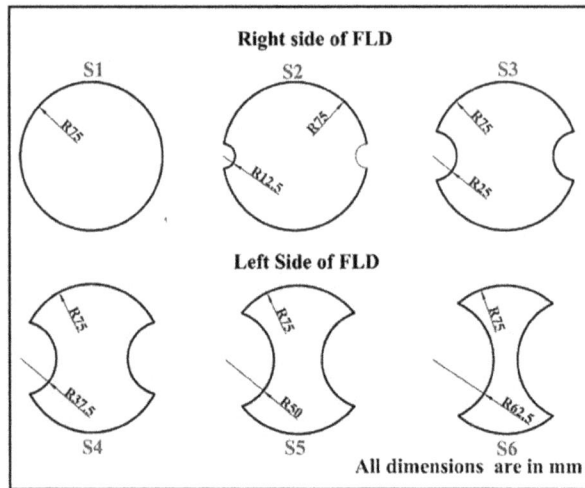

FIGURE 9.4 Schematic of specimen geometry studied for FLDs and FFLDs plots.

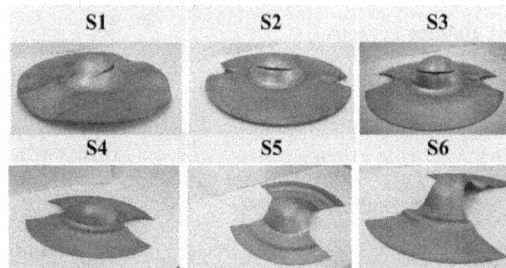

FIGURE 9.5 Representative stretched samples at 300°C for FLD prediction.

and minor diameters of the stretched grid (ellipses) in drawn/stretched blanks were measured with an optical microscope to estimate engineering minor (e_2) and major (e_1) strains using Equation (9.1). Then transferred into corresponding true strains (ε_1 & ε_2) as in Equation (9.2),

$$e_1 = \frac{\text{major axis of deformed ellipse} - \text{grid diameter}}{\text{major axis of the deformed ellipse}}$$

$$e_2 = \frac{\text{minor axis of deformed ellipse} - \text{grid diameter}}{\text{minor axis of the deformed ellipse}} \tag{9.1}$$

$$\varepsilon_{1,2} = \ln\left(1 + e_{1,2}\right) \tag{9.2}$$

9.3 RESULTS AND DISCUSSION

9.3.1 FORMING LIMIT DIAGRAM

Figure 9.6 gives scattered of calculated true minor and major strain values. For S_1-S_6 specimens, different colors were assigned and different symbols were assigned to distinguish necking, failed, and safe ellipses. Without any earlier hint of necking, IN718 specimens failed. Very few necked ellipses

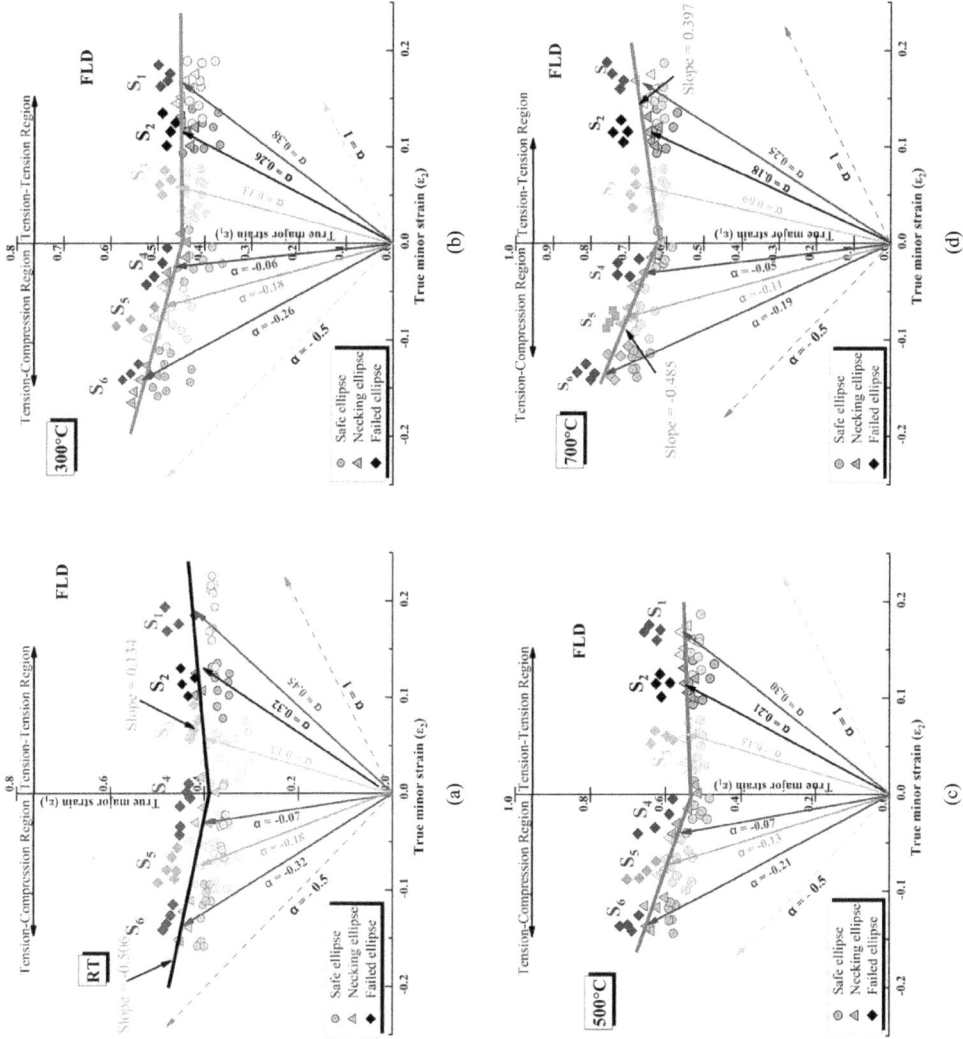

FIGURE 9.6 FLDs of IN718 at (a) RT, (b) 300°C, (c) 500°C, and (d) 700°C.

were observed in deformed regions. No-necking tendency was observed in the T–T region. Strain values are measured for all test conditions. So as to separate maximum safe strain and failed strain values, a reference line is drawn. This sudden failure might be due to the presence of a large quantity of γ' or γ'' phases in IN718 [4]. In Figure 9.6a, the highest major limiting strains at RT in T–C and T–T regions are 0.4555 and 0.4402 respectively. For a particular strain path, an average value of strain ratio $(\alpha = \dfrac{\varepsilon_2}{\varepsilon_1})$ from experimental points on FLC is considered. At plane strain state, limiting strain value 0.374. This value, defines plane strain forming limit (FLD$_0$), indicate a Limiting strain value where FLC intersect to y axis). Obtained limiting strains are well comparable with previous reports [1, 27].

As expected, limiting true strains are enhaned with a rise in temperature for all deformation regions, as shown in Figure 9.6 (b–d) and Figure 9.7 (a–b). This is mainly due to softening of material at higher temperatures. Even, a substantial necking tendency also observed at elevated temperatures. At 300°C, improvement in maximum major strain value is observed around 21% and 4% respectively in T–C and T–T regions with respect to RT. Whereas, at 500°C, 42% and 28% improvement with respect to RT, observed. Necking strain points, as presented in Figure 9.6 (c–d), were increased in all deformation regions at 500°C and 700°C. Limiting strain value at 700°C observed much higher than that at RT. Here, 66.16% and 54.35% improvement with respect to RT, in maximum major strain were found. Improvement in a major strain of IN718 with respect to RT for all temperatures is shown in Figure 9.7(c). Similar behavior showed by FLD$_0$ with rise in temperature which shows dependency on temperature. It is noticed that the FLD$_0$ value is lower

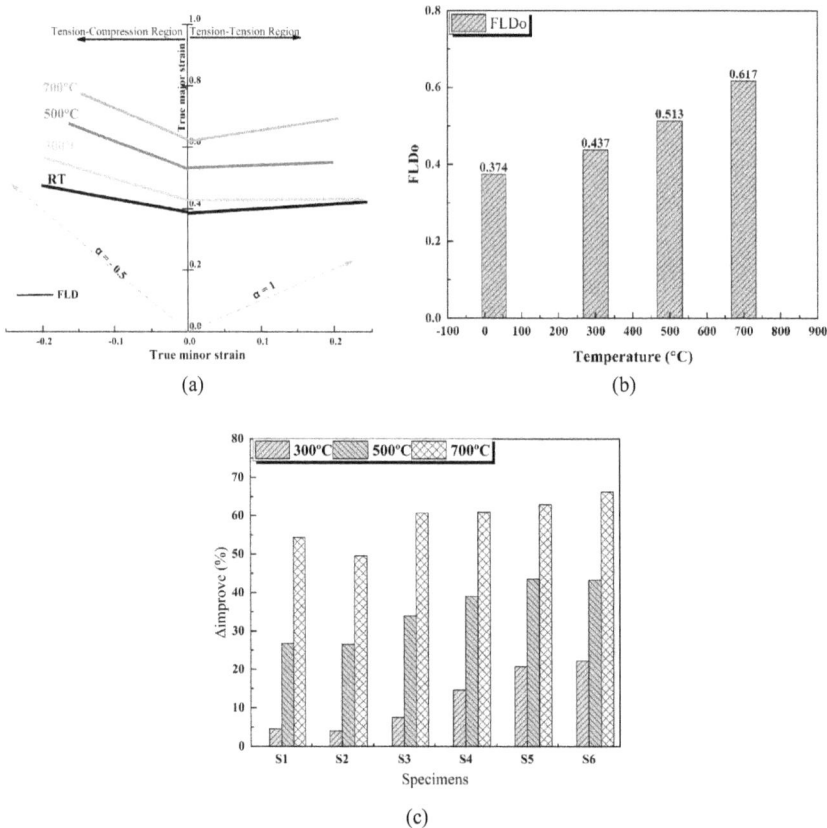

(a)

(b)

(c)

FIGURE 9.7 (a) Influence of temperatures on FLCs, (b) FLD$_0$, and (c) Improvement (%) in maximum major strain values measured for S_1-S_6 specimens (RT values referred as datum).

than maximum limiting strain values observed in T–T and T–C regions for all temperatures. Also absolute value of FLD slope increased significantly in both regions at 700°C, as observed in (Figure 9.7 (a).

9.3.1.1 Strain Distribution

It is important to discuss the fracture location of specimens (S_1, S_2, and S_3) is observed slightly away from the dome apex. High friction coefficient during warm stretching test might be one of possible reasons. Because of friction between workpiece and a tool a significant effect was observed on forming load, die wear, component surface finish, and material deformation. If the friction coefficient controlled properly, it will generate shear stresses necessary to deform sheet workpiece to desired shape. If friction coefficient is not controlled properly, it might also cause occurrence of instability of sheet specimen fracture [18, 28]. In the present study, we used molybdenum-based lubricant (MOLYKOTE) to decrease friction between a tool and workpiece. During hot stretch forming, lubrication is more challenging as it may evaporate. Hence friction coefficient will be much higher than RT. As interface shear stress once exceeds flow shear stress, it may become pointless [10, 29]. Sheet contact region experience more tension because of frictionless contact between sheet specimen and punch than other sheet's regions. As a result, fracturing occurs slightly away from dome apex. Also, strain-rate sensitivity and strain hardening of workpiece also influence fracture location. Variation in interface friction factor and strain-rate sensitivity may precede to substantial variations in predicted strain distribution [30]. Inconel has high value of strain-rate sensitivity and strain hardening exponent. Thus this might be another possible reason for side failure.

Strain signature or surface strain distribution has been plotted for maximum stretched ellipse along the longitudinal direction. Figure 9.8 (a–c) gives representative major and minor true strain distribution curves of three different stretched specimen. Here, true strain values are plotted as a function of the highest distance (pole) in formed specimens. In Figure 9.8 (a–b), both major and minor true strain values are positive which represents biaxial tension induced in specimens. But, minor strains in S_4 are relatively lower magnitude which signifies in-plane strain deformation. While closer positive major and minor strain values in S_1 indicates equi-biaxial tensile deformation. In Figure 9.8(c), specimen S_6 undergoes lateral drawing by negative values (compressive nature) of minor strain and positive values (tensile nature) of major strain. This peak major strain value was detected at a distance of 7 mm approximately, same location where deformed specimen failed. It is also noticed that temperature significantly affects the strain profile for all specimens, as shown in Figure 9.8 (a–c). With the rise in temperature, the strain distribution curve raised from RT to 700°C. This rise in surface strain values is due to thermal activation of the slip system, resulting in easier plastic deformation.

9.3.1.2 Bending Correction on Forming Limit Diagram (FLD)

In the present study, a down-sized hemispherical punch of 50 mm diameter is considered instead of regular standard punch of 101.4 mm diameter, proposed by Hecker [18]. It is downsized to optimize usage and reduce wastage of material. Figure 9.9 a shows bending strain effect observed on blank outer surface. Here sheet specimen enfold around sub-sized punch during stretch forming. Thus, it is necessary to take in account effect of punch curvature on stretch forming limits where sheet specimens were deformed into a convex shape. Strain gradient effect on strain measurement along sheet thickness reported in previous studies [31]. Geometrical factors significantly affect FLD position. Especially for constant sheet thickness, punch curvature (1/R) is directly proportional to limiting strains [31]. It is mathematically expressed as Equation (9.3).

$$\varepsilon_{\text{bending}} = \ln\left(1 + \frac{t_f}{2R_n}\right), \text{where } t_f = t_o - \exp\left(-\varepsilon_1 - \varepsilon_2\right) \tag{9.3}$$

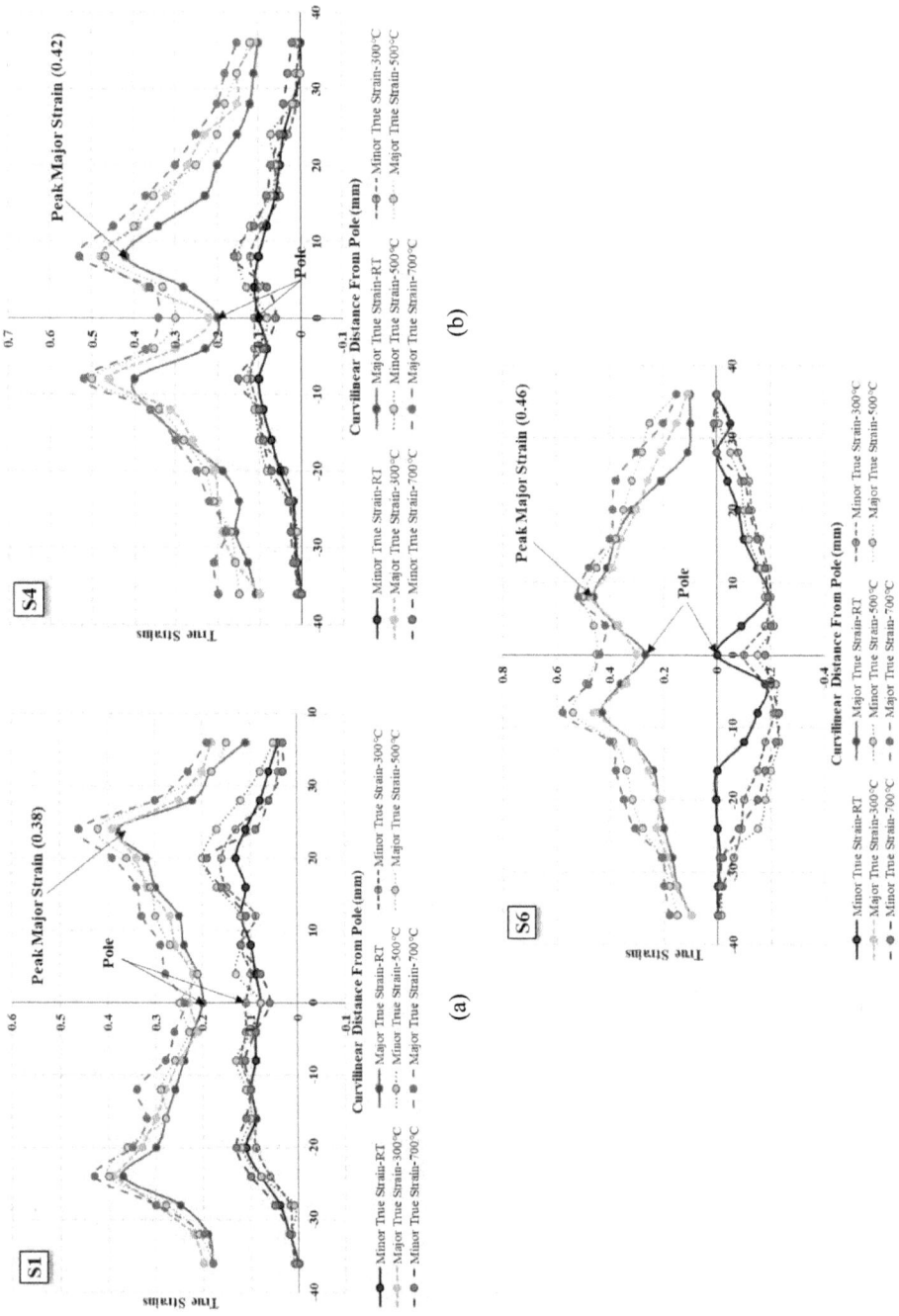

FIGURE 9.8 Strain distribution profiles of specimens (a) S_1, (b) S_4, and (c) S_6 along rolling direction.

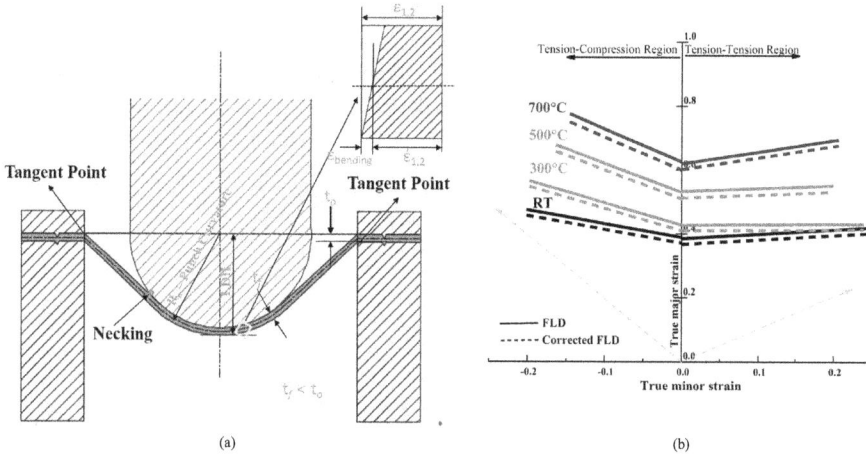

FIGURE 9.9 Schematic diagram of stretch forming tool-setup with punch curvature effect and (b) influence of bending strains on FLCs.

Previously measured surface strain values are actually the combination of stretching and bending strains. For correct limiting surface strain values ($\dot{\varepsilon}_{1n,2n}$) measurement, induced bending strain values are subtracted from measured true strain values, by using Equation (9.4)

$$\varepsilon'_{1n,2n} = \varepsilon_{1,2} - \varepsilon_{bending} \qquad (9.4)$$

Limiting strain values in FLDs are corrected by using Equations (9.3) and (9.4). Bending strain effect on measured and corrected FLDs is illustrated in Figure 9.9 (b). Corrected FLDs moved downwards roughly by 4–5% in all deformation regions for all temperatures.

9.3.1.3 Thickness Distribution

Normalized thickness distribution for S_1 specimens with curvilinear distance from pole is shown in Figure 9.10 (a, b). It is perceived that thickness remains nearly constant initially, and then it starts declining. Minimum thickness is detected at a location where necking or fracture occurred. Then, the thickness is raised gradually till flange part of specimen. Thickness on flat flange part of specimen is nearly equal to original sheet thickness. Comparable nature of curve is observed at tested

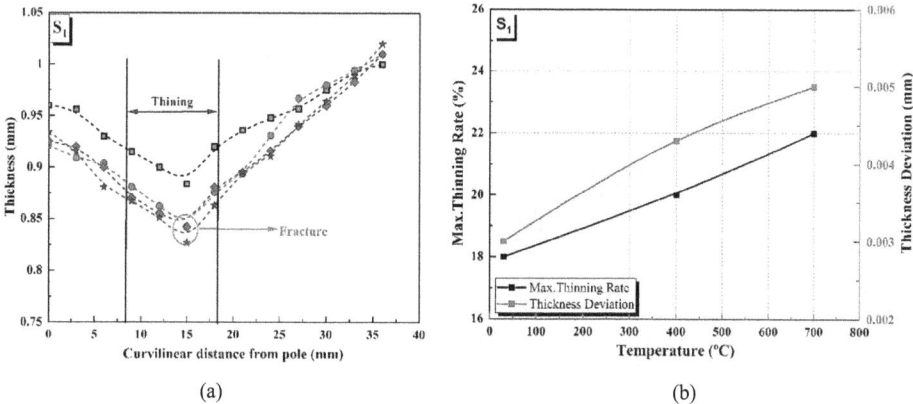

FIGURE 9.10 Thickness distribution (TD) and Maximum thinning rate (MTR) for (a-b) S_1 specimens.

temperatures for different samples. Thickness decreases with increase in temperature because of thermal softening and hence improved ductility at high temperatures. Quality of stretched samples is examined by some qualitative parameters such as Thickness Deviation (TD), Equation (9.5) and Maximum Thinning Rate (MTR), Equation (9.6). Thickness variation is an essential phenomenon in SMF applications. For the preferred formed product, minimum thickness variation is expected. MTR and TD are estimated as

$$\mathrm{TD} = \frac{1}{n-1} \sum_{i=1}^{n} \left(t_i - \overline{t} \right)^2 \tag{9.5}$$

$$\mathrm{MTR} = \left(\left(t_{\mathrm{initial}} - t_{\mathrm{min}} \right) / t_{\mathrm{initial}} \right) \times 100 \tag{9.6}$$

Here, t_i – thickness at any instant deformed specimen, $\overline{t}$ – average thickness, $t_{initial}$ – initial thickness of specimen, t_{min} – minimum thickness of deformed specimen, and n is total number of tested specimens. Figure 9.10 (b, d) shows representative MTR and TD plots for specimens. As expected, the thickness variation of drawn hemispherical cup is affected by temperature. A rise in TD and MTR is observed with increases with temperature. It is noticed that MTR and TD decrease with an increase in sample width as there is a restricted flow of material. Hence fracture occurred much before necking started, which resulted in less deviation of thickness.

9.3.1.4   Limit Dome Height (LDH)

Limiting Dome Height is a most important index that helps to understand the drawability of different width sheet specimens for various temperatures. LDH is measured as the drawn height of formed cup just before fracture occurrence. Variation of LDHs with different temperatures is displayed in Figure 9.11. It is observed that LDH is directly proportional to testing temperature and width of

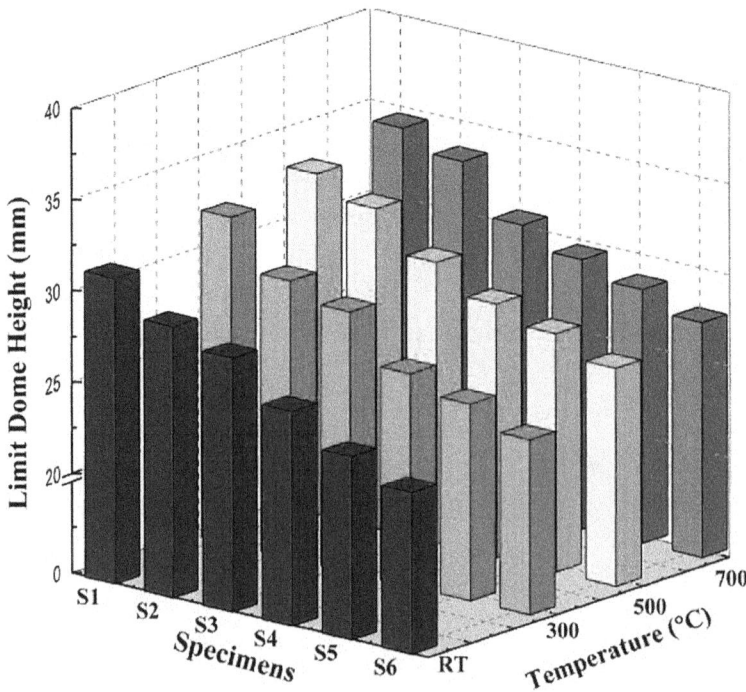

FIGURE 9.11   Variation of LDH at different processing temperatures.

specimen. As a result high LDH is observed. An increase in dome height is might be due to decrease in strain hardening with rise in temperature. Therefore, LDH or total elongation is highly sensitive to temperature.

9.3.2 FRACTURE FORMING LIMIT DIAGRAM

In Figure 9.6 (a–d), solid-color diamond symbol signifies already failed ellipse on surface of deformed specimens. However, these failed strain values didn't describe onset of fracture for sheet specimens. For accurate measure strain values at onset of fracture, by volume constancy expression, as suggested [4]. It is expressed as Equation (9.7),

$$\varepsilon_{1f} + \varepsilon_{2f} + \varepsilon_{3f} = 0 \tag{9.7}$$

There is no significant lateral stretching in blank specimen after occurrence of necking. Because of extreme strain localization, specimens have been thinned along thickness direction. Thus, numerical values of necking strain (ε_{2n}) and minor fracture strain (ε_{2f}) are assumed same. As illustrated in Figure 9.12, each fractured specimen has been wire cut perpendicular to fracture line. Perpendicular distances t_{1f} & t_{2f} from beginning of fracture edge of maximum thinned cross-section, has been measured using an optical microscope. From t_{1f} and t_{2f} fracture thickness values, smallest value has been taken into account for estimation of true thickness fractured strain value (ε_{3f}). By using Equation (9.9), fractured major strain value (ε_{1f}) was computed and further fracture strain state inserted in FLD. Multiple fracture strain points (ε_{1f}, ε_{2f}) were evaluated for individual strain path. Onset of fractured strain points are represented by solid-colored square symbols in all deformed region at RT, as shown Figure 9.13 (a). A straight line was constructed just below scattered fracture points, named as FFLD. Figure 9.13 (b–c) displayed experimental FFLD along various strain paths for IN718 for all temperatures. In Figure 9.14, fracture strain values were rising with rise in temperatures in all deformation regions similar to limiting strains in FLDs. At 700°C, fractured strain values were obtained much higher than that at RT because of thermal softening. Around 65.19% and 68.91% improvement in onset of fracture true strains was observed in T–C and T–T regions at 700°C with respect to RT.

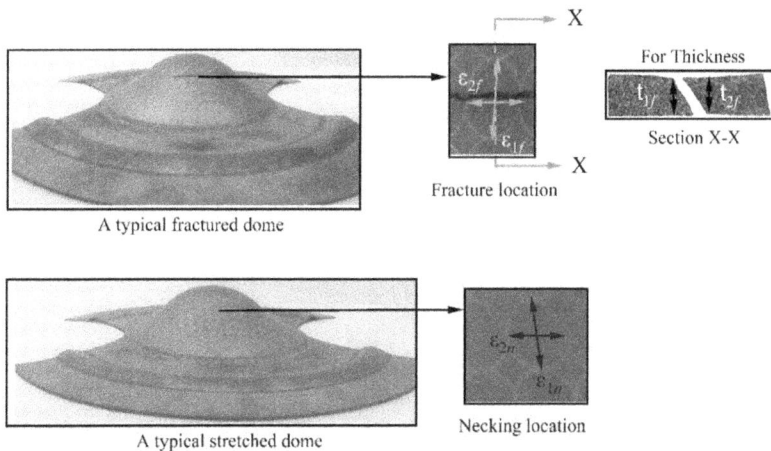

FIGURE 9.12 Fracture measurement technique along thickness direction of fractured specimen.

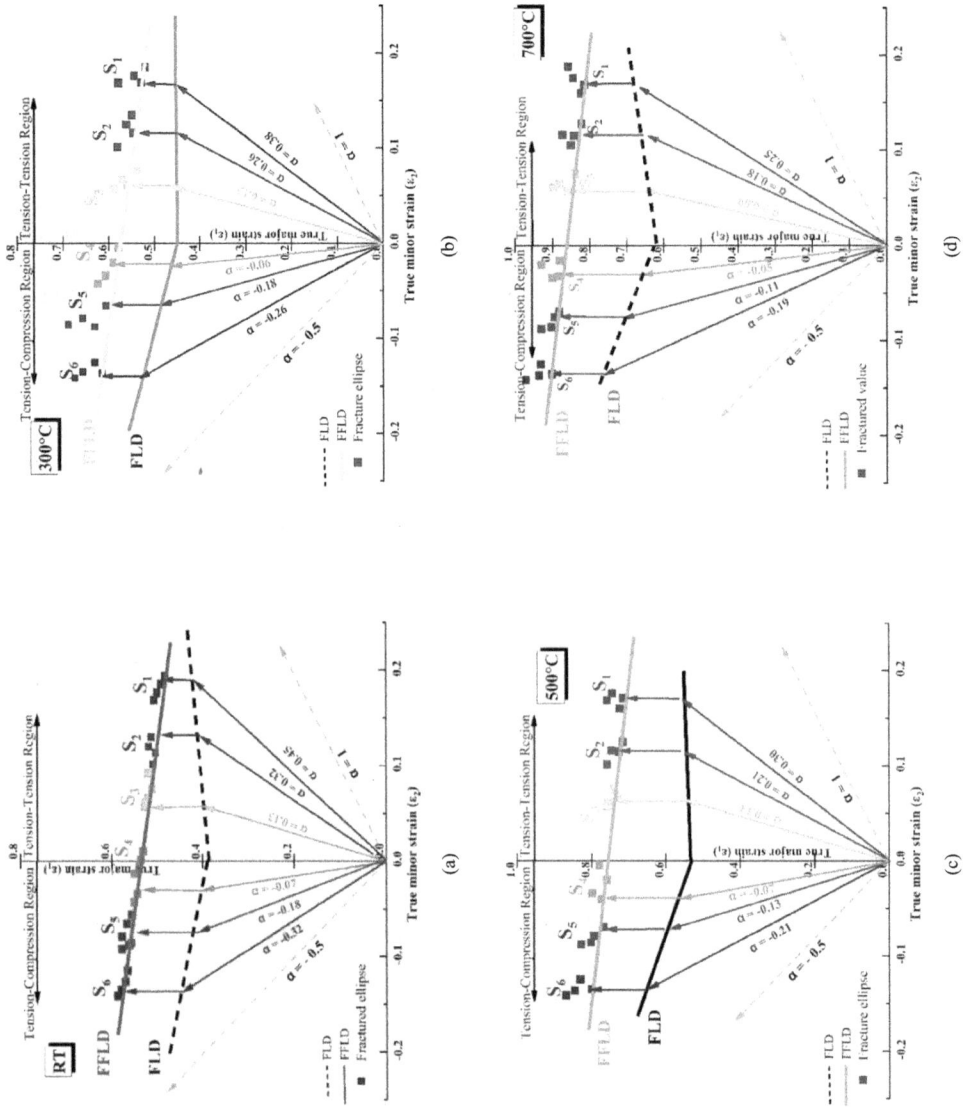

FIGURE 9.13   FFLDs for IN718 at (a) RT, (b) 300°C, (c) 500°C, and (d) 700°C.

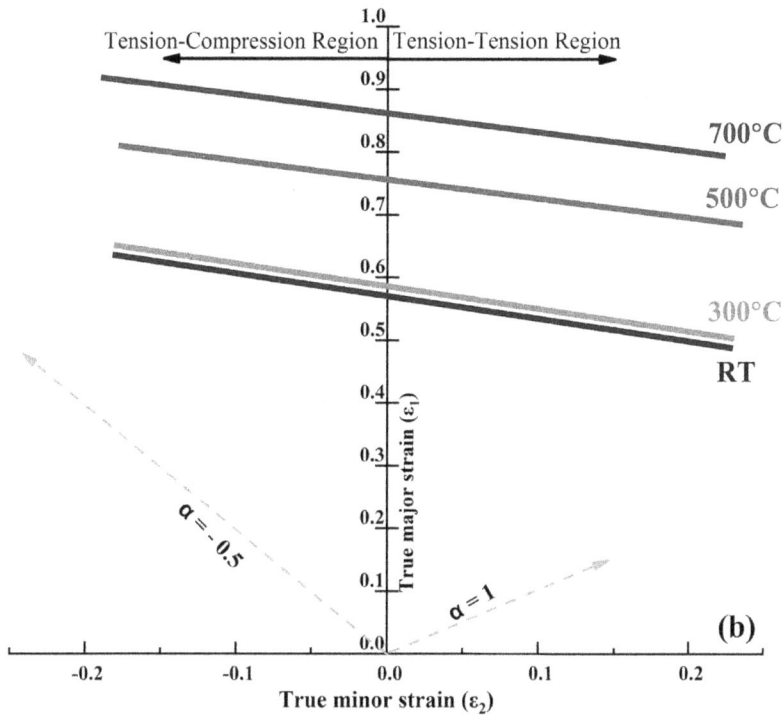

FIGURE 9.14 Forming and Fracture Limit Diagrams.

9.4 CONCLUSION

The key findings are summarized below:

(i) Limiting and fracture true strain values were perceived as increasing with rise in temperature in all deformation regions. Bending strain effect is studied, and FLDs are corrected based on the bending correction factor. Corrected FLDs are shifted downward by approximately 4–5% in all deformation regions.

(ii) Fractured forming limit diagrams (FFLDs) for IN718 at different temperature is also evaluated and significantly influenced by temperatures. Estimated forming process maps of Inconel alloy will provide guidelines for the successful fabrication of space components, especially outer casing of Ni-H$_2$ batteries and high-pressure gas bottles.

ACKNOWLEDGEMENT

The authors would like to thank the Human Resource Development Group, Council of Scientific and Industrial Research, Government of India (File no: 09/1026(0038)/2020.EMR-I), for financial support.

REFERENCES

[1] Roamer, P.; Van Tyne, C. J.; Matlock, D. K.; Meier, A. M.; Ruble, H.; Suarez, F. Room temperature formability of alloys 625LCF, 718 and 718SPF. *Miner. Met. Mater. Soc.*, 1997, 315–329. https://doi.org/10.7449/1997/superalloys_1997_315_329

[2] Mahalle, G.; Kotkunde, N.; Gupta, A. K.; Singh, S. K. Comparative assessment of failure strain predictions using ductile damage criteria for warm stretch forming of IN718 Alloy. *Int. J. Mater. Form.*, 2020, 1–14. https://doi.org/10.1007/s12289-020-01588-3

[3] Kuroda, M.; Tvergaard, V. Forming limit diagrams for anisotropic metal sheets with different yield criteria. *Int. J. Solids Struct.*, 2000, *37* (37), 5037–5059. https://doi.org/10.1016/S0020-7683(99)00200-0

[4] Mahalle, G.; Morchhale, A.; Kotkunde, N.; Gupta, A. K.; Singh, S. K.; Lin, Y. C. Forming and fracture limits of IN718 alloy at elevated temperatures: experimental and theoretical investigation. *J. Manuf. Process.*, 2020, *56*, 482–499. https://doi.org/10.1016/j.jmapro.2020.04.070

[5] Mahalle, G.; Kotkunde, N.; Gupta, A. K.; Singh, S. K. Efficacy of semi-empirical models for prediction of forming limit curve of IN718 alloy at elevated temperatures. *Adv. Mater. Process. Technol.*, 2020, 1–13. https://doi.org/10.1080/2374068X.2020.1792706

[6] Satish, D. R.; Kumar, D. R.; Merklein, M. Effect of temperature and punch speed on forming limit strains of AA5182 alloy in warm forming and improvement in failure prediction in finite element analysis. *J. Strain Anal. Eng. Des.*, 2017, *52*, 258–273. https://doi.org/10.1177/0309324717704995

[7] Wu, Y. T.; Koo, C. H. Effect of temperature on the anisotropic superplasticity of textured Ti-25Al-10Nb alloy. *Scr. Mater.*, 1997, *38* (2), 267–271. https://doi.org/10.1016/S1359-6462(97)00431-4

[8] Kuhlmann-Wilsdorf, D. Theory of plastic deformation: - properties of low energy dislocation structures. *Mater. Sci. Eng. A*, 1989, *113*, 1–41. https://doi.org/10.1016/0921-5093(89)90290-6

[9] Mahalle, G.; Kotkunde, N.; Gupta, A. K.; Singh, S. K. Cowper-symonds strain hardening model for flow behaviour of inconel 718 alloy. *Mater. Today Proc.*, 2019, *18*, 2796–2801. https://doi.org/10.1016/j.matpr.2019.07.145

[10] Bruschi, S.; Altan, T.; Banabic, D.; Bariani, P. F.; Brosius, A.; Cao, J.; Ghiotti, A.; Khraisheh, M.; Merklein, M.; Tekkaya, A. E. Testing and modelling of material behaviour and formability in sheet metal forming. *CIRP Ann. - Manuf. Technol.*, 2014, *63* (2), 727–749. https://doi.org/10.1016/j.cirp.2014.05.005

[11] Huang, L.; Shi, M. forming limit curves of advanced high strength steels: experimental determination and empirical prediction. *SAE Int. J. Mater. Manuf.*, 2018, *11* (4), 409–418. https://doi.org/10.4271/2018-01-0804

[12] Prakash, V.; Kumar, D. R.; Horn, A.; Hagenah, H.; Merklein, M. Modeling material behavior of AA5083 aluminum alloy sheet using biaxial tensile tests and its application in numerical simulation of deep drawing. *Int. J. Adv. Manuf. Technol.*, 2020, *106*, 1133–1148. https://doi.org/10.1007/s00170-019-04587-0

[13] Leu, D. K. A Simplified approach for evaluating bendability and springback in plastic bending of anisotropic sheet metals. *J. Mater. Process. Technol.*, 1997, *66* (1), 9–17. https://doi.org/10.1016/S0924-0136(96)02453-3

[14] Takuda, H.; Enami, T.; Kubota, K.; Hatta, N. Formability of a thin sheet of Mg-8.5Li-1Zn alloy. *J. Mater. Process. Technol.*, 2000, *101* (1), 281–286. https://doi.org/10.1016/S0924-0136(00)00484-2

[15] Venkateswarlu, G.; Singh, A. K.; Davidson, J.; Tagore, G. R. Effect of microstructure and texture on forming limits in friction stir processed AZ31B Mg alloy. *J. Mater. Res. Technol.*, 2013, *2* (2), 135–140. https://doi.org/10.1016/j.jmrt.2013.01.003

[16] Keeler, S.; Backofen, W. Plastic instability and fracture in sheets stretched over rigid punches, Massachusetts Institute of Technology, 1961.

[17] Goodwin, G. M. Application of strain analysis to sheet metal forming problems in the press shop. *SAE Tech. Pap.*, 1970, *62* (8), 767–774. https://doi.org/10.4271/680093

[18] Hecker, S. S. Simple technique for determining forming limit curves. *Sheet Met. Ind.*, 1975, *52* (11), 671–676.

[19] Banabic, D.; Carleer, B.; Comsa, D. S.; Kam, E.; Krasovskyy, A.; Mattiasson, K.; Sester, M.; Sigvant, M.; Zhang, X. *Sheet Metal Forming Processes: Constitutive Modelling and Numerical Simulation*; Springer, Berlin, Heidelberg, 2010.

[20] Nakazima K; Kikuma T; Hasuka K. Study on the formabilityof steel sheets. *JSME Int. Journal*, 1989, *32* (1), 142–148.

[21] Hu, Q.; Li, X.; Chen, J. Forming limit evaluation by considering through-thickness normal stress: theory and modeling. *Int. J. Mech. Sci.*, 2019, *155*, 187–196. https://doi.org/10.1016/j.ijmecsci.2019.02.026

[22] Turetta, A.; Bruschi, S.; Ghiotti, A. Investigation of 22MnB5 formability in hot stamping operations. *J. Mater. Process. Technol.*, 2006, *177* (1–3), 396–400. https://doi.org/10.1016/j.jmatprotec.2006.04.041

[23] Bong, H. J.; Barlat, F.; Lee, M.-G.; Ahn, D. C. Surface roughening of ferritic stainless steel sheets and its application to the forming limit diagram. *Steel Res. Int.*, 2012, 975–978.

[24] Anderson, D.; Butcher, C.; Pathak, N.; Worswick, M. J. Failure parameter identification and validation for a dual-phase 780 steel sheet. *Int. J. Solids Struct.*, 2017, *124*, 89–107. https://doi.org/10.1016/j.ijsolstr.2017.06.018

[25] ASTM E3047-16. Standard test method for analysis of nickel alloys by spark atomic emission spectrometry, 2018. https://doi.org/10.1520/E3047-16

[26] ASTM E2218-15. Standard test method for determining forming limit curves. In *ASTM Book of Standards*, 2015, pp 1–15. https://doi.org/10.1520/E2218-15

[27] Mahalle, G.; Kotkunde, N.; Gupta, A. K.; Singh, S. K. Analysis of hot workability of inconel alloys using processing maps. In *Advances in Computational Methods in Manufacturing*; Springer, 2019; pp 109–118. https://doi.org/10.1007/978-981-32-9072-3_10

[28] Zhang, Z.; Lu, X. Preparation process of magnesium alloys by complex salt dehydration-electrochemical codeposition. *Mater. Manuf. Process.*, 2019, *34* (6), 591–597. https://doi.org/10.1080/10426914.2019.1566616

[29] Güler, B.; Efe, M. Forming and fracture limits of sheet metals deforming without a local neck. *J. Mater. Process. Technol.*, 2018, *252*, 477–484. https://doi.org/10.1016/j.jmatprotec.2017.10.004

[30] Jackson, M. Application of novel technique to examine thermomechanical processing of near β Alloy Ti-10V-2Fe-3Al. *Mater. Sci. Technol.*, 2000, *16* (11–12), 1437–1444. https://doi.org/10.1179/026708300101507433

[31] Charpentier, P. L. Influence of punch curvature on the stretching limits of sheet Steel. *Metall. Trans. A*, 1975, *6*, 1665–1669. https://doi.org/10.1007/BF02641986

10 Tensile Properties and Anisotropy of Cross-Rolled Sheets
An Overview

Murugabalaji V., Matruprasad Rout
National Institute of Technology Tiruchirappalli, Tiruchirappalli, India

Kishore Debnath
National Institute of Technology Meghalaya, Shillong, India

CONTENTS

10.1 INTRODUCTION

Rolling is considered as one of the significant manufacturing processes, and almost all the metals and alloys are subjected to the rolling process at least once before they are made into the finished products [1]. Many products ranging from household utensils to aerospace applications are made from rolled sheets. The recent engineering applications demand lighter and safer structures, which must be manufactured economically. Sheet metal rolling is an excellent manufacturing process that can satisfy all the above-mentioned needs efficiently [2]. The plastic deformation of the sheet during the rolling process occurs between the two rollers rotating in opposing directions, and the gap between the rollers determines the final thickness of the sheet. The frictional force at the contact (between the sheet and the rollers) drag the sheet into the roll gap [3]. Owing to the dimensions of the sheet, plastic deformation during the rolling process is characterized by plane strain deformation, i.e., the amount of reduction in thickness is getting accumulated along the rolling direction (RD) whereas, along the transverse direction (TD), neither elongation nor shortening of material dimension occurs [4]. In general, the required reduction in thickness is achieved in multiple stages, and the continuous deformation along a particular direction introduces anisotropy in the rolled sheet [5].

This anisotropy arises due to the tendency of the grains to align along with preferred orientation which is the result of the plastic deformation along a particular direction. Anisotropy in the mechanical properties of the sheet is desirable for many sheet metal forming processes; however, the degree of anisotropy can be altered by controlling the thermo-mechanical process [6]. Earlier research reflects the importance of RD, i.e., deformation path on the orientation of the grains and subsequent effect on the anisotropy of the material properties [7–10]. In the present chapter, the effect of the change in RD, by 90° about the normal direction (ND), on the tensile properties and anisotropy of the metallic sheets are discussed. The discussion is made by considering the materials with the same crystal structure. Moreover, the discussion includes the highlights of the past research works and the significant results. The effect of change in RD on microstructure and crystallographic orientation (texture) is not included in this chapter. Readers interested in studying the impact of change in RD on microstructure and texture can refer to the available related articles [8, 11–15].

10.2 CROSS ROLLING

The rolling process with a change in RD by 90° about the ND is termed as the cross rolling (CR) process [3]. The change in RD is obtained by rotating the material on the rolling plane (Figure 10.1) and hence, dimensions of the material becomes a constraint for the CR process. Unlike the conventional rolling process where the material is rolled along a particular direction, the effect of CR on mechanical properties can be analyzed by considering the sequence of change in RD [8]. In addition, the parameters like reduction per pass, temperature, roll speed, coefficient of friction, etc. also influence the properties of the cross-rolled sheets.

Depending upon the sequence of change in RD, the CR process can be categorized as follows [15]; two-step cross rolling (TSCR), also known as pseudo cross rolling and multistep cross rolling (MSCR), also known as true cross rolling. In the former, the RD is changed after achieving 50% of the required reduction in thickness, whereas in the latter case, the RD is changed after each pass. On the other hand, rolling with the change in RD by 180° is termed as reverse rolling (RR), and without change in RD is termed as unidirectional rolling (UR) [15]. The schematic representation of different modes of rolling is shown in the Figure 10.2.

The CR process is in use for quite a long time; however, it still remains as a laboratory process with minimal industrial applications. Some of the applications are (i) manufacturing of non-grain oriented electrical sheets by CR process to obtain similar magnetic properties in all directions for their usage in transformers, motors, generators, etc. [12, 9], (ii) manufacturing of stainless steel knife by Bohler-Uddeholm, USA [3], (iii) producing polycrystalline shape memory alloy sheets to achieve in-plane low thermal expansion [16], (iv) manufacturing bio-absorbable Zn alloys to improve their ductility and strength for medical applications [17] etc. It can be noted that the change

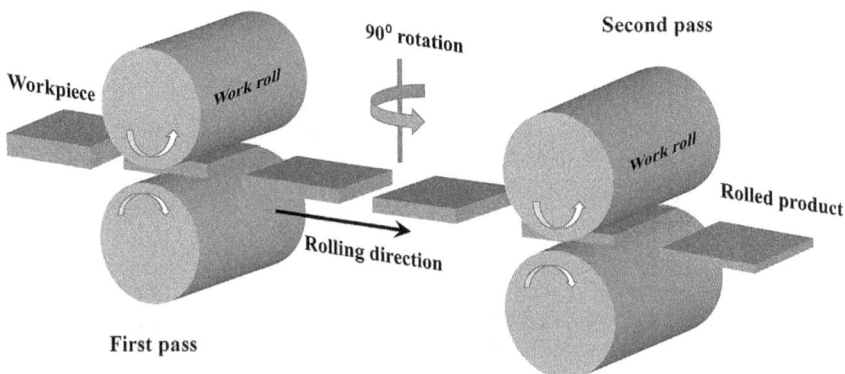

FIGURE 10.1   Schematic representation of cross rolling process.

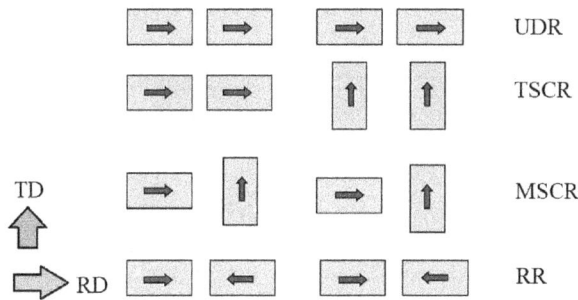

FIGURE 10.2   Schematic representation of different modes of rolling.

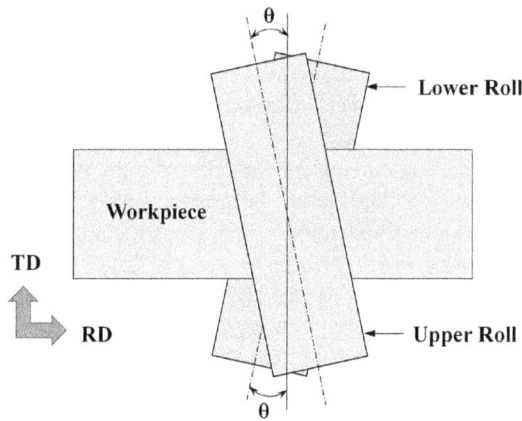

FIGURE 10.3   Schematic representation CR process by tilting of roller axis.

in RD is also employed during the initial passes of flat rolling to obtain the required width of the plate. However, in further passes the continuous deformation along a particular direction reduces the effect of change in RD [18–20]. Few researchers [21–23] have used the term cross rolling to name the rolling process where the rollers are tilted. In this process, the material is rolled by the rollers whose axes are tilted away from the TD in the RD-TD plane (Figure 10.3). However, this type of rolling process is not included in this chapter. The present chapter includes the past research works, where the formability study has been done for the sheets and the sheets are rolled by change in RD obtained through rotating the work material on the RD-TD plane by 90° about ND. The following section gives an overview of the different methods/techniques that has been used by the researchers to study formability.

10.3   METHODS FOR MEASURING FORMABILITY OF SHEETS

The one important property required for any sheet metal manufacturing process is the ability of the sheet to deform upon the application of the load and acquire the required shape without getting fractured. This property is referred to as the formability of sheet metal [24]. The formability of the metal sheets is assessed by deep drawability, stretch formability, stretch-flange formability, bend-ability etc. [25]. The formability of sheets depends upon various factors viz. crystal structure of the material, type of rolling employed, annealing temperature, microstructure, texture, etc. [26]. A sheet metal with optimal formability must distribute strain uniformly and withstand in-plane compressive and shear stresses without wrinkling and fracturing [27]. It should also reach high strain levels without necking or fracturing and at the same time should retain its shape and surface quality [27]. The various techniques like Erichsen test, conical cup test, limit dome height (LDH) test are widely performed by the researchers to measure the formability of sheet metals [28, 29, 30]. In Erichsen test, a sheet specimen is stretched by means of a hemispherical punch until it get fracture. The depth of indentation of the punch in the specimen is expressed in millimeters and is called as Erichsen index, which is considered as the measure of formability [24]. In Conical cup test, hemispherical punch and conical die are used to perform the drawing operation until the specimen get fracture. The formability is measured as the ratio of the bottom diameter of the cone at fracture point to the diameter of the work piece [24]. In LDH method, the rectangular specimens are stretched on a hemispherical punch until the specimen get fracture. The height of the parts are plotted against minimum strains occurring in the specimens and the height corresponding to plane strain condition is considered as formability index, denoted as LDH_0 [24]. In sheet metal forming process, forming limit diagram (FLD) is the widely accepted criterion to predict the failure of the sheet [30]. The sheet specimen is first etched electrochemically with circular-shaped grids, which transforms into an ellipse upon stretching. The specimen is fixed in circular die ring and stretching is done using a hemispherical punch. The engineering strains measured in major and minor axes of the ellipse closest to the fractured part of the specimen is called as major and minor strain, respectively. The plot of major and minor strains with major strain as ordinate and minor strain as abscissa with a curve fitted at strain data points is called as FLD [30].

Earing is an undesired feature occurring in deep drawn cups mainly due to the anisotropy in mechanical properties of the sheet. The normal anisotropy (properties variation along TD) influences drawability of sheets and planar anisotropy (properties variation in plane orientation of sheet) influences the earing formation during deep drawing process [24]. Hence, deep drawing test is considered as a tool to measure the anisotropy of the sheet metal. The formation of undesired wavy rims of the cup during the deep drawing operation is termed as earing. Earing forms due to the difference in radial elongation along different directions of the sheet. In most cases, four ears are formed along the directions $45°+(n \times 90°)$ and $0°+(n \times 90°)$ to the RD, as a result of rolling and recrystallization texture, respectively where "n" is an integer (n = 0, 1, 2,..) [31]. The conventional strategy followed by the researchers to measure the anisotropy in mechanical properties of the sheets, is to conduct the tensile tests along different directions of the sheet, as shown in Figure 10.4. The anisotropy is characterized by the Lankford parameter, R which is the ratio of the strains in width direction (ε_w) to thickness direction (ε_t) [14]. The average R value ($\bar{R}$) and the planar anisotropy ($|\Delta R|$) are determined as follows [32]:

$$\bar{R} = \frac{R_{0°} + 2R_{45°} + R_{90°}}{4} \qquad (10.1)$$

$$|\Delta R| = \frac{|R_{0°} - 2R_{45°} + R_{90°}|}{2} \qquad (10.2)$$

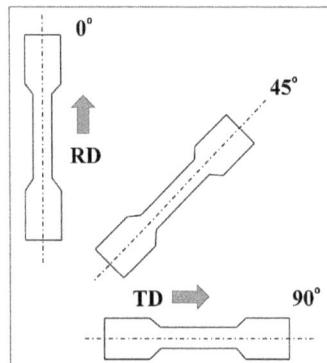

FIGURE 10.4   Tensile specimen orientation to measure tensile properties along different directions.

where, the subscripts 0°, 45° and 90° indicate the tensile specimen direction with respect to RD. The various formability measurements followed by the researchers to access the tensile properties and anisotropy of cross-rolled sheets, for different alloys, are summarized in the following sections.

10.4   TENSILE PROPERTIES AND ANISOTROPY OF CROSS-ROLLED SHEETS

10.4.1   BODY CENTERED CUBIC (BCC) MATERIALS

The high temperature creep strength, low coefficient of thermal expansion and reasonably good Young's modulus make molybdenum a suitable refractory material for industries like lighting, glass, furnace, electronics, nuclear, aerospace, etc. [14]. These versatile applications attracted researchers to analyze the formability of molybdenum sheets under various rolling routes. Oertel et al. [14] analyzed different modes of rolling for molybdenum sheets viz. change in RD by 90° (i.e., CR) at the initial and later stages of hot rolling, no change in RD (i.e., UR) and complex change in RD (about 45°, 90° and 135°). The complex route change was also followed by annealing. The calculated Lankford parameter indicates lowest value for sheet rolled with complex route with subsequent annealing, which indicates good formability and minimal chances for earing formation. In another work, Oertel et al. [33] performed the hot and cold CR and straight rolling (i.e., UR) of pressed and sintered molybdenum powders. They employed Taylor–Bishop–Hill theory of crystal plasticity to calculate Taylor factor (M) and Lankford parameter (R). The full and relaxed constraints models were used for the Taylor factor calculation. The minimum M was obtained for the cold cross-rolled sheets indicating minimum deformation energy. The R values were found to be smaller for cross-rolled sheets compared to the straight-rolled sheets.

The work of Wang et al. [34] includes the hot CR of Mo billets made from Mo powder of size 3 μm. The rolling details used are shown in the Figure 10.5. After rolling, all the plates were annealed for one hour at 800°C. They found an increase in hardness value of plates which was attributed to

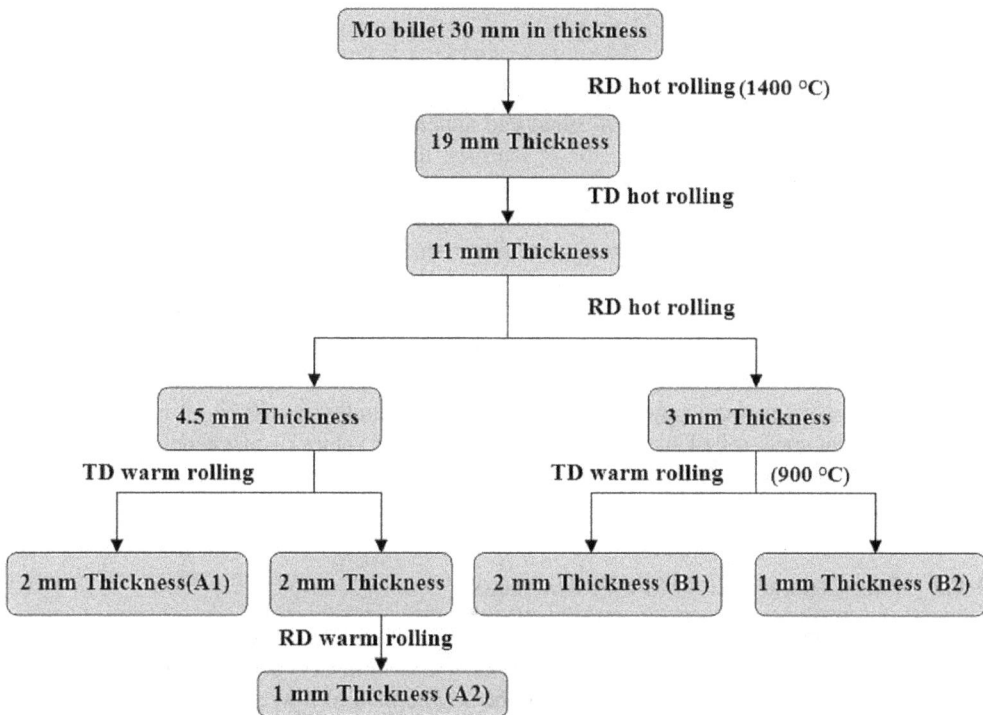

FIGURE 10.5   Routes of CR followed by Wang et al. [34].

the work hardening effect with high degree of deformation. They found the existence of slight anisotropy in all samples with the highest elongation and lowest tensile strength along 45° direction and lowest elongation and highest tensile strength along the TD. They have calculated the in-plane anisotropy (IPA) index, one of the characteristic indexes of anisotropy by adopting the following equation (Equation (10.3)) from the work of Jata et al. [35] and Singh et al. [7].

$$IPA = \left\{ \frac{\left[(N-1)X_{\max} - X_{mid1} - X_{mid2} - LX_{mid(N-2)} - X_{\min} \right]}{\left[(N-1)X_{\max} \right]} \right\} \times 100\% \qquad (10.3)$$

where, N is the number of sampling directions, and X_{max}, X_{min} and X_{mid} are the maximum, minimum, and intermediate values of tensile strength (σ_b) in the sampling direction, respectively. They obtained relatively low IPAs (< 10%) for all samples, indicating that CR can effectively reduce anisotropy of tensile strength in Mo sheets. The IPAs of samples A1 and B2 are comparatively higher due to the higher deformation in the last steps of rolling processes which led to strength enhancement in different directions [34]. The β-Titanium alloys are known for their high strength, corrosion resistance, and plasticity. The plasticity of these alloys can be improved by rolling and recrystallization through heat treatment which aids in the refinement of the β phase [36]. Ma et al. [37] conducted cold UR and TSCR on β-Titanium alloy produced by forging at 950°C followed by the hot rolling at 850°C to produce 2 mm thick sheet. The sheets were solution treated at 800°C (below the β transition temperature 815 ± 10°C) for 30 minutes. The tensile test result of TSCR sheets reveals similar elongation and tensile properties in different directions whereas, a decrease in plasticity with the increase in angle between the RD and tensile direction was observed for the UR sheets [37].

Ferrous alloys are the primary contributing material for vast engineering applications due to their remarkable properties. Pure iron, which has BCC structure, attracted the interest of researchers after its first usage for cardio-vascular applications at the beginning of the 21st century [38]. The cross-rolled pure iron sheets, after annealing at 550°C, exhibit improved ductility than unidirectional rolled sheets as reported by Obayi et al. [39]. They also found enhanced corrosion resistance and bio degradation properties in cross-rolled and annealed samples. Their studies show that CR samples recrystallize at slower rate than UR samples, thereby increasing the ductility of the CR samples after annealing. The BCC ferrous alloys possess magnetic property and are widely used in electrical types of machinery like transformers, generators, motors, etc. The anisotropy of magnetic properties is of major concern in those applications. So, research works [12, 9] are focused on reducing the anisotropy of magnetic properties rather than anisotropy of mechanical properties. Conventionally non-oriented electrical steel with less than 4.5 wt. % of Si are cold rolled, but when Si percentage is higher than 4.5 wt. %, it exhibits poor formability at room temperature. So, Xu et al. (2020) performed the UR and TSCR at 650°C and 450°C on electrical steel with 6.5 wt. % of Si. Further, the rolled sheets were annealed at 1000°C for ten minutes in a 100% nitrogen atmosphere. They reported that cross-rolled sheets possess slightly higher values of magnetic flux density than UR sheets. They also conclude that the anisotropy of the magnetic properties reduces in the samples after the annealing process [9].

10.4.2 FACE CENTERED CUBIC (FCC) MATERIALS

Aluminum alloys are generally classified as heat treatable alloys (e.g., grades 2xxx, 6xxx, and 7xxx) and work hardenable alloys (e.g., grades 1xxx, 3xxx, 4xxx, and 5xxx) depending upon the methods used for strengthening the alloy [40]. The strengthening method like precipitation hardening cannot be employed for work hardenable alloys [40]. On the other hand, the heat treatable alloys of 2xxx and 7xxx grades possess high toughness and fatigue characteristics and hence, are widely used for aerospace applications [40]. The superplastic deformation using the rolling process is an effective option to further increase the strength and formability of the alloys by developing ultrafine grains (UFG) [41]. Literature showing the enhancement of formability of aluminum alloys by the CR process are presented in this section.

Li et al. [42] showed that the warm CR followed by annealing of the AA7075 alloy improves the mechanical anisotropy and corrosion resistance of the alloy. They have worked on an alloy of 10 mm thickness produced by the hot extrusion process and processed it further by reducing its thickness by 70% through the CR process. The extruded billets are preheated to 235°C and held for five minutes before CR. Both the extruded and cross-rolled samples were annealed at different temperatures (300 and 400°C). The hot extruded sample without annealing showed higher anisotropy in tensile properties, whereas the CR with subsequent annealing improved the uniformity of tensile strength along different directions. They found that this improvement is due to the weakened brass texture and strengthened cube texture after the annealing process. They plotted the potentiodynamic polarization curves (E-I behavior of Al alloys) to study the corrosion resistance of different samples and found that annealed samples exhibited better corrosion resistance. They explained that the corrosion resistance in annealed samples was due to the consumption of distortion energy, reduction in the number of crystal defects, and increase in the fraction of low-Σ coincident site lattice (CSL) grain boundaries after annealing [42]. Mondal et al. [43] performed CR of AA7010 alloy using three different modes in which the RD and TD directions are changed once, twice, and thrice. Before the rolling experiment, the samples were homogenized at 465°C for 30 hours and the hot CR was performed after soaking the samples at 430°C for three hours. The rolled samples were solution treated and peak aged separately. Solution treatment was done at 465°C for 1.5 hours whereas, peak aging was done at 100°C for eight hours and at 120°C for 24 hours. They have evaluated the tensile properties of as rolled, solution treated and peak aged samples and found that the peak aged samples possess higher strength due to aging. However, the IPA of peak aged samples was low. They also conclude that the effect of sample orientation on work hardening behavior of the material is minimal. Nayan et al. [44] performed UR and TSCR experiments to study the effect of CR on mechanical properties of Al-Cu-Li (AA2195) alloy. The rolled samples were further heat treated for recrystallization, solutionization, and aging. They found that the cross-rolled samples of higher strength and more isotropic than UR samples. They also studied the influence of precipitates on anisotropy and found that the deterioration in anisotropy is due to the average precipitation strengthening factor due to aging.

As mentioned earlier, earing formation in the deep drawing process is not desirable. Hence, process optimization is required to reduce the ear formation and hence the loss of material. In this context, the earing prediction is of great advantage as it avoids performing a series of mechanical tests. Benke et al. (2020) conducted UR and CR on Al 5056 alloy and developed a method to predict the earing value directly from the {h 0 0} pole figures measured by tilting the X-ray diffractometer. The measured heights of the deep drawn cups and the angle of tilt (CHI) were used to develop the model which gives the qualitative information about the earing [45]. They observed a good agreement between the experimental and predicted results. In comparison to UR sheets, the CR sheets show weaker earing formation [46]. Schweitzer et al. [47] conducted experiments on UR and CR of AW 5056 aluminum alloy sheets. The rolled sheets were further heat treated at 320°C for two hours to obtain a fully recrystallized structure. They predicted the earing using the pole figures and found negligible benefits of CR.

The other FCC material on which the CR process has been implemented is the austenitic stainless steel. This non-magnetic grade of stainless steel possesses superior strength at elevated temperatures and excellent corrosion/oxidation resistance and is widely used in various fields like kitchen utensils, nuclear industries, chemical industries, etc. [48]. Research works [49–51] have been performed on austenitic stainless steels with different routes of rolling. However, most of them are on the study of microstructure and texture evolution. Few works have been performed to study the effect of rolling routes on formability. Rout [8] reported that the variation of tensile properties along different direction can be studied by analyzing the Taylor factor and the number of active slip systems affect the tensile properties of the 304 austenitic stainless steel processed through RR and CR processes. For the 70% reduction in thickness, CR shows lesser anisotropy (in terms of average R and ΔR) in comparison to RR [8]. The work of Raab et al. [52] on different rolling routes for AISI304L austenitic stainless steel sheets subjected to heat treatment for one hour at 1050°C and water cooled revealed that the strength values of samples processed through UR, CR, and RR are relatively close to each other. However, the RR sheets exhibit the highest strength value with strain degree of 70% due to smaller structural elements. On the other hand, the CR sheets showed the highest ductility [52].

10.4.3 HEXAGONAL CLOSED PACK (HCP) MATERIALS

The magnesium alloys have a wide application because of their light weight and good specific strength. However, these alloys exhibit poor formability at room temperature which lead to many researchers to work on these materials to improve their formability [53]. The poor formability at room temperature is mainly due to the presence of a limited number of active slip systems for the HCP structure [54]. Kaya et al. [55] observed significant improvement in the formability of Mg alloys at elevated temperatures compared to room temperature. Similarly, good stretchability and drawability for AZ31 Mg alloys can be observed in the temperature range of 150°C to 300°C, as reported by Chen et al. [30]. Zhang et al. [56] investigated the formability of Mg alloys sheets and found that extrusion and CR are suitable methods for producing high quality Mg sheets. They performed deep drawing tests and found that the extruded Mg alloy sheets exhibited good formability in the temperature range of 250–350°C, whereas extruded and rolled sheets had good formability at 105–170°C [56]. An improvement in formability, at low temperatures, by the CR process has also been reported by Xu et al. [57]. The improvement in the formability of AZ31 Mg alloy was attributed to the grain refinement caused by the CR process.

Chino et al. [28] studied the press formability of AZ31 Mg alloy sheet rolled by UR and CR processes. The material was heat treated at 400°C for 24 hours before rolling and annealed at 400°C for 30 minutes after the rolling process. Erichsen tests carried out at 220°C, 240°C, and 260°C revealed higher Erichsen values for CR samples at 220°C and 240°C. They conclude that the higher Erichsen values for the CR samples was due to the reduction of (0002) texture intensity by the change in RD. At 260°C, the values are almost identical, indicating that texture control has more effect on formability at lower temperatures around 220°C [28]. Their results agree with the relationship developed between Erichsen value and texture of AZ31 Mg alloy sheets by Iwanaga et al. [29]. The Erichsen value is minimum for the sheet with (0002) texture and maximum for the sheet with $(10\bar{1}1)$ texture. In another work, Chino et al. [58] conducted the RR along with the UR and CR for the same material and found that the RR and CR exhibited more press formability. They found that press formability of the sheets does not depend upon the elongation to failure, strain hardening exponent, average R value and planar anisotropy value ΔR. They suggested that the minor texture formation for the RR and CR resulted in reduced anisotropy of strain which results in superior press formability [58]. A comparison of the press formability of AZ31 Mg alloy processed by rotating the sample by 90° about ND (i.e., CR) and by tilting the rolling axis by 7.5° against TD on RD-TD plane was carried out by Chino et al. [59]. The formability of the rolled sheets measured by Erichsen test at temperatures of 433 K, 473 K and 493 K shows higher Erichsen values for the CR samples obtained by tilting the rolling axis. The high press formability of CR samples obtained through the rotation of the sample by 90° about ND was attributed to the reduction in (0002) texture intensity. The CR samples, obtained with tilted axis, produced further enhanced grain refinement in addition to texture intensity reduction [59]. Weaker basal texture and formation of more scattered rotated grains in sheet plane leads to lower anisotropy in mechanical properties for CR than UR samples of Mg–0.6%Zr–1.0%Cd sheets [60]. In addition to the routes like UR and CR, Zhang et al. [11] also studied the effect of change in RD by 45° after each CR path (i.e., RD ➔ TD ➔ 45° to TD). They found that for AZ31 Mg alloy, CR and CR with change in RD by 45° are very effective in yielding weakened basal texture and grain refinement. Also, the Erichsen test results indicated an increase in Erichsen value by 28% and 31%, respectively.

10.4.4 DUAL PHASE MATERIALS

Dual phase materials are characterized by two distinct phases e.g., the dual phase (DP) steel has ferrite as well as martensite present in it. The mechanical properties of DP steel mainly depends on the volume fraction of the martensite [61]. In general, the refinement of ferrite grains, volume fraction and distribution of martensite and heating rate determines the strain hardenability and fracture mechanism of DP steels [62]. Hence, the mechanical properties of the DP steels can be altered by introducing the CR process into the thermo-mechanical processing schedule. Soleimani and Mirzadeh [63] conducted UR and CR experiments, for 45% reduction in thickness, on austenitized and water quenched sheets of a low carbon 0.12 C- 1.11Mn-0.16Si (wt. %) steel. Both UR and CR sheets were inter-critically annealed at 850°C for holding time up to 20 minutes. A different set of samples (both UR and CR) were heated from room temperature to 850°C, at a heating rate of 0.08°C/s and held at that temperature for two minutes. All the samples were water quenched. The hardness and tensile properties study indicated that the CR samples are harder than UR samples owing to the grain refinement due to the activation of more slip systems during CR. The UR and two minutes annealed sample has the highest tensile strength of 470 MPa due to the change in dislocation density of the sample. The cross-rolled and ten-minute-annealed sample was found to have good ductility and isotropic properties due to ferrite grain refinement and homogenous distribution of martensite islands. They established that CR with inter-critical annealing is the best possible way for enhancement of mechanical properties of DP steel [63]. The formability studies on dual phase materials are not explored much. Only a few research works on microstructure and texture studies on duplex stainless steel have been performed by the researchers [64–67].

10.4.5 BIO-COMPATIBLE ALLOYS

Zinc and Titanium based alloys are primarily employed in bio-medical applications due to their significant mechanical properties like high elastic limit to density ratio, corrosion resistance, and bio-compatibility. The bio-medical alloys free from nickel are preferable as nickel leads to cytotoxicity and hypersensitivity to human cells [68]. The stretch formability is one of the desirable properties for these alloys. The research work carried out by Liu et al. [10] on pure titanium sheets by performing rolling in three different routes viz. two UR along RD and TD directions and the CR. They found that CR sheets developed isotropic properties in tensile test when loading along different directions (RD, 45° and TD). The deep drawing test and Erichsen tests also showed satisfactory results for the CR samples when compared to the UR samples. The improved properties in CR are due to the weakened texture during CR. Ramirez-Ledesma et al. [17] investigated the influence of CR on Zinc based bio-absorbable alloys. They found that the CR samples showed an improvement in ductility as the CR allows enormous grain growth. In contrast, the UR provides higher grain refinement and strength, resulting in less isotropic microstructural characteristics than CR samples. Dominguez-Contreras et al. [69] compared changes occurred in mechanical behavior during the 80% rolling reductions using UR and CR of Zn–10.0Ag–1.0Mg alloy sheets. The sheets were reheated for 400°C for ten minutes before each rolling pass to obtain a homogeneous microstructure and to eliminate the presence of residual stresses. The samples were then subjected to tensile test at room temperature. They found that UR samples obtained the highest values of yield strength and ultimate tensile strength, 368.34 ± 38.34 MPa and 442.77 ± 31.46 MPa respectively, and the highest ductility of 31.51 ± 7.82% was obtained by CR. They concluded that the improvement in mechanical properties is due to the solid solution strengthening, precipitation strengthening, and strain hardening occurred during the rolling process.

The above-mentioned CR processes are performed at hot, warm or cold stage which may or may not involve subsequent annealing process. The other rolling process, considering the temperature of the work material, is the cryogenic rolling where the material is rolled at cryogenic temperature. Though cryogenic rolling does not have much industrial applications like hot rolling or cold rolling, still it emerged as an important research area for the researchers working in the metallurgical and materials engineering domain. This process has got much attention as it produces material with UFG. The following section presents the work reported on the CR process where the material is rolled under cryogenic environment.

10.4.6 CRYOGENIC CROSS-ROLLED SHEETS

The UFG structure developed by the cryogenic rolling significantly improves the mechanical strength and hardness of the alloy [70]. The thermo-mechanical processing at cryogenic environment arrests the dynamic recovery leading to refinement of microstructure [71]. Selvan et al. [72] investigated the role of deformation path during cryogenic rolling of Al-4% Cu alloy sheets. The sheets were rolled to two different thickness reduction of 50% and 75% by both UR and CR routes. The samples were placed in liquid nitrogen (LN_2) in a Dewar flask for 30 minutes before rolling and for five minutes between the subsequent passes to retain the cryogenic temperature of -195°C. They found that CR sheets exhibit higher values of mechanical strength in terms of both tensile and yield stresses compared to UR sheets. The CR sample rolled to 50% reduction in thickness showed the highest micro-hardness value of 168 HV. The reduction in mechanical properties, at higher strains, in CR sheets is due to the dynamic recovery whereas, in UR sheets higher strains produced grain refinement [72]. The UR and TSCR of Al-Mg-Sc alloy at room temperature and at cryogenic temperature were carried out by Vigneshwaran et al. [73] to study the formability of the sheet. During tensile deformation, the TSCR sample processed at cryogenic temperature showed a strength value of 423 MPa, whereas the TSCR sample rolled at room temperature revealed 378 MPa. They constructed the combined fracture and forming limit diagram and identified that the TSCR at cryogenic temperature has inferior formability compared to the TSCR at room temperature [73].

10.5 SUMMARY

From the literature, it is evident that CR is a promising technique to alter the formability irrespective of the crystal structure of the material. The key ideas from the literature can be summarized as follows:

- The BCC metals and alloys like pure iron, non-oriented silicon steel and molybdenum sheets are found to exhibit improved biomedical properties, magnetic properties and anisotropy of mechanical properties, respectively due to the CR and subsequent annealing process. There is a significant increase in hardness of Mo sheets due to work hardening effect during CR. Similarly, the β-Titanium alloys are found to exhibit an increase in plasticity after CR and solution heat treatment due to the grain refinement of the β phase.
- The FCC metals particularly aluminum alloys exhibited better anisotropy of mechanical properties due to the formation of randomized textures. Austenitic Stainless steel 304 rolled by CR mode exhibited more elongation in the direction of 45° to RD due to the higher number of slip systems activated. The cryogenic CR of aluminum alloys indicated an increase in mechanical strength and hardness due to the arrest of dynamic recovery and development of UFG. The HCP alloys, mainly Mg alloys are found to exhibit good formability during warm and hot CR compared to room temperature rolling. The improvement in formability was attributed due to basal texture and grain refinement in Mg alloys.
- Bio-compatible alloys are found to possess increased ductility after CR due to the grain growth and also exhibited an increase in isotropy. The DP steels also got beneficial results due to CR. The improvement in ductility and isotropic properties were found to be due to ferrite grain refinement and homogenous distribution of martensite islands.

10.6 CONCLUDING REMARKS

The chapter presents an overview of the work carried out on the tensile properties and anisotropy study of the cross-rolled sheets. The CR process has been studied by the researchers from many decades. However, most of the works are on the development of microstructure and texture. It is evident that the change in rolling path significantly affect the grain orientation and hence the studies made were on texture development. It is also known that the grains orientation affect the formability of the metallic sheets, which brings researchers to work on the formability of the sheets processed by CR. It is in the last few years, significant amount of work has been done in this field to improve formability or strength-ductility ratio of the material. However, further work on accessing the formability of cross-rolled sheets are still necessary as most of the work done till date are based on the tensile test.

REFERENCES

[1] S. Spuzic, K. N. Strafford, C. Subramanian, and G. Savage, "Wear of hot rolling mill rolls: an overview," *Wear*, vol. 176, no. 2, pp. 261–271, 1994, doi:10.1016/0043-1648(94)90155-4.

[2] J. H. Li, D. Y. He, X. T. Zheng, and G. L. Ding, "Annealing effect for cold rolling 6016 aluminum alloy sheet," *Key Eng. Mater.*, vol. 723, pp. 37–43, 2017, doi:10.4028/www.scientific.net/KEM.723.37.

[3] M. Rout, S. K. Pal, and S. B. Singh, "Cross rolling: a metal forming process," pp. 41–64, 2015, doi:10.1007/978-3-319-20152-8_2.

[4] H. Klein, H. J. Bunge, and A. Bocker, "Development of cross-rolling l Ma I," vol. 12, pp. 155–174, 1990.

[5] D. Rahmatabadi, R. Hashemi, M. Tayyebi, and A. Bayati, "Investigation of mechanical properties, formability, and anisotropy of dual phase Mg-7Li-1Zn," *Mater. Res. Express*, vol. 6, no. 9, pp. 0–12, 2019, doi:10.1088/2053-1591/ab2de6.

[6] O. Engler, "Control of texture and earing in aluminium alloy AA 3105 sheet for packaging applications," *Mater. Sci. Eng. A*, vol. 538, pp. 69–80, 2012, doi:10.1016/j.msea.2012.01.015.

[7] R. K. Singh, A. K. Singh, and N. E. Prasad, "Texture and mechanical property anisotropy in an Al-Mg-Si-Cu alloy," *Mater. Sci. Eng. A*, vol. 277, no. 1–2, pp. 114–122, 2000, doi:10.1016/S0921-5093(99)00549-3.

[8] M. Rout, "Texture-tensile properties correlation of 304 austenitic stainless steel rolled with the change in rolling direction," *Mater. Res. Express*, vol. 7, no. 1, 2020, doi:10.1088/2053-1591/ab677c.

[9] H. Xu et al., "Two-stage warm cross rolling and its effect on the microstructure, texture and magnetic properties of an Fe-6.5 wt% Si non-oriented electrical steel," *J. Mater. Sci.*, vol. 55, no. 26, pp. 12525–12543, 2020, doi:10.1007/s10853-020-04861-7.

[10] D. Liu, G. Huang, G. Gong, G. Wang, and F. Pan, "Influence of different rolling routes on mechanical anisotropy and formability of commercially pure titanium sheet," *Trans. Nonferrous Met. Soc. China*, vol. 27, no. 6, pp. 1306–1312, 2017, doi:10.1016/S1003-6326(17)60151-1.

[11] H. Zhang, G. Huang, H. Jørgen, L. Wang, and F. Pan, "Influence of different rolling routes on the microstructure evolution and properties of AZ31 magnesium alloy sheets," *Mater. Des.*, vol. 50, pp. 667–673, 2013, doi:10.1016/j.matdes.2013.03.053.

[12] J. Mishra, S. Sahni, R. Sabat, V. D. Hiwarkar, and S. K. Sahoo, "Effect of cross-rolling on microstructure, texture and magnetic properties of non-oriented electrical steels," *Mater. Res.*, vol. 20, no. 1, pp. 218–224, 2017, doi:10.1590/1980-5373-MR-2016-0437.

[13] S. Mishra, M. Kumar, and A. Singh, "Evolution of rotated Brass texture by cross rolling: implications on formability evolution of rotated Brass texture by cross rolling: implications on formability," 2020, doi:10.1080/02670836.2020.1773036.

[14] C. G. Oertel et al., "Influence of cross rolling and heat treatment on texture and forming properties of molybdenum sheets," *Int. J. Refract. Met. Hard Mater.*, vol. 28, no. 6, pp. 722–727, 2010, doi:10.1016/j.ijrmhm.2010.07.003.

[15] S. Suwas and N. P. Gurao, *Development of Microstructures and Textures by Cross Rolling*, vol. 3, no. 30. Elsevier, 2014.

[16] Q. Li, Z. Deng, Y. Onuki, W. Wang, L. Li, and Q. Sun, "In-plane low thermal expansion of NiTi via controlled cross rolling," *Acta Mater.*, vol. 204, p. 116506, 2021, doi:10.1016/j.actamat.2020.116506.

[17] A. L. Ramirez-Ledesma, L. A. Domínguez-Contreras, J. A. Juarez-Islas, C. Paternoster, and D. Mantovani, "Influence of cross-rolling on the microstructure and mechanical properties of Zn bioabsorbable alloys," *Mater. Lett.*, vol. 279, no. December, 2020, doi:10.1016/j.matlet.2020.128504.

[18] C. H. Moon and Y. Lee, "The effects of rolling method changes on productivity in thick plate rolling process," *J. Mater. Process. Tech.*, vol. 210, no. 14, pp. 1844–1851, 2010, doi:10.1016/j.jmatprotec. 2010.06.018.

[19] J. H. Ruan et al., "3D FE modelling of plate shape during heavy plate rolling 3D FE modelling of plate shape during heavy plate rolling," vol. 9233, 2014, doi:10.1179/1743281213Y.0000000119.

[20] T. Zhang, B. Wang, and Z. Wang, "Side-surface shape optimization of heavy plate by large temperature gradient rolling," *ISIJ International*, vol. 56, no. 1, pp. 179–182, 2016.

[21] D. G. Kim, H. T. Son, D. W. Kim, Y. H. Kim, and K. M. Lee, "Effect of cross-roll angle on microstructures and mechanical properties during cross-roll rolling in AZ31 alloys," *Mater. Trans.*, vol. 52, no. 12, pp. 2274–2277, 2011, doi:10.2320/matertrans.M2011260.

[22] Y. Chino, K. Sassa, A. Kamiya, and M. Mabuchi, "Stretch formability at elevated temperature of a cross-rolled AZ31 Mg alloy sheet with different rolling routes," *Mater. Sci. Eng: A*, vol. 473, pp. 195–200, 2008, doi:10.1016/j.msea.2007.05.109.

[23] Y. Chino, Æ. K. Sassa, and Æ. M. Mabuchi, "Enhanced stretch formability of Mn-free AZ31 Mg alloy rolled by cross-roll rolling," pp. 1821–1827, 2009, doi:10.1007/s10853-009-3248-7.

[24] D. Banabic, "(Engineering Materials) H.J. Bunge, K. Pöhlandt, A.E. Tekkaya, D. Banabic, D. Banabic, Klaus Pöhlandt - Formability of Metallic Materials_ Plastic Anisotropy, Formability Testing, Forming Limits-Sprin.pdf," *Engineering Materials*. pp. xv, 334 p., 2000, [Online]. Available: https://link-springer-com. pbidi.unam.mx:2443/book/10.1007/978-3-662-04013-3%0A; http://link.springer.com/10.1007/978-3-662-04013-3_7%0A; http://link.springer.com/10.1007/978-3-662-04013-3.

[25] H. Inoue, "Simultaneous prediction of bendability and deep drawability based on orientation distribution function for polycrystalline cubic metal sheets," *Mater. Sci. Forum*, vol. 941, pp. 1468–1473, 2018, doi:10.4028/www.scientific.net/MSF.941.1468.

[26] A. Ghosh, A. Roy, A. Ghosh, and M. Ghosh, "Influence of temperature on microstructure, crystallographic texture and mechanical properties of EN AW 6016 alloy during plane strain compression," *Mater. Today Commun.*, vol. 26, no. November 2020, p. 101808, 2021, doi:10.1016/j.mtcomm.2020.101808.

[27] S. L. Semiatin, *Metalworking: Sheet Forming*, vol. 14, 2018.

[28] Y. Chino, J. S. Lee, K. Sassa, A. Kamiya, and M. Mabuchi, "Press formability of a rolled AZ31 Mg alloy sheet with controlled texture," *Mater. Lett.*, vol. 60, no. 2, pp. 173–176, 2006, doi:10.1016/j. matlet.2005.08.012.

[29] K. Iwanaga, H. Tashiro, H. Okamoto, and K. Shimizu, "Improvement of formability from room temperature to warm temperature in AZ-31 magnesium alloy," *J. Mater. Process. Technol.*, vol. 155–156, no. 1–3, pp. 1313–1316, 2004, doi:10.1016/j.jmatprotec.2004.04.181.

[30] F. K. Chen and T. Bin Huang, "Formability of stamping magnesium-alloy AZ31 sheets," *J. Mater. Process. Technol.*, vol. 142, no. 3, pp. 643–647, 2003, doi:10.1016/S0924-0136(03)00684-8.

[31] P. Van Houtte, G. Cauwenberg, and E. Aernoudt, "Analysis of the earing behaviour of aluminium 3004 alloys by means of a single model based on yield loci calculated from orientation distribution functions," *Mater. Sci. Eng.*, vol. 95, no. C, pp. 115–124, 1987, doi:10.1016/0025-5416(87)90503-9.

[32] R. Narayanasamy and C. S. Narayanan, "Forming, fracture and wrinkling limit diagram for if steel sheets of different thickness," *Mater. Des.*, vol. 29, no. 7, pp. 1467–1475, 2008, doi:10.1016/j.matdes.2006.09.017.

[33] C. Oertel, I. Huensche, W. Skrotzki, W. Knabl, A. Lorich, and J. Resch, "Plastic anisotropy of straight and cross rolled molybdenum sheets," vol. 484, pp. 79–83, 2008, doi:10.1016/j.msea.2007.03.107.

[34] D. Z. Wang, Y. X. Ji, and Z. Z. Wu, "Effects of cross rolling on texture, mechanical properties and anisotropy of pure Mo plates," *Trans. Nonferrous Met. Soc. China (English Ed.)*, vol. 30, no. 8, pp. 2170–2176, 2020, doi:10.1016/S1003-6326(20)65369-9.

[35] K. V. Jata, A. K. Hopkins, and R. J. Rioja, "The anisotropy and texture of Al-Li alloys," *Mater. Sci. Forum*, vol. 217–222, no. PART 1, pp. 647–652, 1996, doi:10.4028/www.scientific.net/msf.217-222.647.

[36] P. E. Markovsky, V. I. Bondarchuk, and O. M. Herasymchuk, "Influence of grain size, aging conditions and tension rate on the mechanical behavior of titanium low-cost metastable beta-alloy in thermally hardened condition," *Mater. Sci. Eng. A*, vol. 645, pp. 150–162, 2015, doi:10.1016/j.msea.2015.08.009.

[37] Y. Ma et al., "Effect of cold rolling process on microstructure and mechanical properties of high strength β titanium alloy thin sheets," *Prog. Nat. Sci. Mater. Int.*, vol. 28, no. 6, pp. 711–717, 2018, doi:10.1016/j. pnsc.2018.10.004.

[38] M. Peuster et al., "A novel approach to temporary stenting: degradable cardiovascular stents produced from corrodible metal - Results 6-18 months after implantation into New Zealand white rabbits," *Heart*, vol. 86, no. 5, pp. 563–569, 2001, doi:10.1136/heart.86.5.563.

[39] C. S. Obayi et al., "Influence of cross-rolling on the micro-texture and biodegradation of pure iron as biodegradable material for medical implants," *Acta Biomater.*, vol. 17, pp. 68–77, 2015, doi:10.1016/j. actbio.2015.01.024.

[40] P. Rambabu, N. E. Prasad, and V. V. Kutumbarao, "Aluminium alloys for aerospace applications," 2017, doi:10.1007/978-981-10-2134-3.

[41] S. K. Panigrahi and R. Jayaganthan, "Effect of rolling temperature on microstructure and mechanical properties of 6063 Al alloy," *Mater. Sci. Eng. A*, vol. 492, no. 1–2, pp. 300–305, 2008, doi:10.1016/j. msea.2008.03.029.

[42] Z. Li, L. Chen, J. Tang, W. Sun, G. Zhao, and C. Zhang, "Improving mechanical anisotropy and corrosion resistance of extruded AA7075 alloy by warm cross rolling and annealing," *J. Alloys Compd.*, vol. 863, p. 158725, 2021, doi:10.1016/j.jallcom.2021.158725.

[43] C. Mondal, A. K. Singh, A. K. Mukhopadhyay, and K. Chattopadhyay, "Effects of different modes of hot cross-rolling in 7010 aluminum alloy: Part II. mechanical properties anisotropy," *Metall. Mater. Trans. A Phys. Metall. Mater. Sci.*, vol. 44, no. 6, pp. 2764–2777, 2013, doi:10.1007/s11661-013-1678-y.

[44] N. Nayan, S. Mishra, A. Prakash, S. V. S. N. Murty, M. J. N. V. Prasad, and I. Samajdar, "Effect of cross-rolling on microstructure and texture evolution and tensile behavior of aluminium-copper-lithium (AA2195) alloy," *Mater. Sci. Eng. A*, vol. 740–741, no. August 2018, pp. 252–261, 2019, doi:10.1016/j. msea.2018.10.089.

[45] M. Benke, A. Hlavacs, I. Piller, and V. Mertinger, "Prediction of earing of aluminium sheets from {h00} pole figures," *Eur. J. Mech. A/Solids*, vol. 81, no. September 2019, p. 103950, 2020, doi:10.1016/j. euromechsol.2020.103950.

[46] M. Benke, B. Schweitzer, A. Hlavacs, and V. Mertinger, "Prediction of earing of cross - rolled Al sheets from {h00} pole figures," 2020, doi:10.3390/met10020192.

[47] B. Schweitzer, M. Benke, A. Hlavacs, and V. Mertinger, "Normál- és kereszthengerelt, lágyított AW-5056 Al-lemezek fülesedésének becslése {h00} pólusábrák alapján," *Acta Mater. Transylvanica Magy. kiadás*, vol. 3, no. 1, pp. 38–42, 2020, doi:10.33923/amt-2020-01-07.

[48] X. Fang, K. Zhang, and H. Guo, "Twin-induced grain boundary engineering in 304 stainless steel," vol. 487, pp. 7–13, 2008, doi:10.1016/j.msea.2007.09.075.

[49] D. Gonzalez, J. F. Kelleher, J. Quinta, and P. J. Withers, "Macro and intergranular stress responses of austenitic stainless steel to 90° strain path changes," *Mater. Sci. Eng. A*, vol. 546, pp. 263–271, 2012, doi:10.1016/j.msea.2012.03.064.

[50] M. Nezakat, H. Akhiani, S. Morteza, and J. Szpunar, "Materials characterization electron backscatter and X-ray diffraction studies on the deformation and annealing textures of austenitic stainless steel 310S," *Mater. Charact.*, vol. 123, pp. 115–127, 2017, doi:10.1016/j.matchar.2016.11.019.

[51] Y. Park, J. Kang, and Y. Lee, "The relationship of microstructure and texture to rolling direction in cold-rolled metastable austenitic steel," 2010, doi:10.1007/s11661-010-0467-0.

[52] A. G. Raab et al., "Microstructure and mechanical properties of AISI 304L austenitic stainless steel processed by various schedules of rolling," *J. Phys. Conf. Ser.*, vol. 1688, no. 1, 2020, doi:10.1088/1742-6596/1688/1/012007.

[53] E. Schedin, *Sheet metal forming. Borlänge, Sweden, May 1992*, vol. 13, no. 6. 1992.

[54] Q. Liu, A. Roy, and V. V. Silberschmidt, "Temperature-dependent crystal-plasticity model for magnesium: a bottom-up approach," *Mech. Mater.*, vol. 113, pp. 44–56, 2017, doi:10.1016/j.mechmat.2017.07.008.

[55] S. Kaya, T. Altan, P. Groche, and C. Klöpsch, "Determination of the flow stress of magnesium AZ31-O sheet at elevated temperatures using the hydraulic bulge test," *Int. J. Mach. Tools Manuf.*, vol. 48, no. 5, pp. 550–557, 2008, doi:10.1016/j.ijmachtools.2007.06.011.

[56] S. H. Zhang, Y. C. Xu, G. Palumbo, S. Pinto, L. Tricarico, and Q. L. Zhang, "Formability and process conditions of magnesium alloy sheets," vol. 489, pp. 453–456, 2005, doi:10.4028/www.scientific.net/ MSF.488-489.453.

[57] Y. C. Xu et al., "Improved formability and deep drawing of cross-rolled magnesium alloy sheets at elevated temperatures," vol. 489, pp. 461–464, 2005, doi:10.4028/www.scientific.net/MSF.488-489.461.

[58] Y. Chino, K. Sassa, A. Kamiya, and M. Mabuchi, "Influence of rolling routes on press formability of a rolled AZ31 Mg alloy sheet," vol. 47, no. 10, pp. 2555–2560, 2006, doi:10.2320/matertrans.47.2555.

[59] Y. Chino, K. Sassa, A. Kamiya, and M. Mabuchi, "Enhancement of press formability of rolled Mg alloy sheet by using cross rolling processes," vol. 543, pp. 1615–1619, 2007, doi:10.4028/www.scientific.net/ MSF.539-543.1615.

[60] T. Chen, Z. Chen, L. Yi, J. Xiong, and C. Liu, "Effects of texture on anisotropy of mechanical properties in annealed Mg – 0.6% Zr – 1.0% Cd sheets by unidirectional and cross rolling," *Mater. Sci. Eng. A*, vol. 615, pp. 324–330, 2014, doi:10.1016/j.msea.2014.07.089.

[61] H. Ashrafi, M. Shamanian, R. Emadi, and N. Saeidi, "Correlation of tensile properties and strain hardening behavior with martensite volume fraction in dual-phase steels," *Trans. Indian Inst. Met.*, vol. 70, no. 6, pp. 1575–1584, 2017, doi:10.1007/s12666-016-0955-z.

[62] Y. G. Deng, Y. Li, H. Di, and R. D. K. Misra, "Effect of heating rate during continuous annealing on microstructure and mechanical properties of high-strength dual-phase steel," *J. Mater. Eng. Perform.*, vol. 28, no. 8, pp. 4556–4564, 2019, doi:10.1007/s11665-019-04253-2.

[63] M. Soleimani and H. Mirzadeh, "Enhanced mechanical properties of dual phase steel via cross rolling and intercritical annealing," *Mater. Sci. Eng. A*, vol. 804, no. January, p. 140778, 2021, doi:10.1016/j.msea.2021.140778.

[64] M. Wang, H. Li, Y. Tian, H. Guo, X. Fang, and Y. Guo, "Evolution of grain interfaces in annealed duplex stainless steel after parallel cross rolling and direct rolling," *Materials (Basel).*, vol. 11, no. 5, 2018, doi:10.3390/ma11050816.

[65] M. Wang et al., "Evolution of interface character distribution in duplex stainless steel processed by crossrolling and annealing," *J. Mater. Sci. Technol.*, vol. 34, no. 11, pp. 2160–2166, 2018, doi:10.1016/j.jmst.2018.02.018.

[66] T. R. Dandekar, A. Kumar, R. K. Khatirkar, D. Mahadule, and G. Ayyappan, "Multistep cross rolling of UNS S32101 steel: microstructure, texture, and magnetic properties," *J. Mater. Eng. Perform.*, vol. 30, no. 4, pp. 2916–2929, 2021, doi:10.1007/s11665-021-05510-z.

[67] A. Patel et al., "Strain-path controlled microstructure, texture and hardness evolution in cryo-deformed AlCoCrFeNi2.1 eutectic high entropy alloy," *Intermetallics*, vol. 97, no. March, pp. 12–21, 2018, doi:10.1016/j.intermet.2018.03.007.

[68] J. I. Kim, H. Y. Kim, T. Inamura, H. Hosoda, and S. Miyazaki, "Shape memory characteristics of Ti-22Nb-(2-8)Zr(at.%) biomedical alloys," *Mater. Sci. Eng. A*, vol. 403, no. 1–2, pp. 334–339, 2005, doi:10.1016/j.msea.2005.05.050.

[69] L. A. Domínguez-Contreras, "On the strengthening mechanisms of a cross-rolled Zn-based bioabsorbable alloy," *J. Phys. Conf. Ser.*, 2021, doi:10.1088/1742-6596/1723/1/012003.

[70] P. Nageswara, D. Singh, and R. Jayaganthan, "Effect of post cryorolling treatments on microstructural and mechanical behaviour of ultra fine grained Al e Mg e Si alloy," *J. Mater. Sci. Technol.*, pp. 1–8, 2014, doi:10.1016/j.jmst.2014.03.009.

[71] M. Yadollahpour, H. Hosseini-Toudeshky, and F. Karimzadeh, "The use of response surface methodology in cryrolling of ultrafine grained Al6061 to improve the mechanical properties," *Proc. Inst. Mech. Eng., Part L: J. Mater.: Des. Appl.*, vol. 230, no. 2. pp. 400–417, 2016, doi:10.1177/1464420715574139.

[72] C. C. Selvan, C. S. Narayanan, B. Ravisankar, R. Narayanasamy, and C. T. Valliammai, "The dependence of the strain path on the microstructure, texture and mechanical properties of cryogenic rolled Al-Cu alloy," *Mater. Res. Express*, vol. 7, no. 3, 2020, doi:10.1088/2053-1591/ab7f9a.

[73] S. Vigneshwaran, K. Sivaprasad, and R. Narayanasamy, "Superior strength with enhanced fracture resistance of Al-Mg-Sc alloy through two-step cryo cross rolling," *Metall. Mater. Trans. A*, vol. 50, no. 7, pp. 3265–3281, 2019, doi:10.1007/s11661-019-05253-6.

11 A Review on Process Limitations and Recent Advancements in Single Point Incremental Sheet Forming

Parnika Shrivastava
National Institute of Technology, Hamirpur, India

Puneet Tandon
PDPM Indian Institute of Information Technology Design and Manufacturing Jabalpur, Jabalpur, India

CONTENTS

11.1 INTRODUCTION

Single Point Incremental Forming is a dieless forming operation. Unlike, conventional forming operations the process requires simple ball shaped tool to form sheet metal which is clamped around its periphery. This makes the process viable for small batch production of customized sheet metal components. Irrespective of various advantages like flexibility and time efficiency, there are certain serious issues related to the process, which makes the process economically reasonable only for prototypes and small batch fabrication, thus, hindering the industrial application for mass production (Hussain et al. 2011; Al-Ghamdi and Hussain 2016). In spite of the fact, that the process ensures higher formability at lower setup cost in comparison with conventional forming processes, still there are issues related to forming load, forming limits, thickness distribution, geometrical accuracy, and

DOI: 10.1201/9781003226703-11

surface quality of the SPIF parts (Formisano et al. 2017). The related issues are discussed in detail in the subsequent part of the chapter. As a potential solution to the above-stated problems in SPIF, the process has undergone several advancements in the due course of time. The chapter also discusses the background of the process along with the recent advancements reported in the literature that have been developed to enhance the process characteristics.

11.1.1 BACKGROUND

Single Point Incremental Forming (SPIF) process was started in 1967 when Leszak (1967) obtained the patent named as "Apparatus and Process for Incremental Dieless Forming". The patented process is shown in Figure 11.1. In the same year, another patent by Berghahn led to the formation of sheet metal component by a roller which moved inward along a radial trajectory (Emmens et al. 2010). The proposed methodology by Berghahn is shown in Figure 11.2. The true SPIF can be dedicated to the work done by Mason in 1978, which is shown in Figure 11.3. It was in 2004, when Young and Jeswiet (2004) envisaged the process and named it as "SPIF". Later, they distinguished the process from those conventional forming techniques which utilize progressive dies for bulk forming operations.

It was in 2004, when Young and Jeswiet (2004) envisaged the process and named it as "SPIF". Later, they distinguished the process from those conventional forming techniques which utilize progressive dies for bulk forming operations.

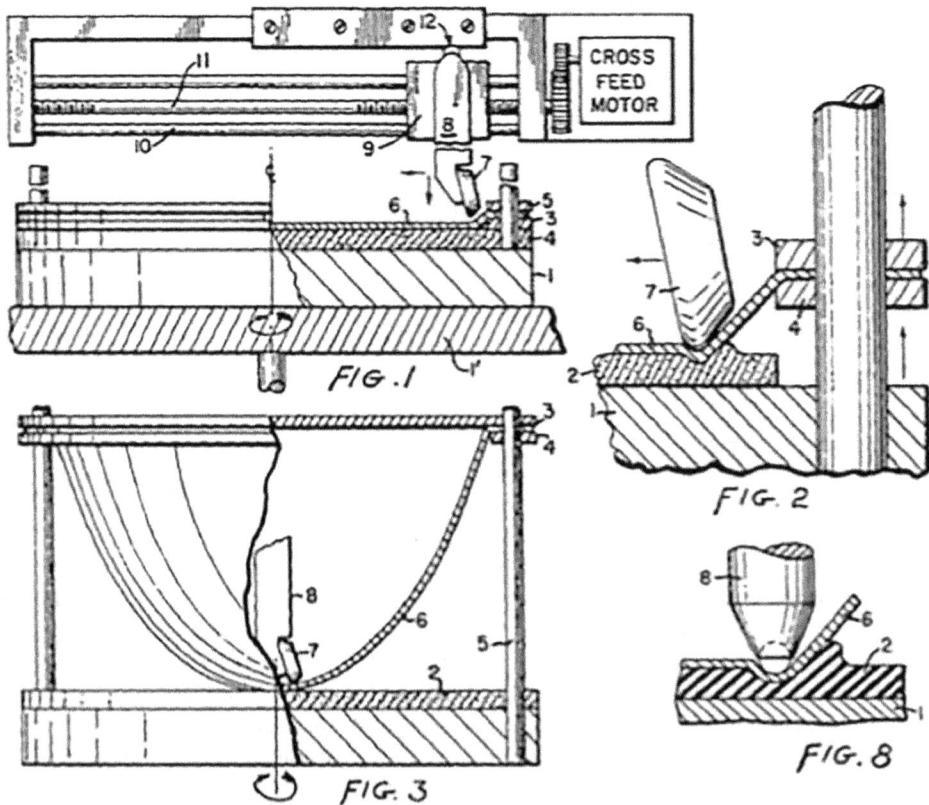

FIGURE 11.1   Process proposed by Leszak [US 3342051] (Leszak 1967).

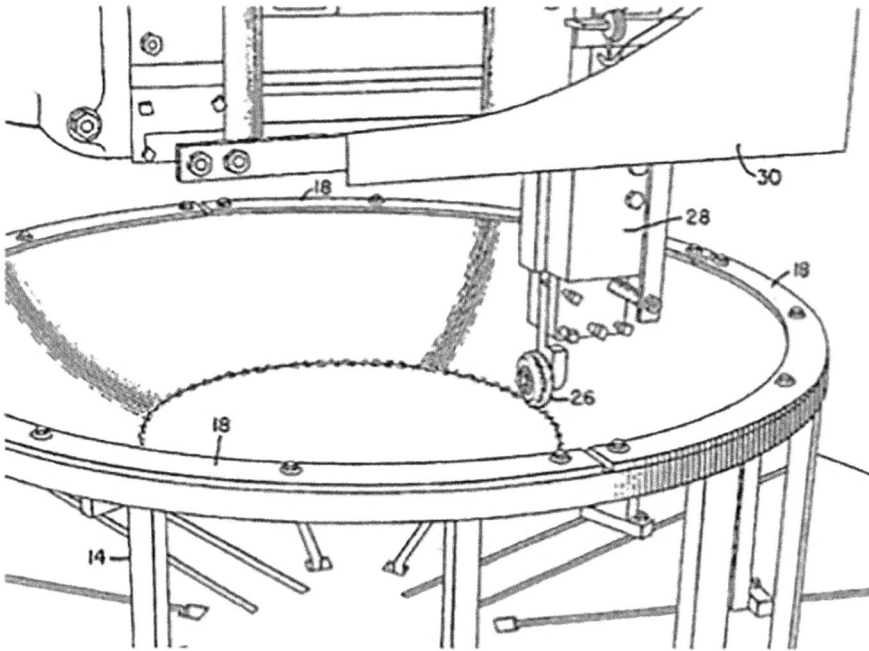

FIGURE 11.2 Process proposed by Berghahn [US 3316745] (Emmens et al. 2010).

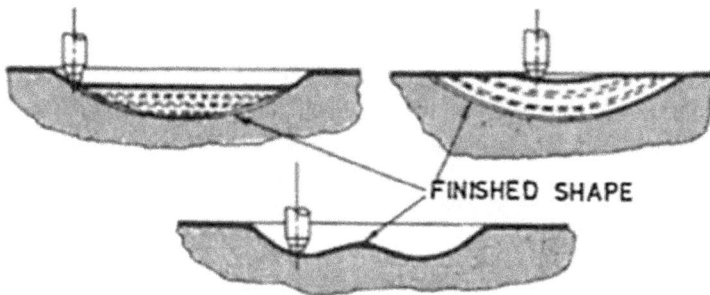

FIGURE 11.3 First true ISF by Mason (Emmens et al. 2010).

11.1.2   BENEFITS OF SINGLE POINT INCREMENTAL FORMING (SPIF)

The process offers various benefits in comparison with conventional forming techniques. Following are the major advantages of SPIF process:

   (i) Flexibility offered by the process, as it merely requires CAD data, conventional CNC milling machine or lathe and universal tooling.
   (ii) The process is dieless and capable of forming asymmetric and complicated shapes. However, it does need at times a backing plate to generate a perfect change of angle at the sheet surface.
   (iii) The process is quick and flexible, particularly for the cases when design needs a change or an up gradation.
   (iv) The process offers higher formability at much lower forming load in comparison with the conventional forming techniques as the deformation is highly localized and incremental in nature.
   (v) There is no process limitation when it comes to size and shape of part to be formed.
   (vi) The process offers lower lead time in comparison with conventional forming processes.
   (vii) The process is relatively noise and clatter free.

11.1.3   APPLICATIONS AND LIMITATIONS OF SPIF

There are various leading areas where highly précised products are required for the efficient and accurate performance. A few typical components formed by SPIF are shown in Figure 11.4. The major industries which are served by SPIF process are as follows:

   • Aerospace industries for the production needs of appliance panel, body panel, passenger seat cover, etc.
   • Automobile industries for the manufacturing of front panel of bullet train, door inner/outer panel, hood panel, engine cover, headlight reflector, noise shield, etc.
   • Medical industries for the rapid prototyping of denture plate, ankle support, metal helmet, prosthetics, dental and cranial implants, etc.
   • Electronics industries for the production of cellular phone body. Integrated Circuit LED frames, hard disk drives, sensors, etc.
   • Architecture sector for the digital fabrication of three dimensional artifacts.

11.1.4   MAJOR LIMITATIONS IN SPIF

This process in comparison with the conventional forming processes carries significant improvements in terms of process flexibility, customization, and complexity of products produced at low industrial costs. However, it has some serious limitations which constraints its implementation on shop floor and hinders the technology in reaching the mainstream (Mulay et al. 2017). The process suffers from high forming time. To circumvent the delay in forming time, high speed forming was performed by increasing the tool feed rate in the range of 1500 to 12,000 mm/min. By this, delay in forming process was eliminated without compromising the feasibility and accuracy (Bastos et al. 2016). However, parts formed by SPIF fail when are subjected to higher forming load during the operation, which is responsible for failure initiation and propagation in SPIF (Ambrogio et al. 2006). The other major limitation of the process is uneven distribution of the wall thickness that limits formability of the parts with steep wall angles (LI et al. 2012). Another major challenge in SPIF is to obtain the geometry of final formed part close to the Computer-Aided Design (CAD) model of the part to be produced. The bending and springback phenomena during the process also lead to the geometrical inaccuracy in the formed products. In addition, the internal and external surface quality

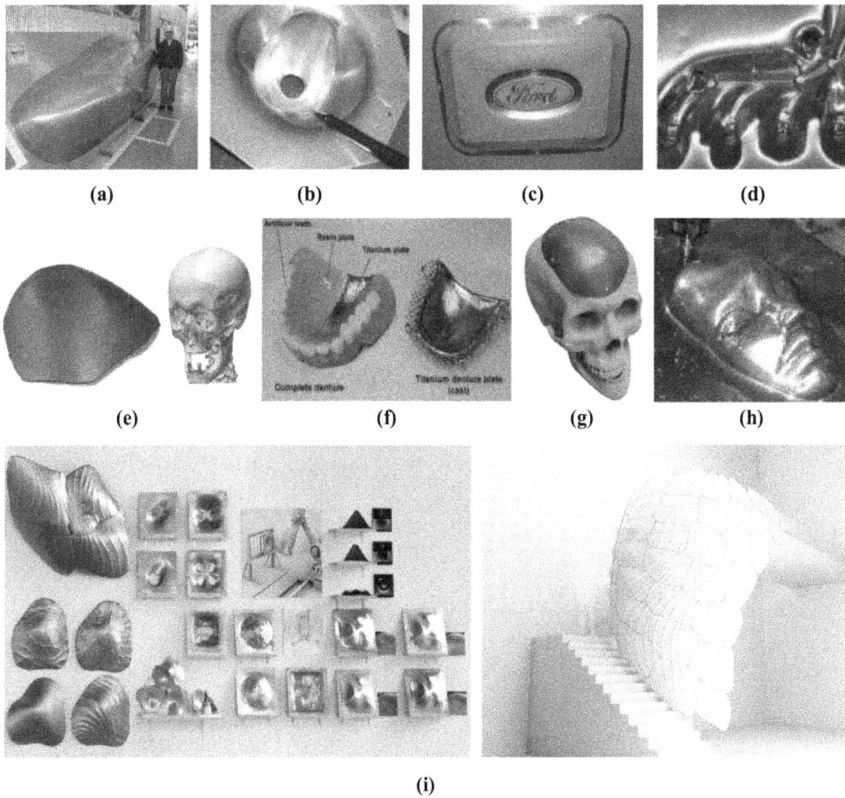

FIGURE 11.4 Components formed by SPIF process (a) 1:8 scaled model of a Shinkansen bullet train (Jeswiet 2002); (b) Head light reflector; (c) Ford badge; (d) Heat shield (Jeswiet and Young 2005); (e) Cranial implant (Duflou et al. 2008); (f) Denture plate (Milutinović et al. 2014); (g) Prosthetic implant (Piccininni et al. 2016); (h) 3D Human Face Mask; (i) 3D artifacts (Kalo and Newsum 2014).

also deteriorates post SPIF operation (Hussain et al. 2011). Further, not only the process allows forming of the ductile material but also offers limited formability when the wall angle is increased above a certain limit.

11.1.4.1 Forming Load in SPIF

One of the major points to ponder in SPIF is the forming load (Shrivastava and Tandon 2015). Various studies were conducted to analyze the effect of parameters like feed rate, tool geometry, incremental step depth, spindle speed, lubrication, and part geometry on forming load during SPIF of parts with constant or variable wall angles (Petek et al. 2009; Mulay et al. 2017). Analysis of forming load is essential as during forming, the occurrence of crack can be predicted and the SPIF process can be controlled online by observing the trend of forming load and keeping track of the peak load. Thus, proper evaluation of forming load helps in designing and safe utilization of machine tools and fixtures for longer duration (Ambrogio et al. 2006). Further, it has been analyzed that forming load and plastic strain limit of the workpiece decides the structural integrity of the formed parts.

Force prediction methods used in SPIF, were either inaccurate or time consuming thus, efficient force prediction strategy was employed by Bansal et al. (2017) based on experimental observations of forming forces. They have developed a model which predicts component thickness, tool-sheet contact area, and forming forces. The model also necessitates usage of fewer computational resources in comparison with FEA. Li et al. (2015) had proposed a tangential force prediction model considering the important deformation modes, i.e., shearing, stretching, and bending.

11.1.4.2   Forming Limits in SPIF

Formability evaluation in SPIF has always been complex and a matter of the concern due to the combined effect of bending, stretching, and shear (Shrivastava and Tandon 2018). In addition, SPIF is limited to formability of ductile materials only and that too when the wall angle is below a critical value. Many efforts have been made to form the parts with steep wall angle; however, the crack occurs in the geometry during the operation.

Correct assessment of forming limits is another issue in SPIF. Due to highly localized nature of deformation in SPIF, which although grows in subsequent tool passes, transition from material instability to failure is comparatively slower than the traditional forming (Shrivastava and Tandon 2018). The slow transition is also due to predominant bending and through-the-thickness shear mechanisms in SPIF. Thus, Forming Limit Curves (FLC) used to predict failure in conventional forming process fails to define the fracture strain in SPIF (Allwood and Shouler 2009; Seong et al. 2014; Benedetti et al. 2017). Thus, Fracture Forming Limit diagrams (FFLD) are the most common and efficient way to predict failure in SPIF (Martins et al. 2008). The typical fracture forming limit diagram obtained in SPIF is shown in Figure 11.5. Alinaghian et al. (2017) had investigated the dependency of forming limits on various process parameters, like tool dimensions, sheet material, sheet thickness, friction between tool and sheet, incremental step depth, and tool rotational speed.

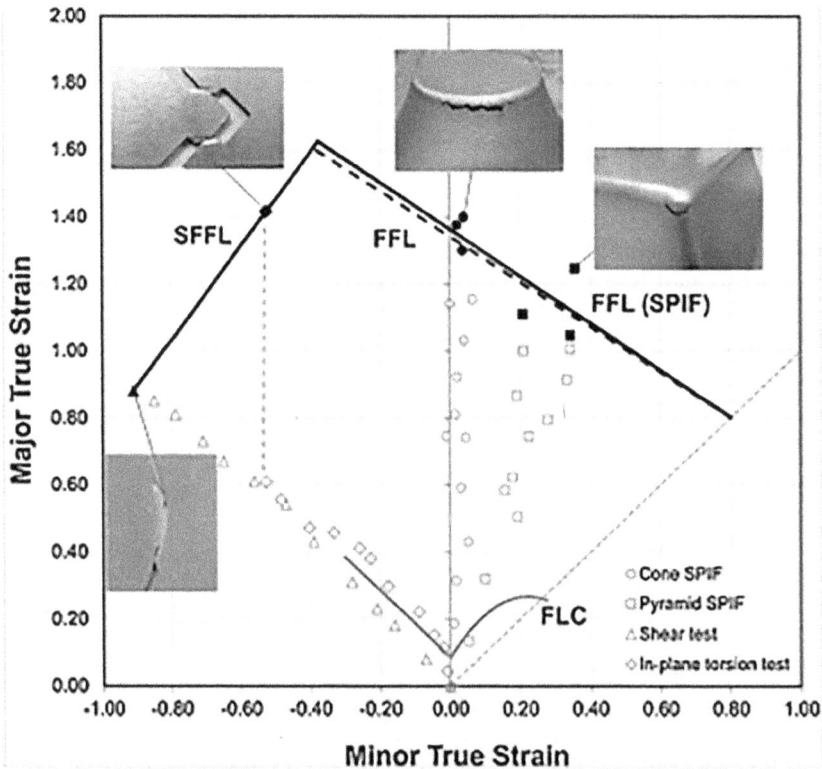

FFL: Fracture Forming Limit obtained by Conventional Forming Test Equipment; FFL (SPIF): Fracture Forming Limit in SPIF; FLC: Forming Limit Curve obtained by Conventional Forming Processes; SFFL: Shear Fracture Forming Limit obtained by Shear Test

FIGURE 11.5 Typical fracture forming loci determined from experimental strains values at fracture corresponding to SPIF, conventional forming tests, shear tests, and conventional sheet forming processes. The solid black markers correspond to strains at fracture and the dashed line denoted as FFL corresponds to the fracture locus that was determined by means of conventional sheet formability tests (Isik et al. 2014).

An analytical model was developed by Ai et al. (2017) to predict critical strain value at which deformation instability arises in SPIF. Further, they also investigated the influence of bending and work hardening on critical strain values. They claimed that fracture in SPIF is solely dependent on deformation stability and ductility of sheet material. Allwood et al. (2007) had claimed that the presence of through the thickness shear increases the forming limits in SPIF. In order to investigate the effect of strain paths on forming limits, SPIF was performed by a rotating ball tool (Shim and Park 2001). They had further observed that formability in SPIF strongly depends on strain path.

11.1.4.3 Thickness Distribution in SPIF

The other major limitation of the process is uneven distribution of the wall thickness in SPIF. This also limits the formability of the parts with steep wall angles. In SPIF, during deformation, stretching accompanies thinning, which acts as a major limitation of the process. Conventionally, sine/cosine law is utilized to predict the thickness of the formed parts. The thinning limit of the sheet material is usually considered to evaluate the forming limits in sheet metal forming operations. The cosine law is expressed with the help of Equation (11.1) as given below (Hussain and Gao 2007; Patidar 2017):

$$t = t_0 * \cos \theta \tag{11.1}$$

where, "t_0" is the thickness of a sheet blank, and "t" and "θ" are the final thickness and slope angle of a part to be formed. The same has been explained in Figure 11.6.

However, in SPIF, shearing and bending phenomena restricts the use of conventional sine/cosine law in predicting the thickness profile of the parts formed with SPIF. In SPIF, studies have proved that residual stresses, bending stresses, and sheet thickness variation are correlated with each other. Various thickness prediction models have been developed for predicting localized thinning in sheet metal parts by Cao et al. (2000) and Thibaud et al. (2012). For improved thickness distribution, a novel hybrid technique has also been employed in conjugation with thickness prediction model to optimize the overall process of improving thickness distribution in SPIF (Zhang et al. 2017). The

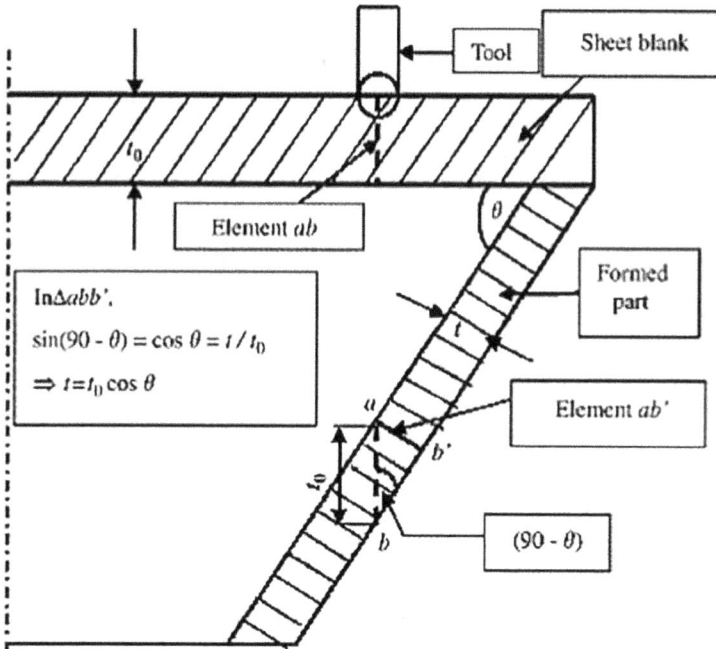

FIGURE 11.6 Depiction of wall thickness by cosine law (Hussain and Gao 2007).

hybrid mechanism involved multipoint forming with a preforming stage followed by incremental sheet forming. Liu et al. (2013) proposed multistage deformation path to control material flow and to improve sheet thickness in SPIF.

11.1.4.4 Geometrical Accuracy in SPIF

Another major challenge in SPIF is to obtain the geometry of the final formed part close to the Computer-Aided Geometry model (Shrivastava et al. 2018). Incremental Sheet Forming is devoid of robust solutions to eliminate geometric inaccuracies, which arise due to sheet bending, sheet spring-back, and pillow effect, to name a few. Efforts had been made to recognize different causes of shape and geometrical errors, as well as its dependency on the most relevant process parameters (Hussain et al. 2011; Jodhave 2015; Panjwani et al. 2017). The elastic recovery tendency of the metals is responsible for the springback in sheet material (Shrivastava and Tandon 2018). Figure 11.7 shows the springback in the sheet component resulted in conventional sheet metal forming operation.

Parts formed by SPIF undergo cyclic deformation and suffer from local and global springback. Bauschinger effect determines the distribution of stress in the sheet metal subjected to plastic deformation. It is a microscopic phenomenon which is associated with dislocation structures that develop during the deformation, especially when the material is subjected to cold forming.. It can be inferred from the figure that under reverse loading, an early re-yielding takes place at the beginning of reversal loading which accompanies with significant change in hardening rate. This lead to permanent softening in the material, as this result in difference in the values of monotonous and reversed flow stresses.

Parameters like tool diameter, step depth, tool rotational speed, lubrication, material properties and part design affect the process mechanism as well as geometrical accuracy of the SPIF parts. For error minimization some strategies like, use of supple support, use of counter force, multiple point incremental forming, back drawing incremental forming and forming with optimized tool trajectory have been discussed (Micari et al. 2007). In addition, to have controlled plastic flow in SPIF, metallic

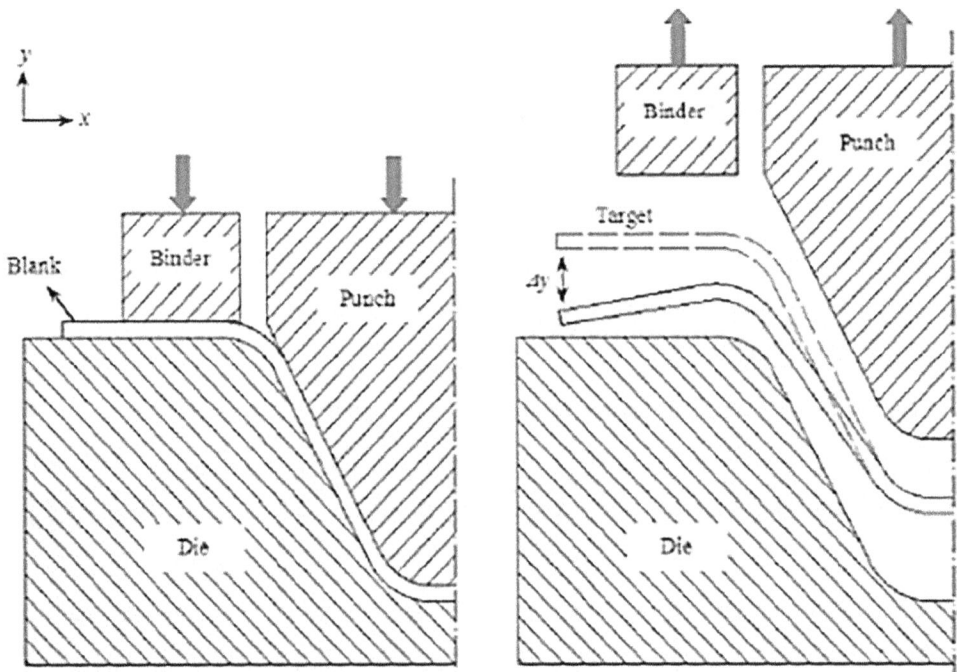

FIGURE 11.7 Deformation due to springback during conventional sheet metal forming operation (Anggono et al. 2012).

foam blocks were also used as flexible die-support. This method has been proved to improve the accuracy of the formed part along with maintaining the flexibility of the process (Min et al. 2018).

The toolpath has been studied as a process parameter influencing the final geometry and forming characteristics. The ISF process has inherent drawbacks such as poor surface finish which includes surface non-uniformity, pillow effect, and springback. The localized deformation and small interaction zone between tool and sheet contribute to these drawbacks. Several methods have been proposed to improve forming characteristics and reduce the adverse effects of interaction between tool and sheet metal or any other material being employed in ISF. Rauch et al. 2009, found the basic CAM toolpaths unsuitable and proposed an Intelligent CAM (ICAM) approach. The purpose was to develop and adopt toolpaths generated through CAM specifically for ISF. Industrial standard accuracy and surface finish was achieved using this approach. Malhotra et al. 2010, exploited the similarities between additive layered manufacturing and ISF to propose a 3D spiral SPIF toolpath. A direct relation between incremental depth, surface finish, and geometrical accuracy was observed. Experiments revealed that the method of rigorous computation in contrast to empirical formulae to generate tool path produced better results. A way to improve geometrical accuracy is a compensation mechanism by predicting forming forces and applying that to CAD model while generating tool path (Asghar et al. 2014). Another novel toolpath strategy, radial toolpath strategy (Grimm and Mears 2020), was developed to obtain better surface uniformity and increased formability. Numerical methods have also been used to generate better tool path strategies or tool path optimization (Azaouzi and Lebaal 2012), a constant scallop height method (Zhu et al. 2011) was simulated numerical and results were identical to the test model. These methods are generally cost and time intensive so a toolpath generation approach in (Tera et al. 2019) aimed at reducing both by making adaptations in softwares meant for simple milling operations.

11.1.4.5   Surface Quality in SPIF

The analysis of the part surface is another vital aspect in SPIF (Shrivastava and Tandon 2019). Studies revealed that surface roughness in SPIF is primarily dependent on the parameters like forming tool geometry, tool path, step size, lubrication, forming angle, and feed rate. Earlier, Kurra et al. (2015) had utilized some optimization techniques, like Artificial Neural Network (ANN) and Support Vector Regression techniques to model surface roughness in SPIF. Parts formed by SPIF process suffers galling, poor surface finish, and ultimate failure of the sheet material (Kurra et al. 2015). Study reports that lubrication is must to ensure minimized friction and better internal surface finish. The surface quality of the part is also affected by other parameters such as tool diameter, wall angle, step depth, feed rate, and sheet material. Some researchers made an effort to develop analytical and empirical models to understand the effect of process parameters on surface roughness. Hagan and Jeswiet (2005) studied the mean surface roughness, RMS roughness, maximum profile height, and mean peak to valley height in SPIF resulted as per different spindle speed and step depth. In conjugation an empirical model was also proposed. Truncated conical and pyramidal frustums were formed to analyze the effect of process parameters on surface roughness of Al5052 and AA1050 material (Bhattacharya et al. 2011; Shrivastava et al. 2015). Recent studies reports that machine learning techniques are more efficient in predicting surface roughness in comparison with analytical models (Li 2011; Hussaini et al. 2014).

Apart from internal surface analysis, external surface of the SPIF parts also needs significant attention. This is because grain-scale roughening or morphological reliefs dominate at the external surface of the formed part, when the tool forms only at one side of the sheet (Hamilton and Jeswiet 2010) (Shrivastava and Tandon 2019). Due to this, a critical kind of external surface gets formed which is termed as orange peel. Orange peel is a kind of microsurface defect which is mainly dependent on the grain size and its distribution and orientations. The effect is undesirable as it deteriorates the surface integrity of the formed parts which adversely affects the quality and reliability of the process.

Effort has been made for online detection of orange peel roughening behavior in stretch forming of Al–Li alloy sheet. Further, they have detected fracture in stretch-forming operation by keeping track of critical values of surface roughness originated due to orange peel effect (Feng et al. 2016). Hamilton and Jeswiet (2010) characterized orange peel for high speed SPIF by evaluating equivalent combinatory roughness, which is the function of shape factor and the incremental step size. They have considered the effects of feed rates and tool rotational speed on orange peel and the characterization of the same was done based on the values of resultant equivalent roughness.

11.1.5 RECENT ADVANCEMENTS IN SPIF

Two-Point Incremental Forming (TPIF) or Positive Dieless Forming process has undergone several advancements in the due course of time (Figure 11.8). As per the fact that ISF process is a variant of spinning process, this category of ISF differs slightly from the spinning process. The process utilizes partial or full die to provide support, while forming tool moves in predetermined trajectory to attain the desired shape. The process lacks flexibility as it requires partial or full die for the forming operation. Figure 11.8(a) provides the schematic representation of the process.

Single Point Incremental Forming (SPIF) also known or Negative Dieless Forming category of ISF is advanced and dieless, as it requires a single forming tool to perform the operation. From the mechanism point of view, SPIF differs from TPIF as the former incorporates the outside in (from clamped edge to the center of the sheet) movement of the tool, while the later includes inside out movement of the forming tool. The schematic representation of the process is shown in Figure 11.8(b).

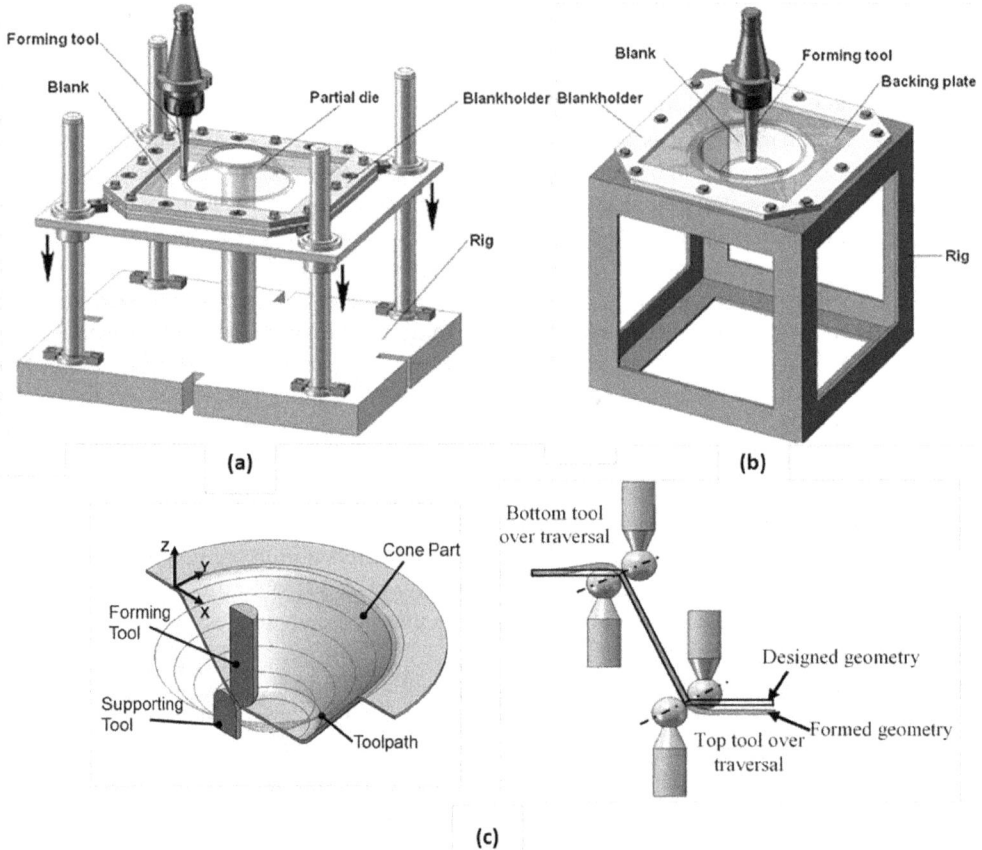

FIGURE 11.8 Variants of ISF process (a) TPIF process, (b) SPIF process (Silva and Martins 2013), and (c) DSIF process (Zhang et al. 2015; Lingam et al. 2016).

Double Sided Incremental Forming (DSIF) category of forming is a recent advancement in the field of incremental sheet forming. The process utilizes two dynamic tools (on each sides of the sheet) at a time. The major forming tool moves in predetermined trajectory to impart the desired shape, while the other tool moves in synchronization to the forming tool to provide the dynamic support. The schematic representations of the process along with the position of forming tools are shown in Figure 11.8(c).

Among them, hybrid techniques like Double Sided Incremental Forming (DSIF) (Figure 11.8c), Robot assisted SPIF (shown in Figure 11.9), Multistage SPIF (MSPIF) (Nirala et al. 2017), and hot/warm SPIF are the recent ones. As a potential solution to the above-stated problems in SPIF, warm forming, a hybrid technique, has been developed in recent years. Warm or hot SPIF can be attained by localized or overall heating of the sheet (Figure 11.10), or tool tip or even both. The process can be differentiated based on the types of heat sources, namely, hot air blowers, heater bands, friction stir, laser, or electricity. Different types of hot/warm incremental forming setups are shown in Figure 11.11.

It was reported that warm forming improves part formability, and reduces forming load as well as reduces springback in SPIF parts (Mohammadi et al. 2016; Wang et al. 2016; Xu et al. 2016). Parts formed by warm forming technique reported enhanced geometrical accuracy, and improved formability with reduction in forming load.

Induction heating assisted hot SPIF process was performed over high strength sheet metal like DP 980 steel, which can hardly form at room temperature. It was reported that improved formability, lesser residual stresses, and forming load were resulted by this hybrid technique (Al-Obaidi et al. 2016). Another warm incremental forming of titanium alloy performed by Khazaali and Fereshteh-Saniee (2018) had reported the ease in forming at elevated temperatures. The reason for the improvement is the reformation of microstructure at elevated temperature which facilitates the dislocations

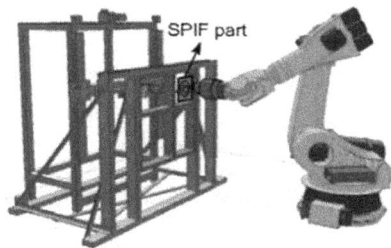

FIGURE 11.9   Robot assisted SPIF (Duflou et al. 2007).

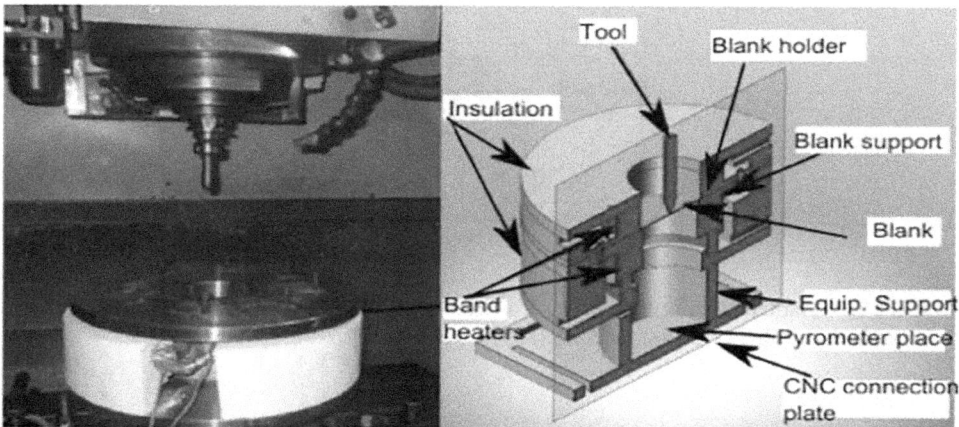

FIGURE 11.10   Overall/global heating setup (Palumbo and Brandizzi 2012).

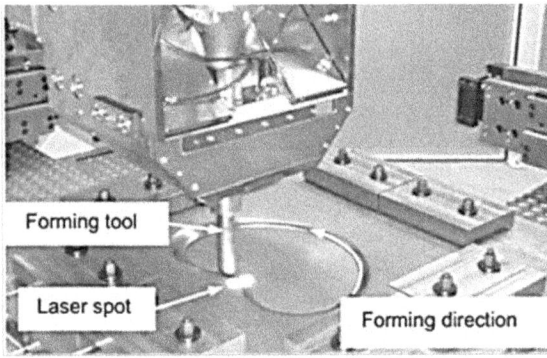

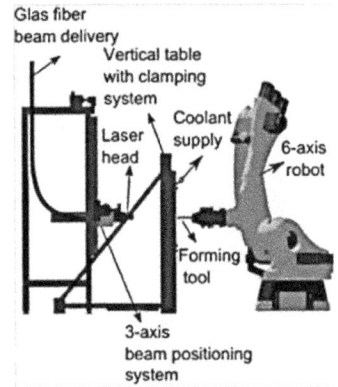

(a)

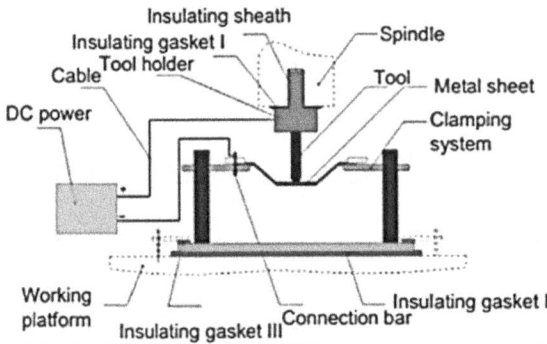

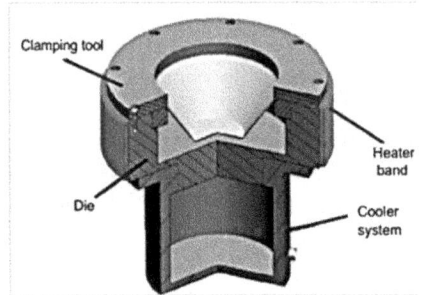

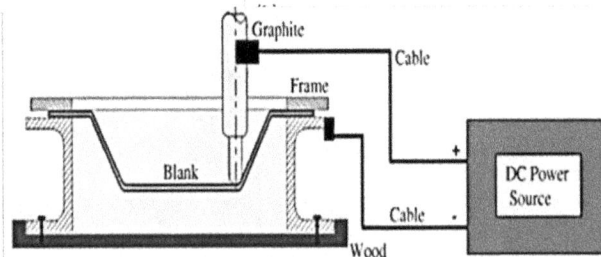

(b)

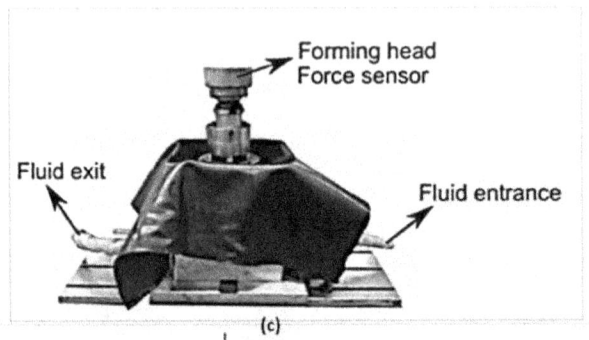

(c)

FIGURE 11.11 Warm/hot SPIF setups (a) laser assisted SPIF, (b) electric assisted SPIF with experimental equipment, and (c) oil based heat assisted SPIF (Palumbo and Brandizzi 2012).

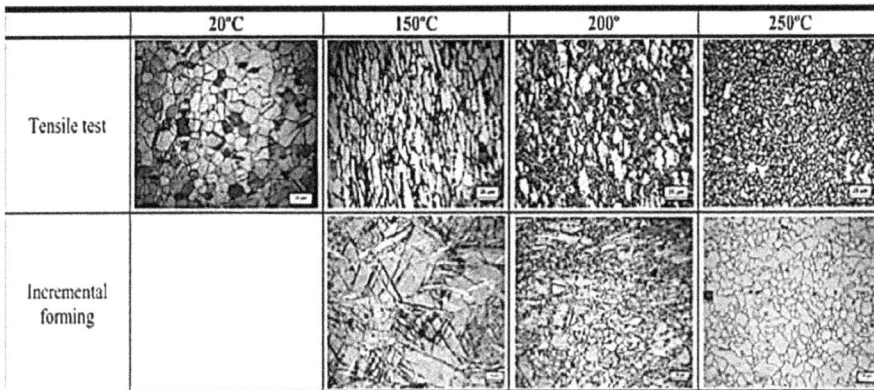

FIGURE 11.12 Microstructure reformation of Mg alloy due to forming at elevated temperature (Ulacia et al. 2014).

movement during forming, which can be elucidated from Figure 11.12. Although, the warm forming process improves geometrical accuracy and formability up to an extent, but it also demands additional setup and skills for its efficient operation. Such investments and extra power consumption is acceptable for hard to form materials, like titanium or magnesium, but not for ductile materials like aluminium, copper, and brass, which although suffer from limited formability and geometrical inaccuracy but still can be formed at room temperature.

11.2   CONCLUSION

The chapter provides an overview and the state of art of Single Point Incremental Forming process along with the major limitations associated with the process characteristics. The chapter also discusses the recent advancements and inventions related to the process that have been technologically developed to improvise the process characteristics. SPIF process in comparison with the conventional forming processes carries significant improvements in terms of process flexibility, customization, and complexity of products produced at low industrial costs. However, parts formed by SPIF fail when subjected to higher forming load during the operation, which is responsible for failure initiation and propagation in SPIF. Uneven distribution of the wall thickness limits formability of the parts with steep wall angles in SPIF. The bending and springback phenomena during the process also lead to the geometrical inaccuracy in the formed products. Further, not only the process allows forming of the ductile material but also offers limited formability when the wall angle is increased above a certain limit. It can be concluded that fracture in SPIF is solely dependent on deformation stability and ductility of sheet material. As a potential solution to the above-stated problems in SPIF, warm forming, a hybrid technique, had been developed in recent years. For error minimization some strategies like, use of supple support, use of counter force, multiple point incremental forming, back drawing incremental forming, and forming with optimized tool trajectory have been proved as a potential solution. SPIF post improvements can be claimed as a potential solution for the production of bio-medical implants and aerospace components.

REFERENCES

Ai, S., Lu, B., Chen, J., Long, H., Ou, H., 2017. Evaluation of deformation stability and fracture mechanism in incremental sheet forming. *Int. J. Mech. Sci.* 124–125, 174–184. https://doi.org/10.1016/J.IJMECSCI.2017.03.012

Al-Ghamdi, K.A., Hussain, G., 2016. Parameter-formability relationship in ISF of tri-layered Cu-Steel-Cu composite sheet metal: Response surface and microscopic analyses. *Int. J. Precis. Eng. Manuf.* 17, 1633–1642. https://doi.org/10.1007/s12541-016-0189-3

Alinaghian, I., Ranjbar, H., Beheshtizad, M.A., 2017. Forming limit investigation of AA6061 friction stir welded blank in a single point incremental forming process. *RSM Approach. Trans. Indian Inst. Met.* 70, 2303–2318. https://doi.org/10.1007/s12666-017-1093-y

Allwood, J.M., Shouler, D.R., 2009. Generalised forming limit diagrams showing increased forming limits with non-planar stress states. *Int. J. Plast.* 25, 1207–1230. https://doi.org/10.1016/J.IJPLAS.2008.11.001

Allwood, J.M., Shouler, D.R., Tekkaya, A.E., 2007. The increased forming limits of incremental sheet forming processes. *Key Eng. Mater.* 344, 621–628. https://doi.org/10.4028/www.scientific.net/KEM.344.621

Al-Obaidi, A., Kräusel, V., Landgrebe, D., 2016. Hot single-point incremental forming assisted by induction heating. *Int. J. Adv. Manuf. Technol.* 82, 1163–1171. https://doi.org/10.1007/s00170-015-7439-x

Ambrogio, G., Filice, L., Micari, F., 2006. A force measuring based strategy for failure prevention in incremental forming. *J. Mater. Process. Technol.* 177, 413–416. https://doi.org/10.1016/J.JMATPROTEC.2006.04.076

Anggono, A.D., Siswanto, W.A., Omar, B., 2012. Algorithm development and application of spring back compensation for sheet metal forming. *Res. J. Appl. Sci. Eng. Technol.* 4, 2036–2045.

Asghar, J., Lingam, R., Shibin, E., Reddy, N.V., 2014. Tool path design for enhancement of accuracy in single-point incremental forming. *Proc. Inst. Mech. Eng. Part B J. Eng. Manuf.* 228, 1027–1035. https://doi.org/10.1177/0954405413512812

Azaouzi, M., Lebaal, N., 2012. Tool path optimization for single point incremental sheet forming using response surface method. *Simul. Model. Pract. Theory* 24, 49–58. https://doi.org/10.1016/j.simpat.2012.01.008

Bansal, A., Lingam, R., Yadav, S.K., Venkata Reddy, N., 2017. Prediction of forming forces in single point incremental forming. *J. Manuf. Process.* 28, 486–493. https://doi.org/10.1016/J.JMAPRO.2017.04.016

Bastos, R., de Sousa, R., Ferreira, J., 2016. Enhancing time efficiency on single point incremental forming processes. *Int. J. Mater. Form.* 9, 653–662. https://doi.org/10.1007/s12289-015-1251-x

Benedetti, M., Fontanari, V., Monelli, B., Tassan, M., 2017. Single-point incremental forming of sheet metals: Experimental study and numerical simulation. *Proc. Inst. Mech. Eng. Part B J. Eng. Manuf.* 231, 301–312. https://doi.org/10.1177/0954405415612351

Bhattacharya, A., Singh, S., Maneesh, K., Reddy, N.V., Cao, J., 2011. Formability and surface finish studies in single point incremental forming. *ASME 2011 Int. Manuf. Sci. Eng. Conf. MSEC 2011*, 1, 621–627. https://doi.org/10.1115/MSEC2011-50284

Cao, J., Yao, H., Karafillis, A., Boyce, M.C., 2000. Prediction of localized thinning in sheet metal using a general anisotropic yield criterion. *Int. J. Plast.* 16, 1105–1129. https://doi.org/10.1016/S0749-6419(99)00091-1

Duflou, J.R., Callebaut, B., Verbert, J., De Baerdemaeker, H., 2007. Laser assisted incremental forming: Formability and accuracy improvement. *CIRP Ann.* 56, 273–276. https://doi.org/10.1016/J.CIRP.2007.05.063

Duflou, J.R., Verbert, J., Belkassem, B., Gu, J., Sol, H., Henrard, C., Habraken, A.M., 2008. Process window enhancement for single point incremental forming through multi-step toolpaths. *CIRP Ann.* 57, 253–256. https://doi.org/10.1016/J.CIRP.2008.03.030

Emmens, W.C., Sebastiani, G., van den Boogaard, A.H., 2010. The technology of incremental sheet forming—a brief review of the history. *J. Mater. Process. Technol.* 210, 981–997. https://doi.org/10.1016/J.JMATPROTEC.2010.02.014

Feng, J.-W., Zhan, L.-H., Yang, Y.-G., 2016. The establishment of surface roughness as failure criterion of Al–Li alloy stretch-forming process. *Metals (Basel).* 6, 13. https://doi.org/10.3390/met6010013

Formisano, A., Boccarusso, L., Capece Minutolo, F., Carrino, L., Durante, M., Langella, A., 2017. Negative and positive incremental forming: Comparison by geometrical, experimental, and FEM considerations. *Mater. Manuf. Process.* 32, 530–536. https://doi.org/10.1080/10426914.2016.1232810

Grimm, T.J., Mears, L., 2020. Investigation of a radial toolpath in single point incremental forming. *Procedia Manuf.* 48, 215–222. https://doi.org/10.1016/j.promfg.2020.05.040

Hagan, E., Jeswiet, J., 2005. Analysis of surface roughness for parts formed by computer numerical controlled incremental forming. 218, 1307–1312. https://doi.org/10.1243/0954405042323559

Hamilton, K., Jeswiet, J., 2010. Single point incremental forming at high feed rates and rotational speeds: Surface and structural consequences. *CIRP Ann.* 59, 311–314. https://doi.org/10.1016/J.CIRP.2010.03.016

Hussain, G., Gao, L., Hayat, N., 2011. Forming parameters and forming defects in incremental forming of an aluminum sheet: Correlation, empirical modeling, and optimization: Part A. Mater. *Manuf. Process.* 26, 1546–1553. https://doi.org/10.1080/10426914.2011.552017

Hussain, G.Ã., Gao, L., 2007. A novel method to test the thinning limits of sheet metals in negative incremental forming 47, 419–435. https://doi.org/10.1016/j.ijmachtools.2006.06.015

Hussaini, S.M., Singh, S.K., Gupta, A.K., 2014. Experimental and numerical investigation of formability for austenitic stainless steel 316 at elevated temperatures. *J. Mater. Res. Technol.* 3, 17–24. https://doi.org/10.1016/J.JMRT.2013.10.010

Isik, K., Silva, M.B., Tekkaya, A.E., Martins, P.A.F., 2014. Formability limits by fracture in sheet metal forming. *J. Mater. Process. Technol.* 214, 1557–1565. https://doi.org/10.1016/J.JMATPROTEC.2014.02.026

Jeswiet, J., 2002. *Rapid proto-typing of a headlight with sheet metal.* Society of Manufacturing Engineers, Dearborn Mich.

Jeswiet, J., Young, D., 2005. Forming limit diagrams for single-point incremental forming of aluminium sheet. *Proc. Inst. Mech. Eng. Part B J. Eng. Manuf.* 219, 359–364. https://doi.org/10.1243/095440505X32210

Jodhave, S., 2015. *Experimental investigations on the effect of clamping parameters and blank shape in incremental sheet forming.* PDPM, IIITDM Jabalpur.

Kalo, A., Newsum, M.J., 2014. An investigation of robotic incremental sheet metal forming as a method for prototyping parametric architectural skins. *Robot. Fabr. Archit. Art Des.* 33–49.

Khazaali, H., Fereshteh-Saniee, F., 2018. Application of the Taguchi method for efficient studying of elevated-temperature incremental forming of a titanium alloy. *J. Brazilian Soc. Mech. Sci. Eng.* 40, 43. https://doi.org/10.1007/s40430-018-1003-1

Kurra, S., Hifzur Rahman, N., Regalla, S.P., Gupta, A.K., 2015. Modeling and optimization of surface roughness in single point incremental forming process. *J. Mater. Res. Technol.* 4, 304–313. https://doi.org/10.1016/J.JMRT.2015.01.003

Leszak, E., 1967. *Apparatus and Process for Incremental-Deless Forming.*

Li, E., 2011. Reduction of springback by intelligent sampling-based LSSVR metamodel-based optimization. *Int. J. Mater. Form.*, 61(6), 103–114. https://doi.org/10.1007/S12289-011-1076-1

Li, J., Li, C., Zhou, T., 2012. Thickness distribution and mechanical property of sheet metal incremental forming based on numerical simulation. *Trans. Nonferrous Met. Soc. China* 22, s54–s60. https://doi.org/10.1016/S1003-6326(12)61683-5

Li, Y., Daniel, W.J.T., Liu, Z., Lu, H., Meehan, P.A., 2015. Deformation mechanics and efficient force prediction in single point incremental forming. *J. Mater. Process. Technol.* 221, 100–111. https://doi.org/10.1016/J.JMATPROTEC.2015.02.009

Lingam, R., Srivastava, A., Reddy, N.V., 2016. Deflection compensations for tool path to enhance accuracy during double-sided incremental forming. *J. Manuf. Sci. Eng.* 138, 091008. https://doi.org/10.1115/1.4033956

Liu, Z., Li, Y., Meehan, P.A., 2013. Vertical wall formation and material flow control for incremental sheet forming by revisiting multistage deformation path strategies. *Mater. Manuf. Process.* 28, 562–571. https://doi.org/10.1080/10426914.2013.763964

Malhotra, R., Reddy, N.V., Cao, J., 2010. Automatic 3D spiral toolpath generation for single point incremental forming. *J. Manuf. Sci. Eng. Trans. ASME* 132, 1–10. https://doi.org/10.1115/1.4002544

Martins, P.A.F., Bay, N., Skjoedt, M., Silva, M.B., 2008. Theory of single point incremental forming. *CIRP Ann.* 57, 247–252. https://doi.org/10.1016/J.CIRP.2008.03.047

Micari, F., Ambrogio, G., Filice, L., 2007. Shape and dimensional accuracy in single point incremental forming: State of the art and future trends. *J. Mater. Process. Technol.* 191, 390–395. https://doi.org/10.1016/J.JMATPROTEC.2007.03.066

Milutinović, M., Lenđel, R., Potran, M., Vilotić, D., Skakun, P., Plančak, M., 2014. Application of single point incremental forming for manufacturing of denture base. *J. Technol. Plast.* 39.

Min, J., Kuhlenkötter, B., Shu, C., Störkle, D., Thyssen, L., 2018. Experimental and numerical investigation on incremental sheet forming with flexible die-support from metallic foam. *J. Manuf. Process.* 31, 605–612. https://doi.org/10.1016/J.JMAPRO.2017.12.013

Mohammadi, A., Vanhove, H., … A.V.B.-I.J. undefined, n.d., 2016. *Towards accuracy improvement in single point incremental forming of shallow parts formed under laser assisted conditions.* Springer.

Mulay, A., Ben, S., Ismail, S., Kocanda, A., 2017. Experimental investigations into the effects of SPIF forming conditions on surface roughness and formability by design of experiments. *J. Brazilian Soc. Mech. Sci. Eng.* 39, 3997–4010. https://doi.org/10.1007/s40430-016-0703-7

Nirala, H.K., Jain, P.K., Roy, J.J., Samal, M.K., Tandon, P., 2017. An approach to eliminate stepped features in multistage incremental sheet forming process: Experimental and FEA analysis. *J. Mech. Sci. Technol.* 31, 599–604. https://doi.org/10.1007/s12206-017-0112-6

Palumbo, G., Brandizzi, M., 2012. Experimental investigations on the single point incremental forming of a titanium alloy component combining static heating with high tool rotation speed. *Mater. Des.* 40, 43–51. https://doi.org/10.1016/J.MATDES.2012.03.031

Panjwani, D., Priyadarshi, S., Jain, P.K., Samal, M.K., Roy, J.J., Roy, D., Tandon, P., 2017. A novel approach based on flexible supports for forming non-axisymmetric parts in SPISF. *Int. J. Adv. Manuf. Technol.* 92, 2463–2477. https://doi.org/10.1007/s00170-017-0223-3

Patidar, S., 2017. *Numerical investigation of blank shape and tool geometry in incremental sheet forming.* PDPM, IIITDM Jabalpur.

Petek, A., Kuzman, K., Kopaè, J., 2009. Deformations and forces analysis of single point incremental sheet metal forming 35.

Piccininni, A., Gagliardi, F., Guglielmi, P., De Napoli, L., Ambrogio, G., Sorgente, D., Palumbo, G., 2016. Biomedical Titanium alloy prostheses manufacturing by means of Superplastic and Incremental Forming processes. *MATEC Web Conf.* 80, 15007. https://doi.org/10.1051/matecconf/20168015007

Rauch, M., Hascoet, J.Y., Hamann, J.C., Plenel, Y., 2009. Tool path programming optimization for incremental sheet forming applications. *CAD Comput. Aided Des.* 41, 877–885. https://doi.org/10.1016/j.cad.2009.06.006

Seong, D.Y., Haque, M.Z., Kim, J.B., Stoughton, T.B., Yoon, J.W., 2014. Suppression of necking in incremental sheet forming. *Int. J. Solids Struct.* 51, 2840–2849. https://doi.org/10.1016/J.IJSOLSTR.2014.04.007

Shim, M.-S., Park, J.-J., 2001. The formability of aluminum sheet in incremental forming. *J. Mater. Process. Technol.* 113, 654–658. https://doi.org/10.1016/S0924-0136(01)00679-3

Shrivastava, P., Kumar, P., Tandon, P., Pesin, A., 2018. Improvement in formability and geometrical accuracy of incrementally formed AA1050 sheets by microstructure and texture reformation through preheating, and their FEA and experimental validation. *J. Brazilian Soc. Mech. Sci. Eng.* 40, 335. https://doi.org/10.1007/s40430-018-1255-9

Shrivastava, P., Roy, J.J., Samal, M.K., Jain, P.K., Tandon, P., 2015. Parameter optimization of incremental sheet metal forming based on taguchi design and response surface methodology, in: *Volume 2A: Advanced Manufacturing*. ASME, p. V02AT02A025. https://doi.org/10.1115/IMECE2015-51227

Shrivastava, P., Tandon, P., 2015. Investigation of the effect of grain size on forming forces in single point incremental sheet forming. *Procedia Manuf.* 2, 41–45. https://doi.org/10.1016/J.PROMFG.2015.07.008

Shrivastava, P., Tandon, P., 2018a. Microstructure and texture based analysis of forming behavior and deformation mechanism of AA1050 sheet during Single Point Incremental Forming. *J. Mater. Process. Technol.* https://doi.org/10.1016/J.JMATPROTEC.2018.11.012

Shrivastava, P., Tandon, P., 2018b. Enhancement of process capabilities and numerical prediction of geometric profiles and global springback in incrementally formed AA 1050 sheets. *Trans. Indian Inst. Met.* https://doi.org/10.1007/s12666-018-1346-4

Shrivastava, P., Tandon, P., 2019. Effect of preheated microstructure vis-à-vis process parameters and characterization of orange peel in incremental forming of AA1050 sheets. *J. Mater. Eng. Perform.* 28, 2530–2542. https://doi.org/10.1007/s11665-019-04032-z

Silva, M.B., Martins, P.A.F., 2013. Two-point incremental forming with partial die: Theory and experimentation. *J. Mater. Eng. Perform.* 22, 1018–1027. https://doi.org/10.1007/s11665-012-0400-3

Tera, M., Breaz, R.E., Racz, S.G., Girjob, C.E., 2019. Processing strategies for single point incremental forming—a CAM approach. *Int. J. Adv. Manuf. Technol.* 102, 1761–1777. https://doi.org/10.1007/s00170-018-03275-9

Thibaud, S., Ben Hmida, R., Richard, F., Malécot, P., 2012. A fully parametric toolbox for the simulation of single point incremental sheet forming process: Numerical feasibility and experimental validation. *Simul. Model. Pract. Theory* 29, 32–43. https://doi.org/10.1016/J.SIMPAT.2012.07.004

Ulacia, I., Galdos, L., Ander Esnaola, J., Larrañaga, J., Larrañaga, L., Arruebarrena, G., De Argandoña, S., Argandoña, A., 2014. Warm forming of Mg sheets: From incremental to electromagnetic forming. *Miner. Met. Mater. Soc. ASM Int.* https://doi.org/10.1007/s11661-014-2322-1

Wang, J., Li, L., Jiang, H., 2016. Effects of forming parameters on temperature in frictional stir incremental sheet forming. *J. Mech. Sci. Technol.* 30, 2163–2169. https://doi.org/10.1007/s12206-016-0423-z

Xu, D., Lu, B., Cao, T., Zhang, H., Chen, J., Long, H., Design, J.C.-M. &, 2016, undefined, n.d. *Enhancement of process capabilities in electrically-assisted double sided incremental forming*. Elsevier.

Young, D., Jeswiet, J., 2004. Wall thickness variations in single-point incremental forming. *Proc. Inst. Mech. Eng. Part B J. Eng. Manuf.* 218, 1453–1459. https://doi.org/10.1243/0954405042418400

Zhang, H., Lu, B., Chen, J., Feng, S., Li, Z., Long, H., 2017. Thickness control in a new flexible hybrid incremental sheet forming process. *Proc. Inst. Mech. Eng. Part B J. Eng. Manuf.* 231, 779–791. https://doi.org/10.1177/0954405417694061

Zhang, Z., Ren, H., Xu, R., Moser, N., Smith, J., Ndip-Agbor, E., Malhotra, R., Cedric Xia, Z., Ehmann, K.F., Cao, J., 2015. A mixed double-sided incremental forming toolpath strategy for improved geometric accuracy. *J. Manuf. Sci. Eng.* 137, 051007. https://doi.org/10.1115/1.4031092

Zhu, H., Liu, Z., Fu, J., 2011. Spiral tool-path generation with constant scallop height for sheet metal CNC incremental forming. *Int. J. Adv. Manuf. Technol.* 54, 911–919. https://doi.org/10.1007/s00170-010-2996-5

12 Single-Stage to Multi-Stage Incremental Sheet Forming Technology
State of the Art

Sameer Vadher and Amrut Mulay
Sardar Vallabhbhai National Institute of Technology, Surat, India

CONTENTS

12.1 INTRODUCTION

Metal forming process is one of the most common and highly used manufacturing processes in many industries with the capability of producing a good quality complex part. However, many conventional forming methods are affected by the high product demand of the customers. Moreover, prototyping is the foremost step in many manufacturing processes which increases the cost and time involved in the final product, especially when specific dies are required. Incremental Sheet Forming (ISF) is the evolved forming method or technique introduced to overcome the drawbacks of conventional forming processes. In the ISF process, the tool moves along the pre-programmed path to deform the metal sheet clamped in the fixture through small increments to achieve the final shape. The metal sheet deformed plastically through small increments provided by the tool along the programmed path which is controlled by the Computer Numerical Control (CNC) controller. The CNC machine tool carries the whole process with the assistance of the Computer-Aided Design/Computer-Aided Manufacturing (CAD/CAM) technique [1]. The toolpath is generated on the basis of the geometry of the required component with the help of appropriate software. The basic elements of the SPIF process include sheet metal blank, fixture, rotating tool, and backing plate are shown in Figure 12.1 [2].

The ISF process can be broadly classified on the basis of the forming method, forming path, tool-path strategy, etc. According to the forming method, the process can be distinguished into single-point incremental forming (SPIF), two point incremental forming (TPIF), double-sided incremental forming (DSIF), and hybrid incremental forming processes. Single Point Incremental Forming (SPIF), also known as the negative incremental forming which deforms the sheet using a single tool

DOI: 10.1201/9781003226703-12

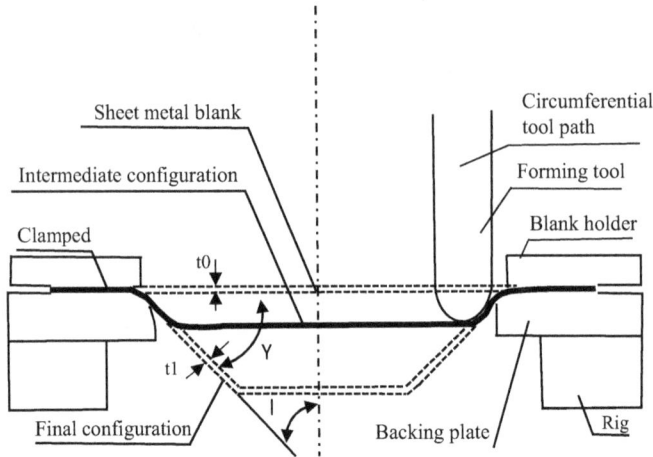

FIGURE 12.1 Schematic representation of a cross-sectional view of SPIF Process [2].

without any involvement of die whereas, in the TPIF process, partial or full die supports the blank providing controlled deformation. In SPIF majorly the applied force act on the tool-blank interface [3]. While in Two Point Incremental Forming (TPIF), also termed as positive incremental forming the forces are applied on two points. Firstly, on the interaction point between the tool and the sheet and secondly at the blank–die interface. Although the setup in TPIF is more complex than in SPIF, greater accuracy can be obtained in TPIF due to the presence of a partial or full die [4]. However, in double-sided incremental sheet forming, two tools are present i.e., one above the clamped sheet and the other below the sheet to support the blank and provide backup forces for better and controlled movement. DSIF provides extra support to the blank like TPIF but without the involvement of any die [5]. Moreover, J. Jeswiet et al. [6] reviewed various papers that study the effect of various process parameters on formability in the SPIF process and presented a detailed discussion on their interactions. Harish and Agrawal [7] studied the residual stress distribution on a geometry formed with the recently developed toolpath namely fractal geometry-based incremental toolpath (FGBIT). Venkata Reddy et al. [8] developed a methodology by combining small deflection and membrane theories to improve the accuracy by compensating the sheet deflection in double-sided incremental sheet forming. In addition, there are different types of ISF processes studied in various studies such as high-speed incremental sheet forming, warm forming or heat-assisted forming, electrical-assisted forming, Roboforming and multi-stage incremental sheet forming, etc. Warm forming is the ISF technology combined with a heating system, also known as heat-assisted forming. The formability of the sheet can noticeably increase using heat-assisted forming as with increasing temperature, tensile strength decreases, and the sheet softness increases which enhances material ability to flow by the external force. Thus hot work reduces yield strength and increases ductility and formability of the material. The heat-assisted forming technology is applicable for hard materials having low formability at room temperature such as titanium. However, in electrical assisted incremental sheet forming the sheet specimen is locally heated by electrical current, and heat is generated locally at tool sheet interface called joule heating effect. In addition, roboforming is another variant which is a die-less ISF Process utilizing two synchronized robots (master robot and slave robot). The each robot is equipped with a sphere-tipped tool. The metal sheet is held between the master robot and slave robot. The master robot moves incrementally on the sheet which deforms the sheet in the shape of the toolpath and the slave robot provides support to the master robot from the opposite face of the sheet. This process is also known as robot assisted incremental sheet forming. Although ISF is widely used in many manufacturing and automobile industries it restricts its usage to the mainstream high-value manufacturing industries due to some production challenges among which accuracy is one of the major factors. To compensate the geometrical deviation (i.e., springback), the most commonly used solution is to modify the toolpath or use the toolpath compensation strategy to implies more deformation on the sheet [9]. Thus the springback effect can be compensated by altering the toolpath. Furthermore, many studies and researches have been done on multi-stage incremental sheet forming as an alternative to SPIF. Kurra Suresh et al. [10] carried out an experimental study in multi-stage forming to achieve maximum wall angle (steep wall). Multi-stage forming shows better thickness distribution compared with single-stage forming, Kai Han et al. [11] studied various toolpath strategies in multi-stage forming to effectively improve thickness distribution. In addition, G. Vignesh et al. [12] reviewed multistage incremental sheet forming and various toolpath strategies which were discussed in the next section. However, multi-stage incremental sheet forming could be an alternative to overcome some drawbacks such as undesired plastic deformations, springback, and non-uniform thickness distribution present in SPIF Process. In multi-stage incremental sheet forming the sheet deforms in the number of stages wherein the sheet deformed in the previous stage is further deformed in the subsequent stages. The formability of the SPIF process can be improved by introducing some intermediate stages into it, this process is termed as Multi-Stage Incremental Sheet Forming (MSPIF) process. More discussion about the multi-stage incremental sheet forming is illustrated here by the authors.

12.2   MULTI-STAGE INCREMENTAL SHEET FORMING

The geometric inaccuracies, excessive thinning of the sheet, and formability are some major challenges faced in the SPIF process. Due to some limitations associated with SPIF the studies were carried out on multi-stage forming technique instead of single-stage. A study about the various causes of the defect is required to improve the geometric accuracy. The springback, undesired plastic deformations, and non-uniform thickness distribution may induce geometric inaccuracy in the SPIF process. The strain hardening effect that appeared in the deformed sheet changes its capacity to deform. Moreover the effect becomes more severe as the deformation goes on increasing. Furthermore, with the increase in wall angle the deformation increases rapidly due to biaxial stretching. With the introduction of some intermediate stages in the ISF process, the strain hardening effect reduces since the overall deformation gets distributed between these stages which ultimately results in increased formability. This is similar to the shaft subjected to axial load where the deformation in the shaft is higher when it is subjected to axial load in the multiple stages. The SPIF process carried out in more than one stage is known as Multi-stage Incremental Sheet Forming as shown in Figure 12.2 [1]. According to the sine law the sheet thickness decreases as the wall angle increases and theoretically becomes zero at maximum wall angle (i.e., 90°). MSPIF can be an alternative to single-stage forming to obtain a large wall angle by redistributing material or shift the material with multi-stages. A several researches have been made in an attempt to find an optimal number of stages to successfully achieve the maximum possible wall angles with satisfactory geometric accuracy of the part. In addition to this, the actual toolpath also has a significant influence over the successful forming of sheet metal with maximum achievable wall angles [13].

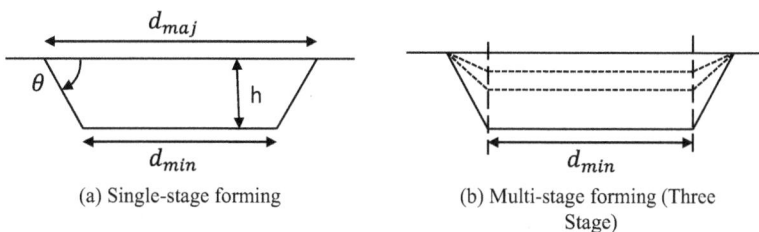

(a) Single-stage forming   (b) Multi-stage forming (Three Stage)

FIGURE 12.2   (a) Single-stage and (b) Multi-stage forming [14].

12.2.1 Toolpath and Forming Strategies in Multi-Stage Incremental Sheet Forming (MSPIF)

In the ISF method, the toolpath is the path or contour along which the forming tool is driven to achieve the targeted shape of the object. The tool moves layer by layer from one contour to another contour. The toolpath is a critical parameter in the ISF process since it specifies the component's dimensional accuracy. Furthermore, the toolpath specifies the necessary forming force which is useful in deciding the ISF machinery requirements. The right selection of the toolpath improves the formability, surface finish, and dimensional accuracy, as well as the processing time of the resulting part [15]. The toolpaths can be divided into two groups namely the profile toolpath and the helical toolpath. In the profile toolpath, also known as the constant Z level toolpath as shown in Figure 12.3 (a) in which the tool moves in the same plane until it reaches the starting position. Once it reaches the starting position step over is given followed by a step down and process repeats until the entire shape has been created. However, in the helical toolpath (continuous toolpath) the tool travels over the peripheral of the finished part in a helical path with pitch equal to step down as shown in Figure 12.3 (b). Many researches have been made to study the effectiveness of profile toolpath and helical toolpath and found that the helical toolpath successfully deforms the sheet by achieving targeted shape of the geometry whereas using constant z-level toolpath, cracks were observed in the sheet before reaching at the targeted depth [7, 8]. Furthermore, the tool force is more stable during the helical toolpath compared with the profile toolpath [9, 10]. The helical toolpath provides forming-force stability with enhancing component surface consistency whereas, the profile toolpath creates stretch marks on the sheet surface due to the force signal instability during the ISF process [6, 9, 10].

In order to form part in MSPIF, different forming strategies can be employed. In MSPIF, the tool-path can be broadly categorized into parallel line type, curved type, and straight-line type with a variable angle according to the forming method. These can be further classified into Out to In and In to Out toolpaths. Two-stage incremental sheet forming process with OI (Out to In) & IO (In to Out) tool-path can be carried out by two methods (down-down (DD) & down-up (DU)). In the DD method, both stages are performed by the OI toolpath. Wherein the DU method, the first stage is performed by the OI toolpath followed by the IO toolpath [12]. The different MSPIF strategies like small corner radius strategy (strategy 1), large corner radius strategy (strategy 2), and In-plane radius strategy (strategy 3) are associated with the pyramidal shape are discussed here (Figure 12.4). The corner radius of the pyramid in the small corner radius strategy is smaller in the first stage than in the final stage while reverse is true in the large corner radius strategy i.e., the corner radius is larger in the first stage than that of the final stage. However, the In-plane radius method cone is formed in the initial stage which is later on converted into a pyramidal shape using the number of stages. In addition, some more strategies

(a) Profile tool path   (b) Helical tool path

FIGURE 12.3   (a) Profile toolpath (Constant Z level Toolpath) and (b) Helical toolpath (Continuous toolpath) [18].

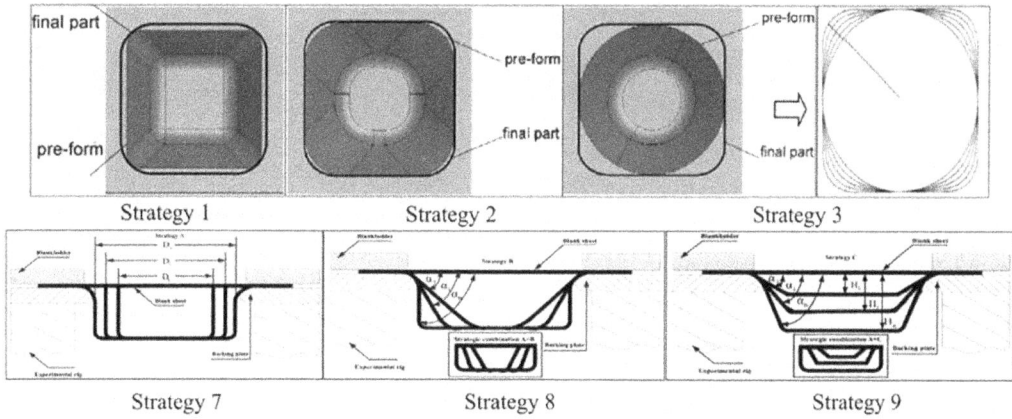

FIGURE 12.4   Forming strategies during MSPIF [12].

were also discussed such as Down, Down, Down, and Down (DDDD) strategy (strategy 4), Down, Down, Down, and Up (DDDU) strategy (strategy 5), Down, Up, Down, and Down (DUDD) strategy (strategy 6), part diameter strategy (strategy 7). The diameter of the cup is increased gradually in each stage in part diameter method (strategy 7) whereas there is a gradual increase in wall angle in each consecutive stage in the incremental part draw angle method (strategy 8). Furthermore, the wall angle and height of the cup are increased gradually and simultaneously in each stage in the incremental part height and draw angle method (strategy 9). Few results of the above strategies are discussed here. The cup formed using DDDU strategy was found without any fracture whereas it resulted in the fracture in the fourth stage while using DUDD due to high thickness strain. It can be observed that the maximum strain was found near the transition region of the cup where the tool was moving upward as more biaxial strains incorporates during upward tool movement than downward tool movement [2]. Using the corner pushing strategy, Cup with 72 degrees wall angle can be formed wherein the material gets squashed near the corner. Moreover in case of flat-bottom method, the cup height increased gradually in each stage by alternating OI and IO tool movements during the process [12].

12.3   PARAMETRIC INVESTIGATION OF MULTI-STAGE INCREMENTAL SHEET FORMING

The performance of the ISF process is significantly affected by various process parameters like step depth, toolpath and spindle speed etc. However, many studies have been done on these process parameters by various researchers in recent times. For example Tera et al. [20] investigated different toolpath strategies to increase part accuracy. Carette et al. [21] used a new methodology to create an automatic toolpath based on feature geometry. The parameters affecting the performance of the MSPIF process are discussed in the next sub-section.

12.3.1   FRICTION AND FORMING TOOL DIRECTION

The presence of little friction can be beneficial for forming as the friction between the tool-sheet interfaces increases the tool pressure and lowers the state of stress in the sheet. Hence, there is a delay in crack occurrence and the formability is improved. Opposite to this, high friction may result in early sheet crack. Kim and Park [22] conducted the study to investigate the effect of friction at tool sheet interface and tool shape for aluminum 1050 sheet of $130 \times 130 \times 0.3$ mm dimensions. The hemispherical head tool and ball tool were used with and without lubrication to study the effects. The value of $\varepsilon_{major} + \varepsilon_{minor}$ was found to be 0.72 and 0.73 with and without lubrication respectively

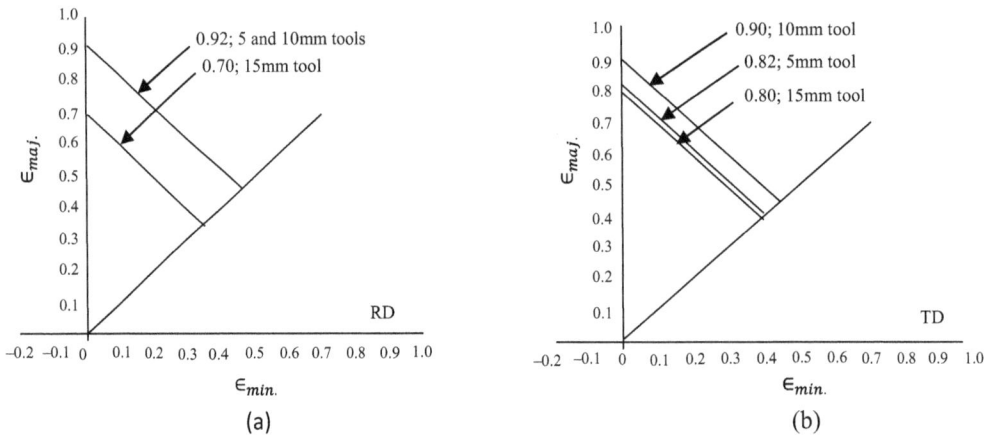

FIGURE 12.5 Forming Limit Curves for different tool size in (a) Rolling Direction (RD) and (b) Transverse Direction (TD) [22].

by using the ball tool. Furthermore, these values were 0.67 and 0.69 with and without lubrication, respectively with the hemispherical head tool. The most ideal condition with improved formability was observed when the ball tool without lubrication was used. It has been concluded that the ball tool was more effective in increasing the formability of the material than the hemispherical head tool. However, the material with plane anisotropy shows different forming abilities in different directions. Hence the forming direction becomes the other important parameter in incremental sheet forming. Kim and Park [22] compared forming of the sheet along the rolling direction and transverse direction with the tool size and found that the value of $\varepsilon_{major} + \varepsilon_{minor}$ was found 0.92 with 5 mm tool diameter, 0.92 with 10 mm tool diameter, and 0.7 with 15 mm tool diameter when the tool was moving parallel to the rolling direction. However, these values tend to 0.82, 0.9, and 0.8, respectively when the tool moves in the transverse direction. The crack was observed at the depth of 6 mm with the tool of 5 mm diameter moving parallel to the rolling direction, at 7.5 mm depth with 10 mm tool diameter moving both in parallel and perpendicular (transverse) to the rolling direction, and at 8.6 mm depth along the transverse direction with 15 mm tool diameter. Furthermore, with the increase in the tool size, the deformation zone also increases with decrease in the strain level as shown in Figure 12.5.

12.3.2 LUBRICANT IN ISF PROCESS

The quality of components and the viability of the ISF process are determined by the lubricant selection. Lubrication is used in the ISF process to minimize friction between the work piece and the tool tip. It can be applied as a paste or spray coating on the blank to achieve self-lubricating results, or it can be applied in liquid form with the forming tool and blank submerged in lubricant. Mulay et al. [23] conducted a detailed study to investigate the effect of different lubricants used in various SPIF processes to achieve the better surface roughness and formability of Al 5052 H32 sheets. The lubricant must be able to withstand forming temperatures and not be squeezed out to ensure the presence of lubricant between the tool and blank during the entire forming process [24]. Lubrication affects the formed component's surface quality, defects, and formability. The right lubricant can minimize friction present between the sheet and the tool, resulting in reduced forming forces, increased heat dissipation, and reduced tool wear and all of which can extend tool life [14–16]. The molybdenum disulfide is one of the widely used lubricants during forming of hard materials such as titanium [28]. The low coefficient of friction (0.03–0.06), with a strong affinity towards metallic surfaces, makes MoS2 more suitable for hard materials. Moreover, it has high stability and effective lubricating properties for cryogenic temperatures in the presence of most solvents. Sornsuwit and Sittisakuljaroen

[29] performed experiments on different grades of stainless steel (SUS 304 and SUS 316L) and Ti-Gr 2 sheets using different lubricants and found that the SUS 316L and SUS 304 grades showed better surface roughness with MoS2 and air blowing lubricants with the acceptable average surface roughness (R_a) value of 0.58 μm and 0.50 μm, respectively for SUS 316L grade and 0.63 μm and 0.35 μm, respectively for SUS 304 grade of stainless steel. Moreover, Ti Gr 2 (Commercially pure titanium) with MoS2 lubricant shows the best surface roughness with a R_a value of 1.12 μm. Further, to investigate the formability behavior of material each sheet was formed up to 15 mm depth or till failure. The SUS 304 and SUS 316L were formed without any noticeable fracture with oil, MoS2, and air as lubricants. However, SUS 304 grade was failed at 12 mm depth when formed without any lubricant. In addition, Ti Gr 2 sheet successfully formed up to the depth of 13 mm, 11 mm, and 12 mm with MoS2, air blowing, and without any lubricant, respectively. As per the results, the stainless steel grades having almost identical properties were successfully formed using the air and MoS2 as a lubricant. However, the titanium sheet was successfully formed up to 13 mm with MoS2 lubricant with an average surface roughness (R_a) value of 1.12 μm. The study shows that the MoS2 lubricant could be suitable for hard materials having low formability such as pure titanium or titanium alloys sheet.

12.3.3 ANGLE INTERVAL (DA) BETWEEN STAGES

The design of multi-stage incremental forming is closely connected with two important factors such as forming stages (n) and angle interval (Da).

Li et al. [30] conducted a detailed analysis to study the influence of the number of forming stages and angle interval between the consecutive stages on section thickness in MSPIF process using both simulation and experimental verification. The DC06 sheet of 150 × 150 × 0.8 mm dimension was used. A more uniform thickness distribution was observed with the increase in the number of stages. However, the increase in the number of stages leads to an increase in the minimum sheet thickness. As shown in Figure 12.6(b) with the increase in angle interval (Da), the thickness increases initially but later on decreases. The maximum and uniform sheet thickness were obtained with the angle interval of 10° between consecutive stages. Further study needs to be done in future to find the optimal value of Da. In addition, more plastic deformation area was obtained with a more uniform thickness which leads to the larger final thickness. Li et al. [31] conducted the simulation analysis on the DC56 sheet of 380 × 380 × 1 mm dimension to study the influence of the number of forming stages (n) and angle interval (Da) between the consecutive stages on the equivalent plastic strain and springback. A conical toolpath was used for forming the sheet material. To analyze the effect of angle interval (Da) and forming stages (n) on the equivalent plastic strain, a graph was plot between

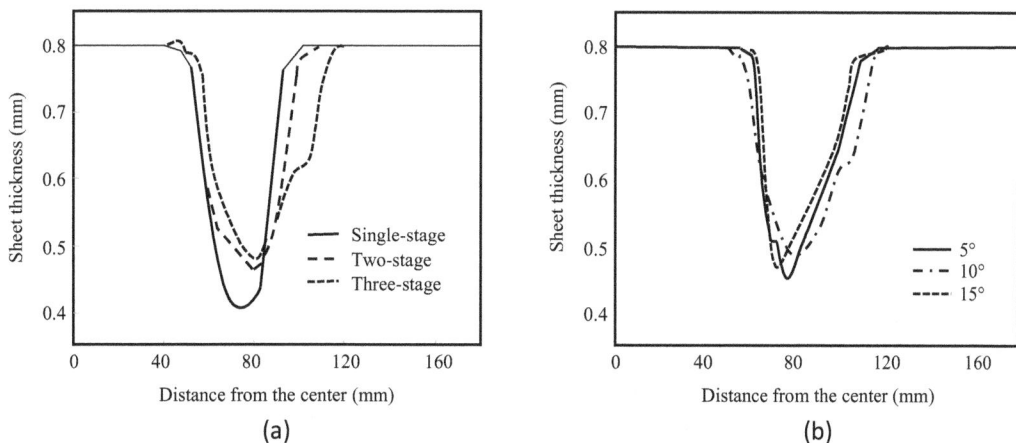

FIGURE 12.6 Thickness distributions with different (a) forming stages and different (b) angle intervals [30].

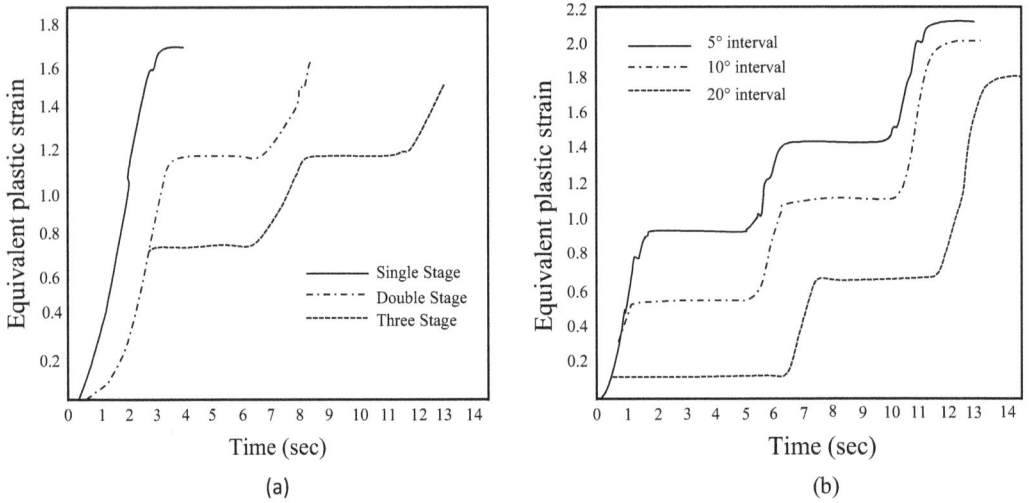

FIGURE 12.7 Equivalent plastic strain (a) At different forming stages and (b) At different angle interval [31].

equivalent plastic strain and time using a node situated on the thinnest area of the deformed part. As shown in Figure 12.7 (a) and (b) with the increase in the forming stages, the equivalent plastic strain and the maximum total strain decreases (depends on wall angle only). Moreover, more uniform deformation is observed with the increase in the forming stages. In addition, the equivalent plastic strain and the final accumulated strain decrease as the angle interval (Da) between stages increases. An angle interval has a negligible effect on springback while for forming stages (n), the cup formed using single-stage forming was found closest to the targeted shape or geometry than the cup formed using multi-stage forming process and this error or deviation goes on increasing as the forming stages increases. However, by modifying the toolpath at each stage, the error observed can be solved or minimized. It is worthwhile to remark that the error near the round corner significantly increases for multi-stage processes.

12.4 CASE STUDY-MSPIF

To study the effect of angle interval between the stages and number of stages on section thickness (STH) simulation analysis has been conducted on commercially pure titanium (Ti Gr 2) of 222 × 222 × 0.5 mm dimension. The hemispherical tool was used in the forming process. A spiral toolpath was generated in MATLAB to form the sheet into a truncated cone shape. A 2-D deformable model of 222 × 222 × 0.5 mm Ti-Gr 2 sheet was made in abaqus explicit. The tool was regarded as an analytical rigid body of 10 mm diameter. The sheet was fixed by the four edges and meshed by linear quadrilateral elements of S4R type with 0.8 mm mesh size. The mechanical properties of Ti Gr 2 are shown in Table 12.1. Moreover, coulomb's friction law was applied using the lubricant having a friction coefficient of 0.045 between the sheet blank and the forming tool.

TABLE 12.1
Mechanical Properties of Titanium Grade 2 [32]

Sr no.	Material	Yield Strength (MPa)	Tensile Strength (MPa)	Strain Hardening Exponent (n value)	Strength Coefficient (MPa) (K value)
1	Titanium grade 2	284	420	0.17	495

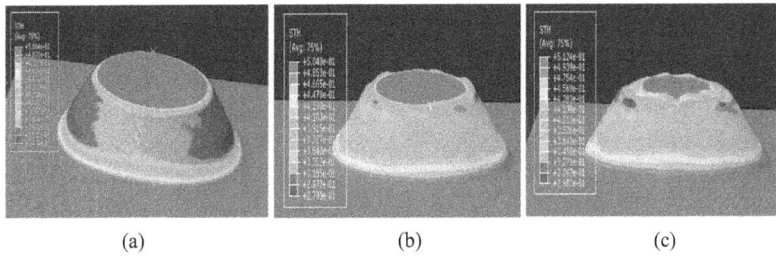

FIGURE 12.8 (a) Single-Stage incremental forming, (b) Two-Stage incremental forming, and (c) Three-Stage incremental forming.

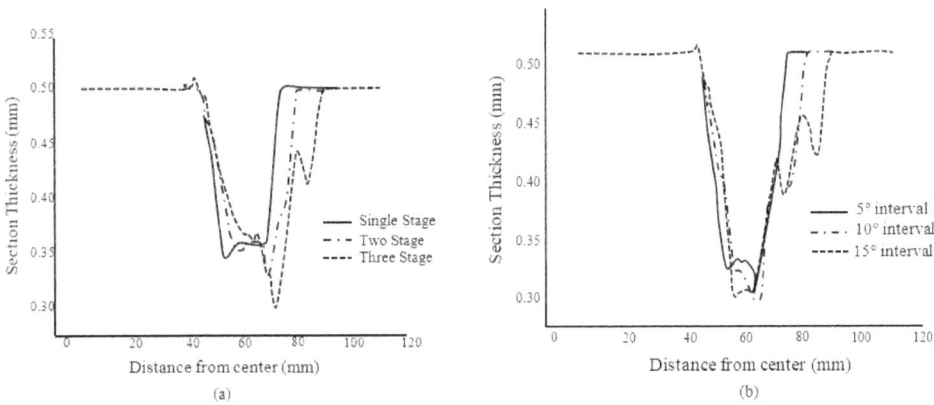

FIGURE 12.9 (a) Effect of forming stages on section thickness and (b) Effect of angle interval on section thickness.

Figure 12.8 (a)–(c) show the simulation results of a cup formed at 40° with 10 mm tool diameter from a Ti Gr 2 sheet of 0.5 mm thickness. The sheet formed in three-stage forming shows more uniform sheet thickness compared with two-stage and single-stage forming. Li et al. [30] carried out a variance analysis of the thickness distribution curves obtained from the simulation result and observed that the variance of three-stage forming was the least compared with single-stage and two-stage forming. Furthermore, the lesser the variance value, the lesser will be the thickness fluctuations. Hence, a more uniform thickness distribution can be observed with an increase in the number of stages. Moreover, it is observed that the gradual and uniform deformation occurred with the increase in intermediate stages. As shown in Figure 12.9 (a) three-stage forming shows a gradual reduction in thickness compared with single-stage and two-stage forming. Moreover, the excessive thinning is delayed or observed later in three-stage forming. Figure 12.9 (b) shows the effect of angle interval on section thickness (STH). The sheet formed with 10° angle interval shows a more uniform section in the middle portion of the cup. Moreover, the sheet formed with a 15° angle interval shows a noticeable increase and decrease of section thickness near the bottom of the cup. This effect may be resulted because of the springback effect or preformed cup in the previous stages. The author agrees that further studies have to perform with the combination of different angle interval in MSPIF.

12.5 CONCLUSION

The multi-stage incremental sheet forming allows precise fabrication of complex geometries with the maximum possible wall angles. With multi-stage, the excessive thinning can be reduced or delayed and better formability and accuracy. Although the time consumption in multi-stage incremental forming is more compared with single-stage incremental forming but better formability and uniform deformation can be obtained. It can be concluded that the angle interval between stages does not have a significant effect on springback. However, the equivalent plastic strain and the final accumulated strain decrease as the angle interval increases. The more uniform deformation is observed with the increase in the number of forming stages. The reduction of surface roughness and better formability of hard materials such as pure titanium (CP Ti) or titanium alloys can be obtained with MoS_2 lubricant. It was noticed that forming along the transverse direction with optimal tool diameter could result in better formability. The simulation result shows that there is a gradual reduction of section thickness in multi-stage forming compared with a single stage. A more uniform section thickness can be obtained in the middle portion of the formed cup with the optimal angle interval between the stages. Further studies with the combination of different angle intervals in multi-stage incremental sheet forming will give more interesting results.

REFERENCES

[1] H. K. Nirala, P. K. Jain, J. J. Roy, M. K. Samal, and P. Tandon, "An approach to eliminate stepped features in multistage incremental sheet forming process: Experimental and FEA analysis," *J. Mech. Sci. Technol.*, vol. 31, no. 2, pp. 599–604, 2017, doi:10.1007/s12206-017-0112-6.

[2] M. Skjoedt, M. B. Silva, P. A. F. Martins, and N. Bay, "Strategies and limits in multi-stage single-point incremental forming," *J. Strain Anal. Eng. Des.*, vol. 45, no. 1, pp. 33–44, 2010, doi:10.1243/03093247JSA574.

[3] M. Tisza, "General overview of sheet incremental forming," *Manuf. Eng.*, vol. 55, no. 1, pp. 113–120, 2012, [Online]. Available: http://www.journalamme.org/papers_vol55_1/55114.pdf.

[4] G. Hirt and M. Bambach, "Incremental sheet forming," *Sheet Met. Form.*, pp. 273–287, 2020, doi:10.31399/asm.tb.smfpa.t53500273.

[5] W. Peng, H. Ou, and A. Becker, "Double-sided incremental forming: A review," *J. Manuf. Sci. Eng. Trans. ASME*, vol. 141, no. 5, pp. 1–12, 2019, doi:10.1115/1.4043173.

[6] T. McAnulty, J. Jeswiet, and M. Doolan, "Formability in single point incremental forming: A comparative analysis of the state of the art," *CIRP J. Manuf. Sci. Technol.*, vol. 16, pp. 43–54, 2017, doi:10.1016/j.cirpj.2016.07.003.

[7] H. K. Nirala and A. Agrawal, "Residual stress inclusion in the incrementally formed geometry using Fractal Geometry Based Incremental Toolpath (FGBIT)," *J. Mater. Process. Technol.*, vol. 279, no. 320, p. 116575, 2020, doi:10.1016/j.jmatprotec.2019.116575.

[8] K. Praveen, R. Lingam, and N. Venkata Reddy, "Tool path design system to enhance accuracy during double sided incremental forming: An analytical model to predict compensations for small/large components," *J. Manuf. Process.*, vol. 58, no. August, pp. 510–523, 2020, doi:10.1016/j.jmapro.2020.08.014.

[9] H. Ren, J. Xie, S. Liao, D. Leem, K. Ehmann, and J. Cao, "In-situ springback compensation in incremental sheet forming," *CIRP Ann.*, vol. 68, no. 1, pp. 317–320, 2019, doi:10.1016/j.cirp.2019.04.042.

[10] K. Suresh, H. R. Nasih, N. V. K. Jasti, and M. Dwivedy, "Experimental studies in multi stage incremental forming of steel sheets," *Mater. Today Proc.*, vol. 4, no. 2, pp. 4116–4122, 2017, doi:10.1016/j.matpr.2017.02.316.

[11] X. Li, K. Han, and D. Li, "Multi-stage two point incremental sheet forming," *J. Phys. Conf. Ser.*, vol. 1063, no. 1, 2018, doi:10.1088/1742-6596/1063/1/012064.

[12] G. Vignesh, C. Pandivelan, and C. Sathiya Narayanan, "Review on multi-stage incremental forming process to form vertical walled cup," *Mater. Today Proc.*, vol. 27, pp. 2297–2302, 2019, doi:10.1016/j.matpr.2019.09.116.

[13] S. Mohanty, S. P. Regalla, and Y. V. D. Rao, "Multi-stage and robot assisted incremental sheet metal forming," pp. 1–5, 2015, doi:10.1145/2783449.2783457.

[14] M. M. Gonzalez et al., "Costing models for capacity optimization in analysis of geometric accuracy and thickness reduction in multistage analysis of geometric accuracy and thickness reduction in multistage incremental sheet," *Procedia Manuf.*, vol. 34, pp. 950–960, 2019, doi:10.1016/j.promfg.2019.06.105.

[15] A. Kumar, V. Gulati, P. Kumar, and H. Singh, *Forming force in incremental sheet forming: a comparative analysis of the state of the art*, vol. 41, no. 6. Springer, Berlin Heidelberg, 2019.

[16] A. Blaga and V. Oleksik, "A study on the influence of the forming strategy on the main strains, thickness reduction, and forces in a single point incremental forming process," vol. 2013, 2013.

[17] S. Thibaud, R. Ben Hmida, F. Richard, and P. Malécot, "Simulation Modelling Practice and Theory A fully parametric toolbox for the simulation of single point incremental sheet forming process: Numerical feasibility and experimental validation," *Simul. Model. Pract. Theory*, vol. 29, pp. 32–43, 2012, doi:10.1016/j.simpat.2012.07.004.

[18] A. Kumar and V. Gulati, "Experimental investigations and optimization of forming force in incremental sheet forming," *Sādhanā*, vol. 43, no. 10, pp. 1–15, 2018, doi:10.1007/s12046-018-0926-7.

[19] H. Arfa, R. Bahloul, and H. Belhadjsalah, "Finite element modelling and experimental investigation of single point incremental forming process of aluminum sheets: Influence of process parameters on punch force monitoring and on mechanical and geometrical quality of parts," pp. 483–510, 2013, doi:10.1007/s12289-012-1101-z.

[20] M. Tera, R. E. Breaz, S. G. Racz, and C. E. Girjob, "Processing strategies for single point incremental forming—a CAM approach," *Int. J. Adv. Manuf. Technol.*, vol. 102, no. 5–8, pp. 1761–1777, 2019, doi:10.1007/s00170-018-03275-9.

[21] Y. Carette, H. Vanhove, and J. Duflou, "Multi-step incremental forming using local feature based toolpaths," *Procedia Manuf.*, vol. 29, pp. 28–35, 2019, doi:10.1016/j.promfg.2019.02.102.

[22] Y. H. Kim and J. J. Park, "Effect of process parameters on formability in incremental forming of sheet metal," *J. Mater. Process. Technol.*, vol. 130, no. 131, pp. 42–46, 2002, doi:10.1016/S0924-0136(02)00788-4.

[23] A. Mulay, S. Ben, and S. Ismail, "Lubricant selection and post forming material characterization in incremental sheet forming lubricant selection and post forming material characterization in incremental sheet forming," 2020, doi:10.1088/1757-899X/967/1/012072.

[24] Q. Zhang et al., "Warm negative incremental forming of magnesium alloy AZ31 Sheet: New lubricating method," *J. Mater. Process. Technol.*, vol. 210, no. 2, pp. 323–329, 2010, doi:10.1016/j.jmatprotec.2009.09.018.

[25] "A general overview of tribology of sheet metal forming A General Overview of Tribology of Sheet," no. March, 2020.

[26] B. H. Lee, Y. T. Keum, and R. H. Wagoner, "Modeling of the friction caused by lubrication and surface roughness in sheet metal forming," *J. Mater. Process. Technol.*, vol. 130–131, pp. 60–63, 2002, doi:10.1016/S0924-0136(02)00784-7.

[27] N. G. Azevedo, J. S. Farias, R. P. Bastos, P. Teixeira, J. P. Davim, and R. J. Alves de Sousa, "Lubrication aspects during Single Point Incremental Forming for steel and aluminum materials," *Int. J. Precis. Eng. Manuf.*, vol. 16, no. 3, pp. 589–595, 2015, doi:10.1007/s12541-015-0079-0.

[28] S. Imoa and S. Imoa, "Molybdenum-sulfur compounds in lubrication," pp. 1–4, 2021.

[29] N. Sornsuwit and S. Sittisakuljaroen, "The effect of lubricants and material properties in surface roughness and formability for single point incremental forming process," vol. 979, pp. 359–362, 2014, doi:10.4028/www.scientific.net/AMR.979.359.

[30] J. C. Li, F. F. Yang, and Z. Q. Zhou, "Thickness distribution of multi-stage incremental forming with different forming stages and angle intervals," *J. Cent. South Univ.*, vol. 22, no. 3, pp. 842–848, 2015, doi:10.1007/s11771-015-2591-x.

[31] J. Li, P. Geng, and J. Shen, "Numerical simulation and experimental investigation of multistage incremental sheet forming," no. October 2013, doi:10.1007/s00170-013-4870-8.

[32] G. Yoganjaneyulu and C. Sathiya Narayanan, "A comparison of fracture limit analysis on titanium grade 2 and titanium grade 4 sheets during single point incremental forming," *J. Fail. Anal. Prev.*, vol. 19, no. 5, pp. 1286–1296, 2019, doi:10.1007/s11668-019-00721-y.

13 Multi-Response Optimization of Process Parameters to Minimize Geometric Inaccuracies in the Single Point Incremental Forming Process

Narinder Kumar and Anupam Agrawal
Indian Institute of Technology, Ropar, India

CONTENTS

13.1 INTRODUCTION

The contemporary manufacturing sector is still dedicated to providing high-quality items with great geometric precision and surface texture in large quantities. The sheet metal processes range from traditional manual operations to automated systems. The main factors essential for selecting a particular sheet metal forming process are the number of parts to be produced and their accuracy level. There is also a growing demand for alternative manufacturing methods to meet global challenges. Therefore, a new manufacturer needs to use new manufacturing techniques to succeed in the global market. The basic tools must be produced quickly to reduce the production time for the new parts. The production time can also be reduced by adopting rapid manufacturing techniques. These techniques can use standard tools and materials. Due to the high cost of equipment and tools, the conventional technology of sheet metal forming requires large-scale production. Many contemporary manufacturing companies now employ forming processes to produce high-profitable sheet metal products. These techniques need high initial estimates and prolonged planning durations,

DOI: 10.1201/9781003226703-13

with specific passes for each item, especially when parts with complicated forms necessitate a small arrangement, as in the case of aerodynamic and automobile parts. As a result, flexible innovation is required for small-medium parts/products.

Incremental Sheet Forming (ISF) is a new forming technique that is ideal for producing small quantities or prototypes of sheet metal parts. The ISF technique is broadly categorized into:

(a) Single-point incremental forming (SPIF) process.
(b) Two-point incremental forming (TPIF) process.

Another approach is to combine incremental sheet forming along with stretch forming. SPIF process is highly developed and offered flexibility with adequate precision for creating symmetrical, asymmetrical, and random forms.

ISF is the forming technique in which the thin layer of structure or sheet metal is gradually deformed according to a given tool trajectory on the computer numerical controlled machine (CNC) with a solid hemispherical-shaped single-point tool, as shown in Figure 13.1.

ISF sheet metal processes are well-known methods that are flexible and accurate in developing symmetrical, asymmetric, and random shapes. The modes of deformation in the ISF process are mentioned below (Singh and Agrawal 2016):

(a) In incremental forming, the tool moves horizontally as well as vertically in each increment according to the generated toolpath.
(b) The deformation at the beginning and co3mpletion of each step is bi-axial as the tool plunges through a sheet.
(c) The deformation when the tool travels horizontally is plane-strain stretching.
(d) As the forming depth increases with each passing increment, the deformation turns more into biaxial stretching.

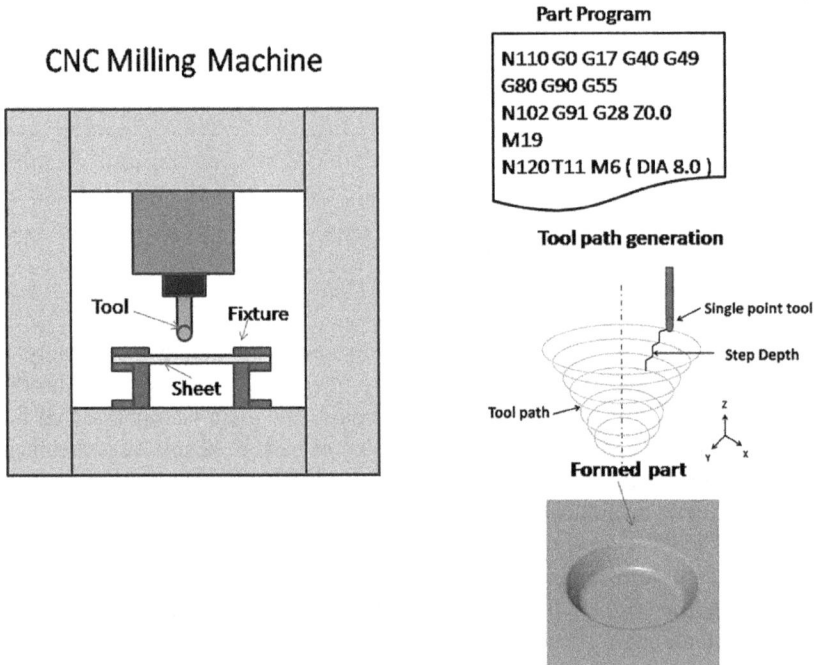

FIGURE 13.1   Elements of the incremental forming process.

13.1.1 Geometric Accuracy

A product designer specifies the tolerances within which a part has to be manufactured. Then, a manufacturer develops a part within these constraints. In the case of the SPIF process, there are three types of geometric errors occurred in the finished part when the tool retract from the surface of the part as mentioned below (Figure 13.2):

(i) Sheet bending: When the sheet is fixed and the tool is moved over the sheet surface, part of the sheet is not directly in touch with the tool. Because of this, rather than the forming, sheet bending has been developed. However, a backup plate can address this problem.

(ii) Sheet springback: This phenomenon occurred when the tool moved away from the sheet after completing the process. The sheet lifts to a certain level which creates a geometric error in the formed part.

(iii) Pillow effect: This occurred at the minor base of the part, as shown in Figure 13.2. Here, a concave surface in the middle portion of the part reflects this type of error.

The geometric error described above can be measured by comparing the ideal CAD profile with the real part profile. Here, in this chapter, an emphasis is put on reducing the geometric error by using a support, optimization method, and extending the tool path.

In the present scenario, geometric accuracy is one of the main drawbacks and needs to be further investigated. Several approaches have been proposed to address this issue, as shown in Table 13.1.

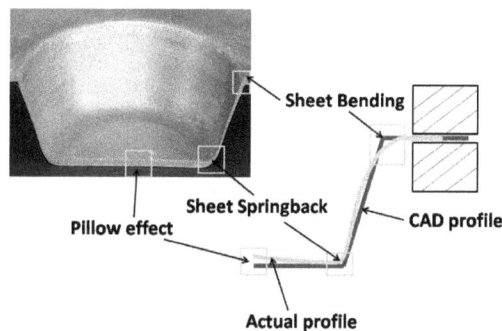

FIGURE 13.2 Geometrical inaccuracies during SPIF process.

TABLE 13.1
Techniques to Improve Accuracy in ISF Process

S. No	Authors Name and Year	Findings
1.	Malhotra et al. (2011)	Squeezing toolpath strategy
2.	Wei et al. (2011)	Forming error compensation method
3.	Han et al. (2013)	PSO-ANN algorithm
4.	Ambrogio et al. (2013)	Decremental slope trajectory
5.	Khan et al. (2014)	Intelligent process model (IPM) based on cloud computing
6.	Yao et al. (2017)	Contour compensation strategy
7.	Shrivastava et al. (2018)	Preheating of sheet blank
8.	Taherkhani et al. (2018)	Group method of data handling (GMDH)
9.	Nirala and Agrawal (2018)	Fractal geometry-based incremental toolpath (FGBIT)
10.	Singh and Agrawal (2018)	Quantification of geometrical errors

13.2   EXPERIMENTAL DETAILS

13.2.1   PROCESS VARIABLES OF SINGLE POINT INCREMENTAL FORMING PROCESS (SPIF)

Based on a review of the literature related to the SPIF process, geometric inaccuracy was obtained at a wall angle of 60° (Li et al. 2015). Furthermore, past research indicates that the following process variables have the greatest influence on the formed part geometrical accuracy by using the SPIF process were: (i) step depth, (ii) tool diameter, (iii) feed rate, and (iv) spindle speed. These four process variables were taken into account in the current study, and the process window for manufacturing the component with a wall angle of 60° was proposed. Table 13.2 shows the process variables that were analyzed at four levels (Kumar and Belokar 2019). As the test geometry in this study, a conical profile with a major diameter of 70 mm, forming depth of 15 mm, and wall angle of 60° was used as test geometry.

The trials were categorized in accordance with the orthogonal array that was chosen. While identifying a certain orthogonal array, the following inequality should be fulfilled.

$$\text{total DOF}_{OA} \geq \text{total DOF}_{experiment} \text{ where DOF: degrees of freedom.}$$

A total of 12 DOF trials are conducted in this study with four parameters at four levels. For this combination of variables, the L16 orthogonal array has thus been selected as previously utilized (Echrif and Hrairi 2014).

The influence of chosen process parameters is investigated on the following SPIF response characteristics:

(a)   Average Radial Error (mm)
(b)   Springback (°)
(c)   Pillow Effect (mm)

Average Radial Error, Springback, and Pillow effect were taken as "Lower-the-Better" type. This study aims to obtain with a multi-response technique based on Taguchi's method and utility concept. Table 13.3 shows the observed response parameter values.

13.2.2   UTILITY CONCEPT

Optimization of a process refers to identifying conditions in which the process results in the desired outcome. The desired outcome depends upon the product specification as well as the practical requirement of higher productivity. The optimization exercise helps the end-user to decide on the selection of process parameters for certain objectives. The industrial usage of every process can only be justified if the process parameters are optimally specified.

TABLE 13.2
Range of Forming Parameters

Parameters	Units	Level 1	Level 2	Level 3	Level 4
Step depth (A)	mm	0.2	0.4	0.6	0.8
Tool diameter (B)	mm	8	10	12	14
Feed rate (C)	mm/min.	1000	1500	2000	2500
Spindle speed (D)	rpm	1500	2000	2500	3000

TABLE 13.3

Experimental Results of Various Response Characteristics

Exp. No.	Average Radial Error (mm)			S/N Ratio (dB)	Springback°			S/N Ratio (dB)	Pillow Effect (mm)			S/N Ratio (dB)
	N1	N2	N3	(dB)	N1	N2	N3	(dB)	N1	N2	N3	(dB)
1	0.52	0.57	0.53	0.54	2.86	3.12	2.91	2.96	0.24	0.31	0.29	0.28
2	0.63	0.65	0.62	0.63	3.14	2.96	2.99	3.03	0.28	0.21	0.23	0.24
3	0.73	0.67	0.68	0.69	3.33	3.07	3.24	3.21	0.17	0.24	0.22	0.21
4	0.71	0.68	0.73	0.71	3.17	3.31	3.36	3.28	0.15	0.14	0.19	0.16
5	0.57	0.6	0.61	0.59	2.71	2.86	2.96	2.84	0.36	0.29	0.34	0.33
6	0.64	0.63	0.6	0.62	3.01	3.21	3.15	3.12	0.27	0.29	0.34	0.3
7	0.73	0.71	0.74	0.73	3.41	3.56	3.44	3.47	0.26	0.21	0.22	0.23
8	0.78	0.72	0.77	0.76	3.65	3.53	3.69	3.62	0.26	0.19	0.21	0.22
9	0.55	0.6	0.57	0.57	2.69	2.8	2.72	2.74	0.41	0.34	0.36	0.37
10	0.63	0.58	0.62	0.61	2.83	2.95	2.97	2.92	0.36	0.31	0.35	0.34
11	0.75	0.84	0.77	0.79	3.72	3.82	3.74	3.76	0.35	0.28	0.3	0.31
12	0.86	0.8	0.84	0.83	3.95	3.92	4.07	3.98	0.28	0.23	0.27	0.26
13	0.7	0.66	0.69	0.68	3.16	3.31	3.34	3.27	0.46	0.44	0.39	0.43
14	0.76	0.72	0.75	0.74	3.64	3.49	3.51	3.55	0.44	0.42	0.37	0.41
15	0.75	0.82	0.77	0.78	3.49	3.44	3.36	3.43	0.43	0.35	0.36	0.38
16	0.82	0.85	0.87	0.85	3.98	4.14	4.03	4.05	0.29	0.34	0.33	0.32

An optimization strategy is needed in situations where more than one response variable is to be studied. This optimization strategy should provide a single criterion to show the overall optimal settings of input variables concerning the response characteristics (Kumar et al. 2020). This type of problem falls under the category of multi-optimization techniques. Most of the literature published incorporates the Taguchi technique to optimize the quality features, while process efficiency is typically described by a set of quality features (Walia et al. 2006). It is a very hard task to select the optimal settings when multiple responses are considered, as optimizing one quality characteristic may hamper the other quality characteristic that will eventually lead to loss to society. The preference number (Ni) has been found by using the following equation (Dubey 2009):

$$N_i = K \times \log\left(\frac{V_i}{V_i'}\right) \tag{13.1}$$

where

V_i = value of any response variable i
V_i' = just acceptable value of response variable i
K = constant

Here, K can be predicted if $V_i = V^*$ (Where V^* is at the optimum level), then $N_i = 9$
Therefore,

$$K = \frac{9}{\log\dfrac{V*}{V_i'}} \tag{13.2}$$

The following equation can be used to compute the utility value corresponds to each response variable:

$$T = \sum_{i=1}^{n} E_i N_i \qquad (13.3)$$

Subject to: $\sum_{i=1}^{n} E_i = 1$

Where E_i = weight assigned to the attribute i. The aggregate of weights would have to be equal to one for all the attributes.

For the utility function, the Higher-the-Better type of quality characteristic is selected. The step-wise procedure for a multi-response improvement procedure demonstrates the utility concept (Figure 13.3). In the previous sections, single response parameters such as average radial error, springback, and pillow effect are considered. Based on the objective of the response, the optimal level for each response is determined by using S/N ratio plots. Hence, the Taguchi method deals with only single response optimization problems. In this section, the utility concept has been implemented in which these multiple responses are combined into a single response, and then the optimal conditions are determined. The individual optimal level summary for each response parameter is presented in Table 13.4.

TABLE 13.4

Optimal Value of Individual Response Parameter

Response Parameters	The Optimal Level of Process Parameters	Significant Process Parameters	Predicted Optimal Value
Average radial error	$A_1, B_1, D_4,$	A, B, D	0.51 mm
Springback	$A_1, B_1, C_4, D_4,$	A, B, C, D	2.55 degree
Pillow effect	A_1, B_4	A, B	0.16 mm

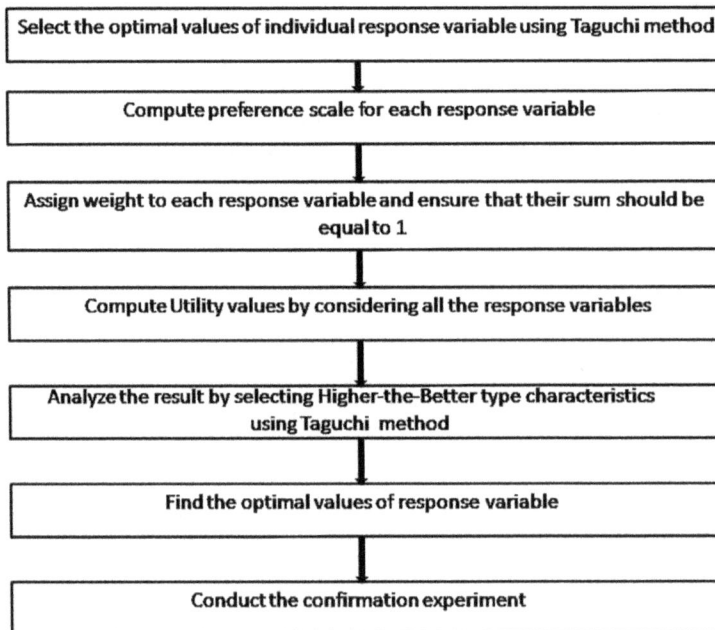

FIGURE 13.3 Utility concept flow diagram.

13.3 RESULTS AND DISCUSSIONS

13.3.1 PREFERENCE SCALE

The method for converting experimental data into utility data is outlined below in a step-by-step manner.

(I) Preference scale construction
(a) Average radial error (P_{RE}):

V* = Predicted optimal value of average radial error =0.51 (refer to Table 13.4)
V_i' = Minimum acceptable value of average radial error = 0.88 (All the observed values of RE are lesser than 0.88)

Using these values and Equations (13.2) and (13.3), the following preference scale for average radial error has been found;

$$N_{RE} = -38.13 \times \log\left(\frac{V_{RE}}{0.88}\right)$$

(b) Springback (P_{SB}):

V* = Predicted optimal value of springback = 2.55 (refer to Table 13.4)
V_i' = Minimum acceptable value of springback = 4.15 (All the observed values of SB are lesser than 4.15)

Using these values and Equations (13.2) and (13.3), the following preference scale for springback has been found;

$$N_{SB} = -42.85 \times \log\left(\frac{V_{SB}}{4.15}\right)$$

(c) Pillow effect (P_{PE}):

V* = Predicted optimal value of pillow effect = 0.16 (refer to Table 13.4)
V_i' = Minimum acceptable value of pillow effect = 0.47 (All the observed values of PE are lesser than 0.5)

Using these values and equations two and three, the following preference scale for pillow effect has been found;

$$N_{PE} = -19.23 \times \log\left(\frac{V_{PE}}{0.47}\right)$$

13.3.2 UTILITY VALUE CALCULATION

A minimized geometric error is an important condition in a component formed by the incremental forming process. Thus the measurement of these errors is therefore of major practical importance

TABLE 13.5
Utility Values Based on Experimental Results

Exp No.	Step Depth (mm)	Tool Diameter (mm)	Feed Rate (mm/min.)	Spindle Speed (rpm)	Utility				
					Y_1	Y_2	Y_3	Avg.	S/N Ratio
1.	0.2	8	1000	1500	6.90	5.19	6.18	6.09	15.51
2.	0.2	10	1500	2000	4.87	5.85	5.80	5.51	14.73
3.	0.2	12	2000	2500	5.12	5.11	4.96	5.06	14.08
4.	0.2	14	2500	3000	5.90	6.09	4.76	5.59	14.78
5.	0.4	8	1500	2500	5.57	5.59	4.86	5.34	14.50
6.	0.4	10	1000	3000	5.13	4.66	4.60	4.80	13.59
7.	0.4	12	2500	1500	3.80	4.31	4.15	4.08	12.19
8.	0.4	14	2000	2000	3.05	4.56	3.65	3.75	11.14
9.	0.6	8	2000	3000	5.44	5.27	5.56	5.43	14.68
10.	0.6	10	2500	2500	4.77	5.42	4.67	4.95	13.84
11.	0.6	12	1000	2000	2.33	2.17	2.58	2.36	7.40
12.	0.6	14	1500	1500	1.85	2.84	1.91	2.20	6.39
13.	0.8	8	2500	2000	2.88	3.07	3.11	3.02	9.58
14.	0.8	10	2000	1500	1.74	2.41	2.51	2.22	6.57
15.	0.8	12	1500	3000	2.12	2.29	2.69	2.36	7.35
16.	0.8	14	1000	2500	1.97	1.11	1.22	1.43	2.34

that determines its usefulness and application. All three qualitative traits were seen to be equally important; hence their weights were equal. However, no weight restriction exists, and depending on the scenario or the user's needs, it can be any value from zero to one.

The utility function was computed using the following relationship based on the experimental trials:

$$T(n,r) = N_{RE}(n,r) \times E_{GE} + N_{SB}(n,r) \times E_{SB} + N_{PE}(n,r) \times E_{PE}$$

Where $E_{RE} = \dfrac{1}{3}$; $E_{SB} = \dfrac{1}{3}$; $E_{PE} = \dfrac{1}{3}$

n = trial number (n = 1,2,3,....,16) and r = repetition number (r = 1,2,3). Table 13.5 represents the computed utility values.

13.3.3 ANALYSIS S/N RATIO OF UTILITY

For the S/N ratio, the data (utility values) were evaluated. Higher-the-Better (HB) attribute is selected for utility value. Figure 13.4 (a–d) and Figure 13.5 (a–d) illustrate the average and major response in terms of utility values and S/N ratio. Table 13.6 shows the average response and ranking of process parameters. Figure 13.4 (a–d) shows the highest value of the utility and the S/N ratio within the experimental space for the first step depth level (A1), the first the tool diameter level (B1), the fourth feed rate level (C4), and fourth spindle rate level (D4).

ANOVA table for utility data is shown in Table 13.7. Table 13.7 shows that all of the input parameters are significant at a 95% confidence level. The optimal values of output variables under investigation are anticipated at the optimal levels of statistically significant parameters.

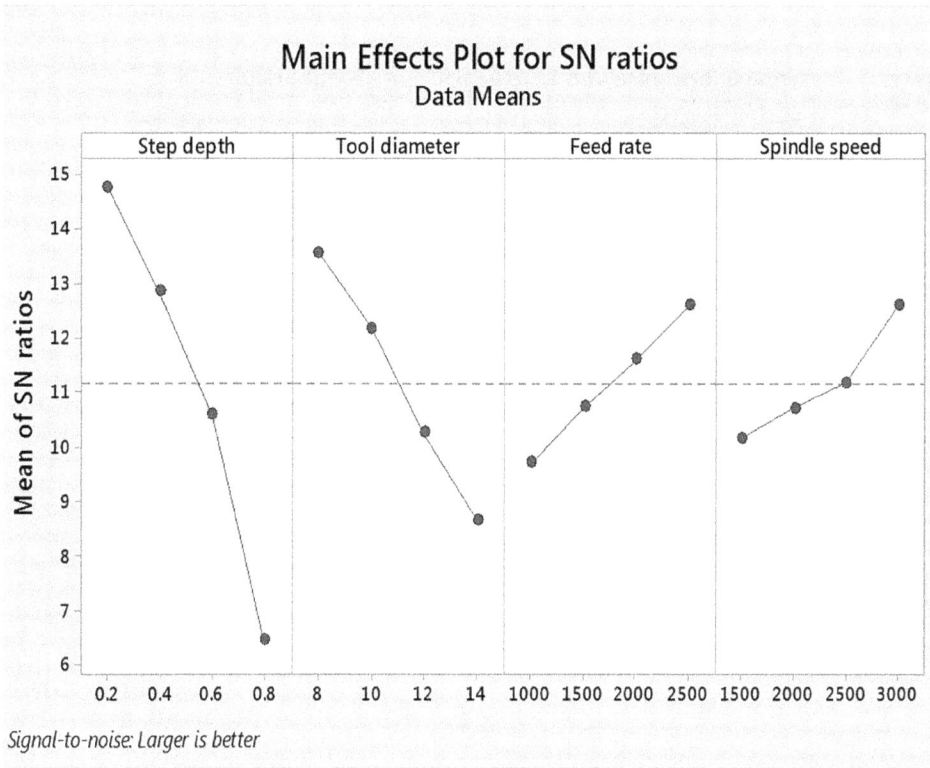

FIGURE 13.4   S/N ratio plot for utility value.

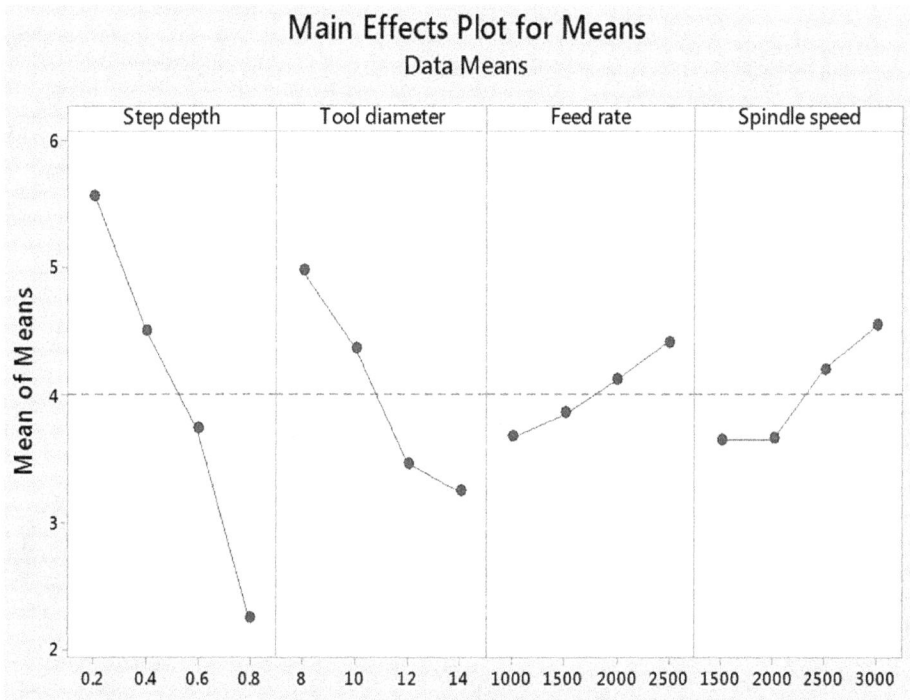

FIGURE 13.5   Main effect plot for the mean of utility value.

TABLE 13.6
Average Response and Ranking of Process Parameters

Level	Step Depth	Tool Diameter	Feed Rate	Spindle Speed
1	14.776	13.568	9.710	10.164
2	12.852	12.183	10.740	10.711
3	10.578	10.253	11.618	11.190
4	6.459	8.662	12.597	12.601
Delta	8.317	4.907	2.887	2.437

TABLE 13.7
ANOVA for Utility Value

Source	Degrees of Freedom	Sum of Squares	Variance	F-ratio	P-Value
Step depth	3	69.385	23.1284	90.19	0.000
Tool diameter	3	23.190	7.7300	30.14	0.000
Feed rate	3	3.730	1.2435	4.85	0.006
Spindle speed	3	6.861	2.2871	8.92	0.000
Error	35	8.975	0.2564		
Total	47	112.142			

13.3.4 ESTIMATION OF MEAN FOR UTILITY

Table 13.4 shows the mean values of all the response characteristics concerning the utility function at the optimum levels of significant parameters. Optimum values of the mean (μ) of various response parameters can be determined using the following equation:

$$\mu = \overline{A1} + \overline{B1} + \overline{C4} + \overline{D4} - 3T \tag{13.4}$$

Where

$\overline{A1}$ - First level of step depth

$\overline{B1}$ - First level of tool diameter

$\overline{C4}$ - Fourth level of feed rate

$\overline{D4}$ - Fourth level of spindle speed

$\overline{T}$ - Overall mean.

Confidence Interval CI_{CE} for the confirmation experiments

$$CI_{CE} = \pm\sqrt{F_\alpha\left(1,f_e\right)V_e\left[\frac{1}{n_{eff}} + \frac{1}{R}\right]} \tag{13.5}$$

Where
$F_\alpha(1,f_e)$ = F value from table
V_e = pooled error mean square/variance
R = sample size for the confirmation experiment.
n_{eff} = effective number of observations

Confidence Interval for Predicted Mean

$$CI_{pop} = \pm\sqrt{\frac{F_\alpha\left(1, f_e\right)V_e}{n_{eff}}}$$

(13.6)

(a) Average radial error:

$$\mu_{RE} = \overline{A1} + \overline{B1} + \overline{C4} + \overline{D4} - 3\overline{T} = 0.49$$

Where $\overline{A1} = 0.64$, $\overline{B1} = 0.60$, $\overline{C4} = 0.68$, $\overline{D4} = 0.67$ (Table 13.3):

$\overline{T} = 0.70$ (Table 13.3)

The ANOVA yielded the following results (Kumar and Belokar 2019):

$$N = 48, f_e = 35; v_e = 0.000988, n_{eff} = 4.8, R = 3, F_{0.05}\left(1,35\right) = 4.12$$

From Equation (13.5), $CI_{CE} = \pm0.05$
From Equation (13.6), $CI_{POP} = \pm0.03$
The optimum range for a geometric error (for confirmatory test runs of three) is predicted as mentioned below:

$CI_{CE} : 0.44 < \mu_{RE} > 0.54$
$CI_{POP} : 0.46 < \mu_{RE} > 0.52$

(b) Springback:

$$\mu_{SB} = \overline{A1} + \overline{B1} + \overline{C4} + \overline{D4} - 3\overline{T} = 2.55$$

Where $\overline{A1} = 3.12$, $\overline{B1} = 2.95$, $\overline{C4} = 3.23$, $\overline{D4} = 3.14$ (Table 13.3):

$\overline{T} = 3.33$ (Table 13.3)

The ANOVA yielded the following results:

$$N = 48, f_e = 35; v_e = 0.013, n_{eff} = 3.69, R = 3, F_{0.05}\left(1,35\right) = 4.12$$

From Equation (13.5), $CI_{CE} = \pm0.18$
From Equation (13.6), $CI_{POP} = \pm0.12$
The optimum range for springback (for confirmatory test runs of three) is predicted as mentioned below:

$CI_{CE} : 2.37 < \mu_{SB} > 2.73$
$CI_{POP} : 2.43 < \mu_{SB} > 2.67$

(c) Pillow effect:

$$\mu_{PE} = \overline{A1} + \overline{B1} + \overline{C4} + \overline{D4} - 3T = 0.26$$

Where $\overline{A1} = 0.22$, $\overline{B1} = 0.35$, $\overline{C4} = 0.29$, $\overline{D4} = 0.30$:
$\overline{T} = 0.30$ (Table 13.3)
The ANOVA yielded the following results:

$$N = 48, f_e = 35; v_e = 0.001096, n_{eff} = 6.85, R = 3, F_{0.05}(1,35) = 4.12$$

From Equation (13.5), $CI_{CE} = \pm0.05$
From Equation (13.6), $CI_{POP} = \pm0.03$
The optimum range for a pillow effect (for confirmatory test runs of three) is predicted as mentioned below:

CI_{CE} : $0.21 < \mu_{PE} > 0.31$
CI_{POP} : $0.23 < \mu_{PE} > 0.29$

13.3.5 CONFIRMATION EXPERIMENT

This is the final stage in confirming the results of Taguchi's utility data parameter design technique. The experimental trials of significant parameters are conducted at optimum conditions. The result is compared with the estimated values. The findings of three confirmatory studies utilizing the optimal levels are reported in Table 13.8. The findings are within 95% CI_{CE} of the estimated optimum of the performance parameters.

The confirmation experiments show that there is an improvement in the springback as it decreases from 2.55° to 2.52°. The average radial error, on the other hand, remains unchanged. Furthermore, smaller tool diameter is used as proposed by the utility concept, the pillow effect increases from 0.16 mm to 0.24 mm. The small tool diameter may induce greater bending of the sheet than the larger tool diameter, produced by normal stress and buckling produced by in-plane forces (Isidore et al. 2016). Small tool diameter tends to concentrate strain under the tool in the deformation zone, whereas bigger tool diameter tends to spread strain more uniformly on the sheet, as in conventional sheet stamping. As a result, a smaller tool diameter enhances the pillowing of formed parts.

TABLE 13.8
Observed Values of Quality Attributes

Response Parameters	Predicted CI($\alpha = 0.05$)	Overall Average
Average radial error	CI_{CE} : $0.44 < \mu_{RE} > 0.54$ CI_{POP} : $0.46 < \mu_{RE} > 0.52$	0.51 mm
Springback	CI_{CE} : $2.37 < \mu_{SB} > 2.73$ CI_{POP} : $2.43 < \mu_{SB} > 2.67$	2.52°
Pillow effect	CI_{CE} : $0.21 < \mu_{PE} > 0.31$ CI_{POP} : $0.23 < \mu_{PE} > 0.29$	0.24 mm

13.4 SUMMARY

In this chapter, the detailed experimentation has been explained to explore the potential application of the Taguchi method based robust design for an optimal parametric combination of the developed SPIF process that can ensure the quality of this process through effective control over the process variables. The influence of SPIF process parameters, i.e., step depth (A, mm), tool diameter (B, mm), feed rate (C, mm/min.), and spindle speed (D, rpm) on response characteristics such as average radial error (mm), springback (°) and pillow effect (mm) were analyzed through experimentation. The ANOVA, S/N Ratio (dB) were employed and identified the most significant parameter as well as optimized the process parameter during the forming of the AA1200 H14 aluminum alloy sheet. The parameters' percentage contributions to achieving a higher value of the utility function are listed in decreasing order: Step depth (61.87%) > Tool diameter (20.68%) > Spindle speed (6.12%) > Feed rate (3.33%). It has been observed that the first level of step depth (A1:0.2mm), the first level of tool diameter (B1:8mm), the fourth level of feed rate (C4:2500 mm/min), and fourth level of spindle speed (D4:3000 rpm), yield a maximum value of the utility and S/N ratio within the experimental space.

REFERENCES

Ambrogio, G., F. Gagliardi, and L. Filice. 2013. "Robust design of incremental sheet forming by Taguchi Method." *Procedia CIRP* 12: 270–275. https://doi.org/10.1016/j.procir.2013.09.047

Dubey, A. K. 2009. "Multi-response optimization of electro-chemical honing using utility-based Taguchi Approach." *International Journal of Advanced Manufacturing Technology* 41 (7–8): 749–759. https://doi.org/10.1007/s00170-008-1525-2

Echrif, Salah B. M., and Meftah Hrairi. 2014. "Significant parameters for the surface roughness in incremental forming process." *Materials and Manufacturing Processes* 29 (6): 697–703. https://doi.org/10.1080/10426914.2014.901519

Han, Fei, Jian Hua Mo, Hong Wei Qi, Rui Fen Long, Xiao Hui Cui, and Zhong Wei Li. 2013. "Springback prediction for incremental sheet forming based on FEM-PSONN technology." *Transactions of Nonferrous Metals Society of China (English Edition)* 23 (4): 1061–1071. https://doi.org/10.1016/S1003-6326(13)62567-4

Isidore, B. B. Lemopi, G. Hussain, S. Pourhassan Shamchi, and Wasim A. Khan. 2016. "Prediction and control of pillow defect in single point incremental forming using numerical simulations." *Journal of Mechanical Science and Technology* 30 (5): 2151–2161. https://doi.org/10.1007/s12206-016-0422-0

Khan, Muhamad S., Frans Coenen, Clare Dixon, Subhieh El-Salhi, Mariluz Penalva, and Asun Rivero. 2014. "An intelligent process model: Predicting springback in single point incremental forming." *The International Journal of Advanced Manufacturing Technology* 76 (9–12): 2071–2082. https://doi.org/10.1007/s00170-014-6431-1

Kumar, Ajay, Vishal Gulati, and Parveen Kumar. 2019. "Experimental investigation of forming forces in single point incremental forming." *Lecture Notes in Mechanical Engineering* 10 (1): 423–430. https://doi.org/10.1007/978-981-13-6412-9_41

Kumar, Narinder, and Rajendra M Belokar. 2019. "Experimental investigation of geometric accuracy in single point incremental forming process of an aluminum alloy." *International Journal of Materials Engineering Innovation* 10(1): 46–59. https://doi.org/10.1504/IJMATEI.2019.097914

Kumar, Narinder, Arshpreet Singh, and Anupam Agrawal. 2020. "Formability analysis of AA1200 H14 aluminum alloy using single point incremental forming process." *Transactions of the Indian Institute of Metals* 73 (7): 1975–1984. https://doi.org/10.1007/s12666-020-02014-7

Li, Yanle, Haibo Lu, William J. T. Daniel, and Paul A. Meehan. 2015. "Investigation and optimization of deformation energy and geometric accuracy in the incremental sheet forming process using response surface methodology." *International Journal of Advanced Manufacturing Technology* 79: 2041–2055. https://doi.org/10.1007/s00170-015-6986-5

Malhotra, Rajiv, Jian Cao, Feng Ren, Vijitha Kiridena, Z. Cedric Xia, and N. V. Reddy. 2011. "Improvement of geometric accuracy in incremental forming by using a squeezing toolpath strategy with two forming tools." *Journal of Manufacturing Science and Engineering, Transactions of the ASME* 133 (6): 061019. https://doi.org/10.1115/1.4005179

Nirala, Harish K., and Anupam Agrawal. 2018. "Fractal geometry rooted incremental toolpath for incremental sheet forming." *Journal of Manufacturing Science and Engineering, Transactions of the ASME* 140 (2): 1–9. https://doi.org/10.1115/1.4037237

Shrivastava, Parnika, Pavan Kumar, Puneet Tandon, and Alexander Pesin. 2018. "Improvement in formability and geometrical accuracy of incrementally formed AA1050 sheets by microstructure and texture reformation through preheating, and their FEA and experimental validation." *Journal of the Brazilian Society of Mechanical Sciences and Engineering* 9: 1–15. https://doi.org/10.1007/s40430-018-1255-9

Singh, Arshpreet, and Anupam Agrawal. 2016. "Comparison of deforming forces, residual stresses and geometrical accuracy of deformation machining with conventional bending and forming." *Journal of Materials Processing Technology* 234: 259–271. https://doi.org/10.1016/j.jmatprotec.2016.03.032

Singh, Arshpreet, and Anupam Agrawal. 2018. "Investigation of parametric effects on geometrical inaccuracies in deformation machining process." *Journal of Manufacturing Science and Engineering, Transactions of the ASME* 140 (7). https://doi.org/10.1115/1.4039586

Taherkhani, Abolfazl, Ali Basti, and Nader Nariman-Zadeh. 2018. "Achieving maximum dimensional accuracy and surface quality at the shortest possible time in single-point incremental forming via multi-objective optimization." *Proceedings of the Institution of Mechanical Engineers, Part B: Journal of Engineering Manufacture* 233 (3): 900–913. https://doi.org/10.1177/0954405418755822

Walia, R. S., H. S. Shan, and P. Kumar. 2006. "Multi-response optimization of CFAAFM process through Taguchi method and utility concept." *Materials and Manufacturing Processes* 21 (8): 907–914. https://doi.org/10.1080/10426910600837814

Wei, Hongyu, Wenliang Chen, and Lin Gao. 2011. "Springback investigation on sheet metal incremental formed parts." *World Academy of Science, Engineering and Technology* 55 (1): 285–289.

Yao, Zimeng, Yan Li, Mingshun Yang, Qilong Yuan, and Pengtao Shi. 2017. "Parameter optimization for deformation energy and forming quality in single point incremental forming process using response surface methodology." *Advances in Mechanical Engineering* 9 (7): 1–15. https://doi.org/10.1177/1687814017710118

14 A Review on Formability of Tailored Sheets in Incremental Forming

Kuntal Maji and Gautam Kumar
National Institute of Technology Patna, Patna, India

CONTENTS

14.1 INTRODUCTION

In the recent years, the requirement of lightweight components of vehicles to reduce fuel consumption and overall cost reduction in automotive industries motivated the industrialists to the growing utilization of Tailor sheets characterized by different material or different thicknesses produced through different joining and laminating processes. Incremental sheet forming (ISF) is an emerging flexible forming process suited for rapid prototyping and small batch production of sheet material components. Higher formability of produced components in ISF compared to that of conventional forming make it more suitable to make complex parts with higher formability. Tailored blanks are made of two sheets with different material properties or geometries which have been widely used in various industries like automobiles, aerospace, marine for weight reductions, fuel efficiency without affecting safety, lower product cost, and others. The increasing demand of customized lightweight sheet metal parts with high strength to weight ratio drives the researchers to investigating fabrications and processing of tailored blanks. Compared with monolithic metallic sheets, tailored blanks made of metal-polymer sheets have potential applications as lightweight materials for structural parts due to lower density, higher specific flexural stiffness, better vibration damping characteristics, excellent strength/weight ratio, ease of production etc. There are different types of tailored sheets/blanks such as tailor welded blanks (TWBs), tailor laminated blanks (TLBs), tailored bimetallic sheets (TBS) etc. made by different fabrication methods depending on the sheet material combinations and geometries, i.e., welding, lamination, rolling as shown in Figure 14.1.

Incremental sheet forming (ISF) is a flexible sheet metal forming process, in which the sheet is deformed gradually according to the desired toolpath controlled by a CNC machine or robotic arm. Incremental forming has better formability contrasted with the traditional forming and it is affected by several process parameters namely sheet thickness, tool diameter, tool geometry, incremental depth, feed rate, rotational speed and direction, and their interactions. Sheet metal can be deformed to limited extent without any necking or failure known as formability which very much important and essential to know for successful formed parts. This chapter presents deformation behaviors and formability of tailored blanks made of metallic and non-metallic sheet materials prepared by welding,

DOI: 10.1201/9781003226703-14

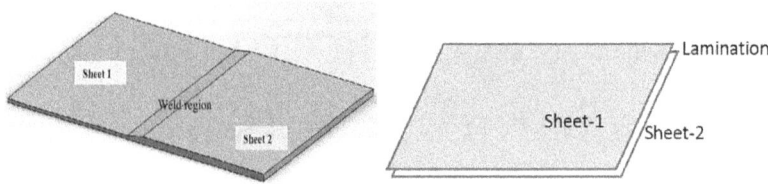

FIGURE 14.1 Tailored sheets made by welding and lamination processes.

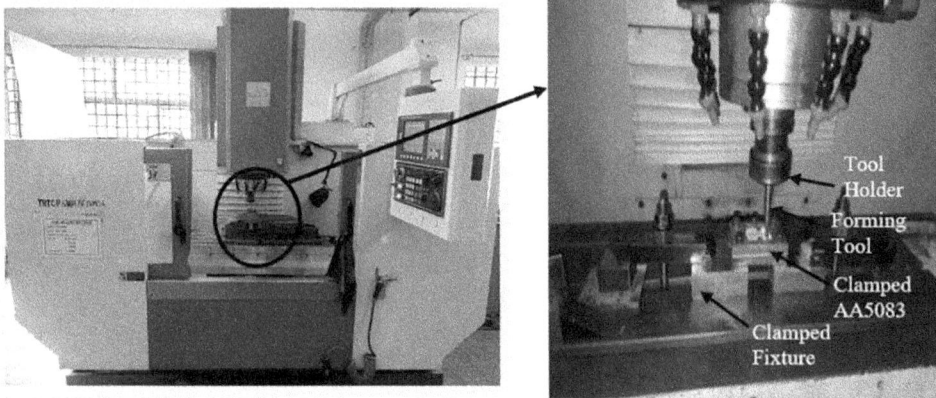

FIGURE 14.2 Experimental setup for incremental sheet forming in a CNC milling machine.

lamination, and rolling processes in incremental forming. Experimental setup for single point incremental forming (SPIF) process in a vertical CNC milling machine is shown in Figure 14.2.

14.2 TAILOR WELDED SHEET

Fabrication techniques, characterization and applications of tailored sheets were presented by Merklein et al. (2014) Tailor welded blanks made of aluminium alloys by FSW have higher formability in comparison to that obtained by the fusion welding. Silva et al. (2009) showed that FSWed blanks of aluminium alloy AA1050-H111 sheets can be formed to complex shaped parts with high forming depths by SPIF. Liu et al. (2020) found that formability of AA7075-O in terms of maximum wall angle and fracture FLD was affected by different draw angles, part height, step-down sizes, toolpath etc. It was also shown that the trend of forming force could indicate the failure in SPIF. Alinaghian et al. (2017) investigated formability of FSWed AA6061 sheet in SPIF using RSM. Tayebi et al. (2019) showed that selection of suitable process parameters and tool design could produce higher formability and strength in the FS-TWB of AA5083 and AA6061 compared to that of the base AA6061 sheet. The microhardness was found to be decrease after FSW but increased after SPIF in the welded region due to the combined effects of frictional heating and work hardening. Fracture modes of the FS-TWB in tensile testing and SPIF were seen to be of ductile, and mixed ductile/brittle types, respectively. Singh et al. (2021) observed a delay the onset of fracture or improved formability of AA1050 sheet in SPIF at elevated temperature compare to that at room temperature. Uniform thickness distribution and good surface quality was observed for deformed samples at elevated temperature.

Incremental forming of monolithic aluminium alloy sheets, i.e., AA5083 and AA7075, and friction stir welded AA5083 and AA7075 sheets, was investigated by Kumar (2021). The monolithic and tailor welded sheets were incrementally formed in a vertical CNC milling machine after holding the sheet in a fixture as shown in Figure 14.2. The sample sheets were formed into truncated square pyramid shapes and the forming wall angles were increased gradually till fracture. The experiments were conducted with the process parameters as forming tool diameter of 6 mm, feed rate of 500 mm/min, spindle speed of 2000 rpm, and incremental depth of 0.25 mm. The wall angle at the onset of fracture was taken as the maximum forming wall angle as an indicator of formability. The maximum forming wall angles for AA5083, AA7075 and the tailor welded sheet were found as 60°, 55° and 42.5°, respectively as shown in Figure 14.3. Therefore the tailor welded sheet was found to have lower formability compared to that of the monolithic sheets in terms of maximum forming wall angle.

Forming limit curves (FLCs) for monolithic and tailor welded blanks were also determined by Kumar (2021) through incremental sheet forming line tests. The sheet samples were marked with circular grids of 1 mm size using screen printing method, and the sheets are deformed by gradually increasing the depth of deformations till fracture. The grid marks on the deformed samples become elliptical as shown in Figure 14.4, and deformed grid marks near the fracture zones are measured using optical method and image analyis technique.

Major and minor diameters (d_1 and d_2) of the elliptical grid marks near the fracture zone on the deformed sheets were measured using the ImageJ software and corresponding major and minor strain values were calculated as $\varepsilon_1 = \ln(d_1/d)$ and $\varepsilon_2 = \ln(d_2/d)$, respectively where$\varepsilon_1$ and ε_2 represents major and minor strains, the symbols d, d_1, and d_2 denotes circular grid diameter before deformation, major and minor diameters of the grid marks after deformation, respectively. The calculated true major and minor strains ($\varepsilon_1, \varepsilon_2$) were plotted as FLCs for the monolithic and tailor welded sheets as shown in Figure 14.5 indicating lower formability of the TWB compared to the base sheets due to strain hardening effect of the FSW process and subsequent incremental forming process.

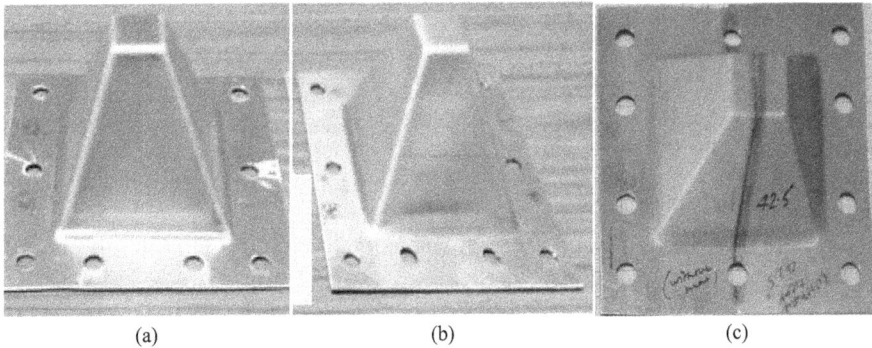

FIGURE 14.3 Maximum forming wall angles in SPIF of (a) AA5083, (b) AA7075, and (c) Friction stir welded AA5083 & AA7075 sheets.

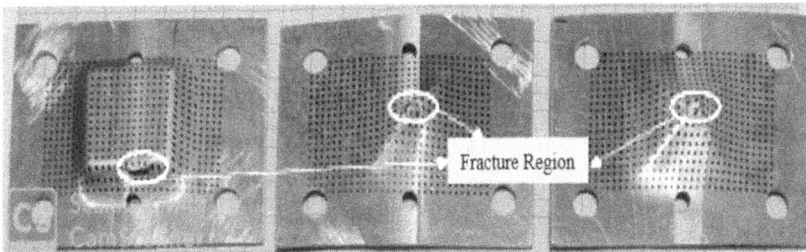

FIGURE 14.4 Grid marks on the deformed samples of TWBs in ISF tests.

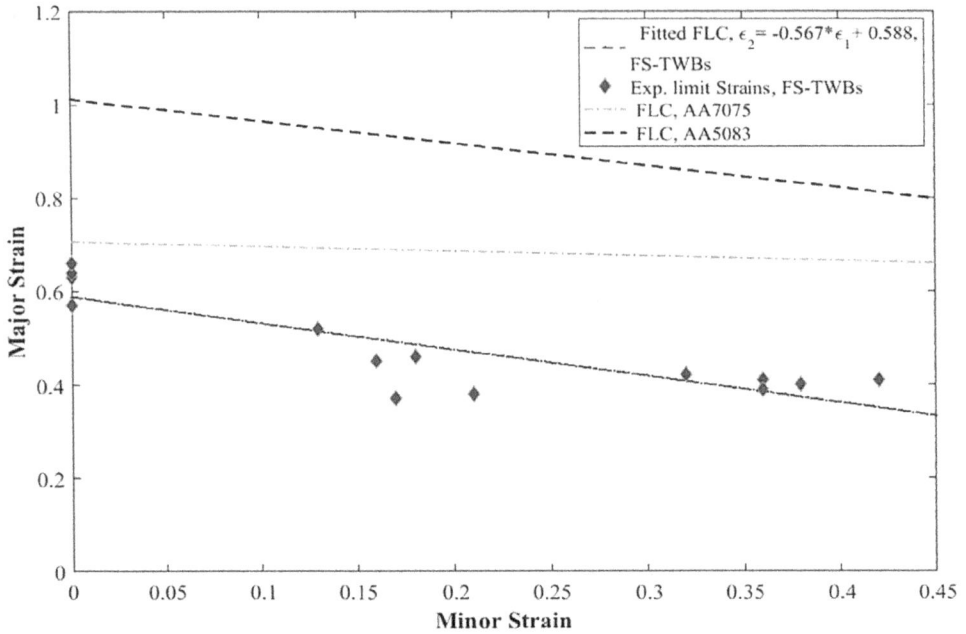

FIGURE 14.5   Experimental forming limit curves (FLC) for AA5083, AA7075 and friction stir welded sheets.

14.3   TAILOR BIMETALLIC SHEET

Tailor bimetallic sheets (TBS) are composed of two dissimilar sheets fulfilling requirement of very different properties on two different sides of the sheet, and these sheets are able to offer advantages of both the sheets and alleviate disadvantages of the individual sheets. Bimetallic sheets are fabricated by mechanical bonding or using adhesives having high strength to weight ratio and have numerous applications in components of aerospace, cars, ships, buildings, sports equipment, automotive and biomedical implants. A schematic representation of the single point incremental forming of bimetal sheet is shown in Figure 14.6 indicating the different factors such as forming tool diameter, step down, sheet thickness, tool rotational speed, feed rate, toolpath, forming angle as summarized in Table 14.1.

The toolpath affects the formability, surface quality, and dimensional accuracy in incremental forming of Al6061/DC04 bimetallic sheet. Tera et al. (2012) conducted finite element analysis

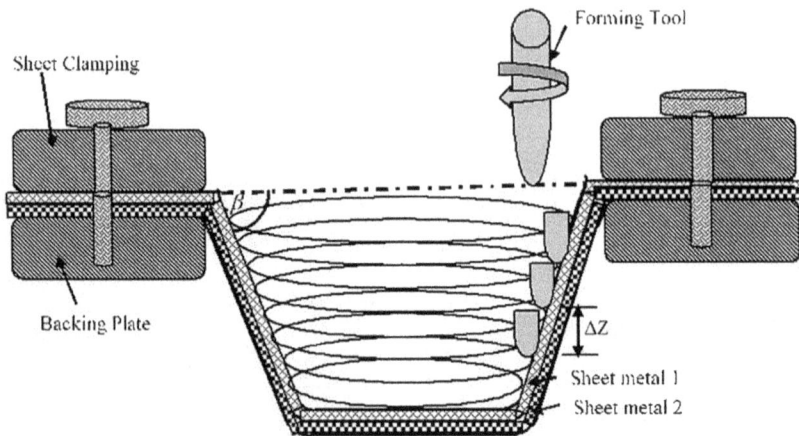

FIGURE 14.6   Incremental forming of tailor bimetal sheet.

TABLE 14.1
Study on Incremental Forming of Bimetal Sheets

SL. No.	Tailor Bimetal Sheets	Findings	References
1.	Al6061/DC04	Maximum major strains spiral toolpath	Tera et al. (2012)
2.	Al/Cu	Anisotropic mechanical properties	Uscinowicz (2013)
3.	Ti/Steel	Forming force & thickness distributions	Abdelkader et al. (2020)
4.	Al/Steel	Simultaneous forming & bonding	Li et al. (2020)
5.	Steel/steel	Delamination	Hassan et al. (2020)
6.	Cu/Al	Layer arrangement effect formability	Liu & Li (2019)
7.	Al/Cu	Optimal process parameters	Gheysarian (2018)
8.	Al/SUS	Formability & failure	Ali et al. (2019)
9.	Steel/Ti	Layer arrangement effect	Sakhtemanian (2018)
10.	Cu/Al	Bulge defect control	Quin et al. (2021)

and showed that the maximum values of major strains were found along the Archimedean spiral toolpath on the side wall of the cone frustum of the bimetallic sheet formed by ISF process. Tera et al. (2014a) shows incremental forming of a hemispherical shape from the Al6061/DC04 TBS. Thickness reductions are uniformly distributed on the surface of the incrementally formed blank. Forming limit curves in incremental forming of bimetallic and multilayer sheets were estimated by Tera et al. (2014b). The rolled Al/Cu bimetal sheet was seen to have highly directional structure and anisotropic mechanical properties as experimentally observed by Uscinowicz (2013). Multiresponse optimization in incremental forming of Al/Cu bimetal sheet made by explosion welding was performed by Honarpisheh et al. (2018a) considering forming process parameters such as tool diameter, step down, rotational speed, and sheet arrangement. Finite element analysis was done to estimate forming force, thickness variation, and stress distributions for the optimal parameters. Higher values of step down, lower values of tool diameter and rotational speed with the tool in contact to the Cu-sheet results higher formability of the bimetallic sheet in terms of higher fracture depth and lower thickness reduction. Honarpisheh et al. (2018b) carried out finite element simulation on incremental forming of explosive welded Al/Cu bimetal sheet into truncated pyramid shape revealed that increase in tool diameter and vertical pitch increased forming force but decreased wall thickness. Layer arrangement affects deformation behavior in incremental forming of bimetal sheet. Sakhtemanian et al. (2018) carried out finite element simulations and experimental study effects of layer arrangement on forming force in SPIF of steel/titanium (St/Ti) bimetal sheet. Lower forming force and better surface quality of the bimetal sheet are obtained for St/Ti compared to that for Ti/St sheet arrangement. Optimization of process parameters in incremental forming of explosive welded Al/Cu bimetal sheet was carried out by Gheyserian and Honarpisheh (2019a). Optimal input parameters settings were found to be different for optimization considering different objectives like forming force, surface roughness, thickness variation etc. Influencing process parameters in SPIF of Al/Cu bimetal sheet were identified and multiresponse optimization was carried out by Gheyserian and Honarpisheh (2019b). Formability and failure modes of Al/SUS bimetal sheet in SPIF was analysed by Ali et al. (2019) using finite element simulation and experiments considering layer arrangement, tool diameter, and step down. Layer arrangement was found the most significant factor affecting formability and forming force. Fracture always occurs in steel side and delamination occurs for SUS/Al arrangement only. Experimental investigations and finite element modeling of SPIF of roll bonded Al/Cu bimetal sheet carried out by Liu and Li (2019) revealed that higher formability and forming force is obtained for Al/Cu arrangement compared to that for Cu/Al one. Hassan et al. investigated delamination in steel/steel bilayer sheet in single point incremental forming using numerical simulations and experiments. Strength of the TBS was found to increase for higher rolling thickness reduction and delamination occurred beyond a certain depth of deformation.

Li et al. (2020) showed that simultaneous forming and bonding of Al/steel sheet could be obtained by friction-stir-assisted incremental forming. Formability and bonding quality are good for Steel/Al layer arrangement and five factors were identified as influential in decreasing order of significance on bonding and forming quality as wall angle, step down, stacking sequence, rotational speed, and ratio of rotational speed to feeding rate. Abdelkader et al. (2020) carried out finite element simulations to study the effects of process parameters on forming forces and thickness distributions in incremental forming of Ti/Steel TBS. The forming wall angle was seen to be the most significant process parameter affecting forming force and thickness distribution. Qin et al. (2021) studied bulge defect in incremental forming of Cu/Al bimetal sheet. It was seen that the stress inconsistency between lower and upper sheets results into bulge defect and the bulge height was mostly affected by the step down factor. Chang and Chen (2020) observed an increase in formability of AA2024 and AA7075 sheets in case of three sheet incremental forming (TISF) using membrane analysis and numerical simulation. Better surface quality was obtained for deformed sheet by TISF compared to that achieved in conventional incremental forming process due to indirect contact of sheet and tool.

14.4   TAILOR LAMINATED BLANK

Formability of tailor laminated blanks made of different combinations of metals and polymer sheets was studied by researchers in past is discussed here. Jackson et al. (2008) studied SPIF of sandwich sheets made of different materials such as mild steel and polypropylene (MS/PP/MS), aluminium and polypropylene core (Al/PP/Al) and stainless-steel and a stainless-steel fibre (SS/SS fibre/SS). Sheet thinning and through-thickness strains were observed for the sandwich panels deformed by SPIF. Liu et al. (2013) investigated formability of AA5052/polyethylene/AA5052 sandwich sheet using Nakazima tests and numerical simulation. Formability for the sandwich sheet was observed to be better than that of the monolithic AA5052 sheet and it increased with increase of polyethylene core layer thickness. Davarpanah and Malhotra (2018) examined the effects of step depth, metal and polymer thicknesses on formability and failure modes of metal-polymer laminated sheets during SPIF. The mode of failure was found to be dependent on the metal and polymer sheet thicknesses. Delamination and metal tearing work observed for the laminated sheet with thinner metal laminates, and galling of polymer with delamination and metal tearing were seen for thicker metal sheet. Ambrogio et al. (2018) investigated SPIF of thermo-plastic composite consists of short glass fibers and Polyamide 6 matrix. The experiments were performed to evaluate part soundness, thinning, fibre distortion, and polymer degradation in the SPIF. The material structure was not observed to differ much for the process. Miranda et al. (2019) investigated formability of sandwich sheet of two interstitial free steel sheets and a polymeric core through experiments and simulation. Premature fracture on the skin layer from the punch side was observed in the hole expansion test. Formability of metal-polymer laminated sheets was also studied by Kumar (2021). Aluminium alloy AA5083 and polycarbonate (PC) sheets were laminated using neoprene adhesive (3 M Scotch-weld 1300) to produce tailored laminated sheets into two configurations of metal/polymer laminates, i.e., AA5083/PC and PC/AA5083. Surfaces of both sheet materials were cleaned using methyl ethyl ketone and after that, the adhesive was applied on both the cleaned surfaces before the lamination. The sheets after application of adhesives were kept under uniform load for about 24 hours to get the materials joined properly. Forming wall angle tests were conducted to estimate the depths of deformations at the onset of fracture for the monolithic and laminated sheets. For both monolithic AA5083 and PC sheets, the depth of deformation was seen to be equal to 37 mm without occurrence of any fracture at 55° angle. However the laminated AA583/PC and PC/AA5083 sheets were found to fracture at depths of 7.5 mm and 8.5 mm, respectively for a forming wall angle of 55° as shown in Figure 14.7.

FLC were also predicted for AA5083, PC and PC/AA5083 sheets though ISF line tests. The AA5083/PC arrangement laminated sheet was seen to have lower formability compared to that of the individual sheets as seen from the ELCs in Figure 14.8.

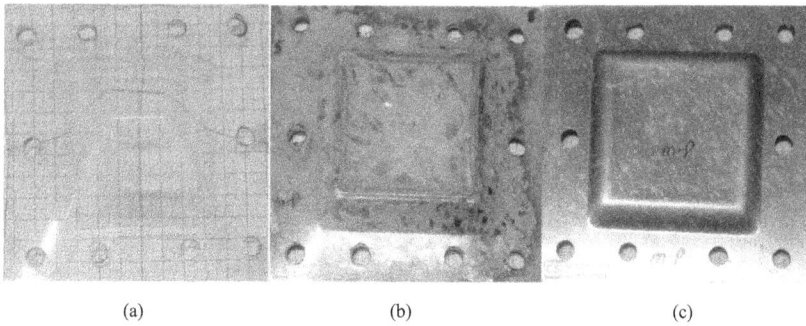

FIGURE 14.7 Deformed Sheets in SPIF of (a) PC and laminated, (b) AA583/PC, and (c) PC/AA5083 sheets.

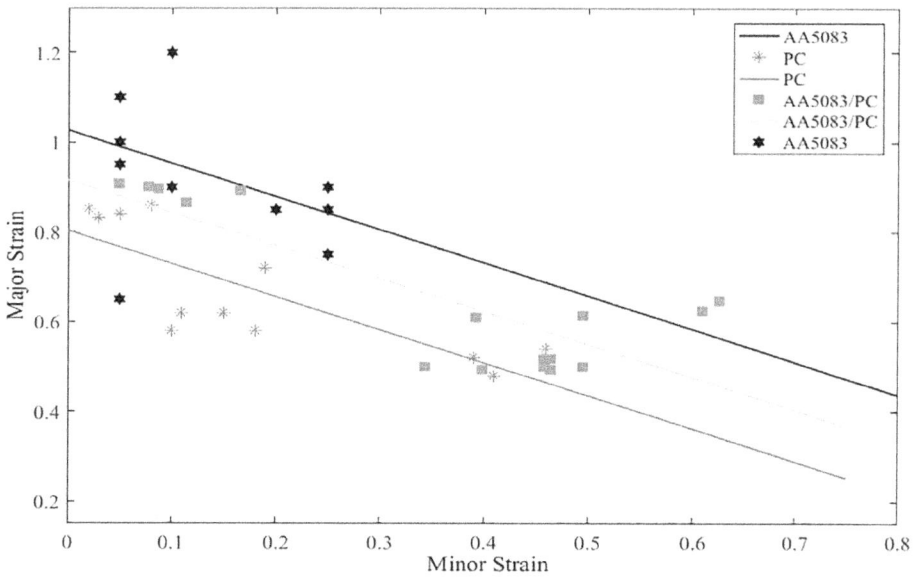

FIGURE 14.8 Experimental forming limit curves (FLC) for AA5083, PC and AA5083/PC sheets in SPIF.

14.5 FUTURE SCOPES

This chapter described single point incremental forming of tailored sheets made by friction stir welding, lamination processes considering deformation behaviors, formability, and failure modes. Incremental forming of complex geometries from tailored sheets by other variations of the process namely two point incremental forming and double sided incremental forming may be studied in future to exploit the potential of this process for broader applications. Moreover, incremental forming of other types of tailored sheets, i.e., tailored blank with non-uniform thickness induced by rolling, additively manufactured blank etc. may be investigated in future.

REFERENCES

Abdelkader, W.B. et al. Numerical investigation of the influence of some parameters in SPIF process on the forming forces and thickness distributions of a bimetallic sheet CP-titanium/low-carbon steel compared to an individual layer, *Procedia Manufacturing* 47 (2020) 1319–1327.

Ali, R. et al. Formability and failure analyses of Al/SUS bilayer sheet in single point incremental forming, *The International Journal of Advanced Manufacturing Technology* 105 (2019), 2785–2798.

Alinaghian, I. et al. Forming limit investigation of AA6061 friction stir welded blank in a single point incremental forming process: RSM approach, *Transactions of the Indian Institute of Metals* 70 (9) (2017), 2303–2318.

Ambrogio, G. et al. A new approach for forming polymeric composite structures, *Composite Structures* 204 (2018), 445–453.

Chang, Z. and Chen, J. Investigations on the deformation mechanism of a novel three-sheet incremental forming, *Journal of Materials Processing Technology*. 281 (2020), 116619.

Davarpanah, M.A. and Malhotra, R. Formability and failure modes in single point incremental forming of metal-polymer laminates, *Procedia Manufacturing* 26 (2018), 343–348.

Gheyserian, A. and Honarpisheh, M. Process parameter optimization of the explosive-welded Al/Cu bimetal in the incremental sheet forming process, *Iranian Journal of Science and Technology, Transactions of Mechanical Engineering* 43 (2019a), 945–956.

Gheyserian, A. and Honarpisheh, M. Investigation of fracture depth of Al/Cu bimetallic sheet in single point incremental forming process, *Iranian Journal of Materials Forming* 6 (1) (2019b), 2–15.

Hassan, M. et al. Delamination analysis in single point incremental forming of steel/steel bi-layer sheet metal, *Archives of Civil and Mechanical Engineering* 20 (39) (2020), 1–14.

Honarpisheh, M. et al. Multi-response optimization on single point incremental forming of hyperbolic shape Al-1050/Cu bimetal using response surface methodology, *The International Journal of Advanced Manufacturing Technology* 96 (2018a), 3069–3080.

Honarpisheh, M. et al. Numerical and experimental study on incremental forming process of Al/Cu bimetals: Influence of process parameters on the forming force, dimensional accuracy and thickness variations, *Journal of Mechanics of Materials and Structures* 13(1) (2018b), 35–51.

Jackson, K.P. et al. Incremental forming of sandwich panels, *Journal of Materials Processing Technology* 204 (2008), 290–303.

Kumar, G. Formability analysis of tailored blanks in single point incremental forming, PhD Thesis, *National Institute of Technology Patna, India* 2021.

Li, M. et al. Experimental investigation on friction-stir-assisted incremental forming with synchronous bonding of aluminum alloy and steel sheets, *Journal of Materials Engineering and Performance* 29 (2020), 750–759.

Liu, Z. and Li, G. Single point incremental forming of Cu-Al composite sheets: A comprehensive study on deformation behaviors, *Archives of Civil and Mechanical Engineering* 19 (2019), 484–502.

Liu, Z. et al. Experimental investigation of mechanical properties, formability and force measurement for AA7075-O aluminum alloy sheets formed by incremental forming, *International Journal of Precision Engineering and Manufacturing* 14–11 (2013a), 1891–1899.

Liu, Z.et al. Forming limit diagram prediction of AA5052/polyethylene/AA5052 sandwich sheets, *Materials & Design* 46 (2013b), 112–120.

Merklein, M. et al. A review on tailored blanks—Production, applications and evaluation, *Journal of Materials Processing Technology* 214 (2014), 151–164.

Miranda, S.S. et al. Characterization and formability analysis of a composite sandwich metal-polymer material. In: Silva L. (eds) *Materials Design and Applications II. Advanced Structured Materials*, vol 98 (2019). Springer, Cham.

Qin, Q. et al. Control and optimization of bulge defect in incremental forming of cu-Al bimetal, *International Journal of Material Forming* (2021), https://doi.org/10.1007/s12289-020-01605-5.

Sakhtemanian, M.R. et al. Numerical and experimental study on the layer arrangement in the incremental forming process of explosive welded low-carbon steel/CP-titanium bimetal sheet, *The International Journal of Advanced Manufacturing Technology* 95 (2018), 3781–3796.

Silva, M.B. et al. Single point incremental forming of tailored blanks produced by friction stir welding, *Journal of Materials Processing Technology* 209 (2009), 811–820.

Singh, S.A. et al. Comparative study of incremental forming and elevated temperature incremental forming through experimental investigations on AA1050 sheet, *Journal of Manufacturing Science and Engineering* 143(6) (2021), 064501–064507.

Tayebi, P. et al. Formability analysis of dissimilar friction stir welded AA 6061 and AA 5083 blanks by SPIF process, *CIRP Journal of Manufacturing Science and Technology* 25 (2019), 50–68.

Tera, M. et al. Study of incremental deep-drawing of bimetallic sheets, *Proceedings in Manufacturing Systems* 7(4) (2012), 235–240.

Tera, M. et al. Contributions regarding incremental forming of bimetallic sheets, *Applied Mechanics and Materials* 657 (2014a), 178–182.

Tera, M. et al. Theoretical and experimental researches regarding multilayer materials used for incremental forming, *Applied Mechanics and Materials* 555 (2014b), 413–418.

Uscinowicz, R. Experimental identification of yield surface of Al-Cu bimetallic sheet, *Composites: Par B* 55 (2013), 96–108.

Index

For Product Safety Concerns and Information please contact our EU
representative GPSR@taylorandfrancis.com
Taylor & Francis Verlag GmbH, Kaufingerstraße 24, 80331 München, Germany